DSP FIRST:

A MULTIMEDIA APPROACH

James H. McClellan

Ronald W. Schafer
School of Electrical and Computer Engineering
Georgia Institute of Technology
Atlanta, GA 30332–0250

Mark A. Yoder
Department of Electrical and Computer Engineering
Rose-Hulman Institute of Technology
Terre Haute, IN 47803–3999

MATLAB® CURRICULUM SERIES

PRENTICE HALL
Upper Saddle River, NJ 07458

Library of Congress Cataloging-in-Publication Data

McClellan, James H.,
    DSP first: a multimedia approach / J. H. McClellan, R. W. Schafer, M. A. Yoder
       p.   cm.
    Includes index.
    ISBN 0–13–243171–8
    1. Signal processing—Digital techniques., 2. Signal processing—Mathematics.
    I. Schafer, Ronald W.,   II. Yoder, M. A. (Mark A.),   III. Title.
TK5102.M388 1998   97–36447
621.382'2'078553042—dc21   CIP

Publisher: *Tom Robbins*
Editor-in-Chief: *Marcia Horton*
Production Coordinator: *Wanda España–WEE DESIGN GROUP*
Vice President Director of Production and Manufacturing: *David W. Riccardi*
Managing Editor: *Bayani Mendoza deLeon*
Cover Designer: *Design Source*
Manufacturing Buyer: *Donna Sullivan*
Editorial Assistant: *Nancy Garcia*
Composition: *PreTEX, Inc.*

Reprinted with corrections June, 1999

© 1998 by Prentice-Hall, Inc.
Upper Saddle River, NJ 07458

TRADEMARK INFORMATION

MATLAB is a registered trademark of the MathWorks, Inc.

    THE MATHWORKS, INC.
    24 Prime Park Way
    Natick, Massachusetts 01760-1500
    Phone: (508) 647-7000
    Fax: (508) 647-7001
    Email: info@mathworks.com
    http://www.mathworks.com

Microsoft Internet Explorer Logo is a trademark of Microsoft Corp.

Quick Time and the Quick Time Logo are trademarks used under license.

Printed in the United States of America

10 9 8 7 6 5

ISBN 0-13-243171-8

Prentice-Hall International (UK) Limited, *London*
Prentice-Hall of Australia Pty. Limited, *Sydney*
Prentice-Hall Canada Inc., *Toronto*
Prentice-Hall Hispanoamericana, S.A., *Mexico*
Prentice-Hall of India Private Limited, *New Delhi*
Prentice-Hall of Japan, Inc., *Tokyo*
Prentice-Hall Asia Pte. Ltd., *Singapore*
Editora Prentice-Hall do Brasil, Ltda., *Rio de Janeiro*

To Carolyn, who transformed my life.
JHMc

To Dorothy, who will always be First.
RWS

To Sarah, for her helping me to know better every day the World's
Greatest Engineer, who is truly time-invariant and not at all linear.
MAY

Contents

► **Appendix C   Laboratory Projects**   415

Preface

Signal processing is concerned with processing signals. That obvious statement is often overlooked in digital signal processing (DSP) textbooks, where the concentration is on the mathematical techniques and methods needed to develop DSP algorithms. Such emphasis is probably correct in graduate courses, but this text and the accompanying CD-ROM were developed to introduce undergraduate students to DSP and real signals. The CD-ROM contains many actual signals so that students can implement and see the results of actual processing. In our effort to define an introductory course based on DSP, we have concentrated on examples such as sound and image signals.

We have used the term *multimedia* to describe this approach, because sound and images are integral to multimedia. More important, however, is the fact that sound and images are manipulated using the methods of DSP. So if one agrees that these signals are known as multimedia, then DSP is concerned with the infrastructure of multimedia because it provides the tools for developing and implementing that infrastructure. Perhaps the foremost example is signal compression. The movement of large audio and image files over networks demands data compression. The "signal" nature of audio and images can be exploited through filtering and transforms to obtain very compact representations such as the JPEG standard. Although these signal compression algorithms are often based on sophisticated mathematical descriptions and demand considerable expertise to design, we believe that it is possible in a first course to teach the fundamentals of DSP that underlie multimedia systems. Furthermore, we believe that working with the actual signals provides a high level of motivation to learn more about the theory and potential applications of DSP.

Many DSP researchers and engineers have designed lowpass filters, but how many have "seen" or "heard" the effects of a lowpass filter? By that we mean, how many have processed sound through a lowpass filter and listened to the result? Nowadays, it is not a difficult task to implement a lowpass filter and process audio, but it is rarely done in college courses. Why? One likely reason is that the facilities and materials to do so have not been readily available until now. The *DSP First* CD-ROM addresses the availability issue because it contains many sound files, both raw and processed, that can be viewed with a Web browser or imported into MATLAB. When used in conjunction with MATLAB, these sounds can be processed in many different ways. New filters can be designed and then applied, so that the concept of frequency response takes on a new meaning when the human auditory sense is involved. Likewise for visual stimulation where the effects of lowpass and highpass filtering of photographs can be seen. Questions such as how highpass filtered speech sounds or what a lowpass filter does to an image can be answered.

We provide a CD-ROM bundled with this textbook, and hope that the readers and their teachers value the CD-ROM material as more important than the written text. If so, then a first step in a new direction will have been taken to enhance learning. The CD-ROM contains four types of materials: demos, labs, exercises, and homework problems with solutions. Of these, the first two are the key items. Each chapter of the text has associated demos. These were developed for classroom use, but have now been preserved so they can be "browsed" outside of class. In this form, the demos have two uses: First, they define some key presentation that should be made during lecture time; second, they provide a resource for self-study. Over time, the database of good demos will grow, and we hope that the present CD-ROM is just a beginning for this sort of resource. At Georgia Tech, we will host a web site containing the material in its evolving form.[1]

We have also written a text. It is a conventional book not unlike any other on the subject of DSP, although, as our title *DSP First* suggests, the distinguishing feature of the text (and the CD-ROM) is that it presents DSP at a level consistent with an introductory ECE course, i.e., the sophomore level in a typical U.S. university. The list of topics in the book is standard, but since we must combine DSP with some introductory ideas, the progression of topics may strike some teachers as unconventional. Part of the reason for this is that DSP has typically been treated as a senior or first-year graduate level course in electrical engineering, for which a traditional background of circuits and linear systems is assumed. We believe strongly that there are compelling reasons for turning this order around, since the early study of DSP affords a perfect opportunity to show electrical and computer engineering students that mathematics and digital computation can be the key to understanding familiar engineering applications. Furthermore, this approach makes the subject much more accessible to students in other majors such as computer science and other engineering fields. This point is increasingly important because non-specialists are beginning to use DSP techniques routinely in many areas of science and technology.

When you think of the basic ideas needed to understand signals and systems, and also DSP, a few key ones must be included: Frequency response is one; sampling is another, filtering is a third. Beyond that, most other topics in our text are needed because they support those key concepts. For example, to explain frequency response, one must know about sinusoids because the frequency response is really concerned with the sinusoidal response. Then it is convenient to represent sinusoids in a complex number representation, so we treat phasors. In a conventional (circuits-based) curriculum, these basic notions are covered under the umbrella of AC circuits, so in effect we have developed an alternative to that path. This idea is not ours, but

---

[1] http://www.ece.gatech.edu/~dspfirst

one that we have adapted from Prof. Ken Steiglitz (Princeton University), whose books on introductory DSP have blazed this trail. Indeed much of our work has been inspired by two of Steiglitz's books.[2,3]

This project owes a considerable debt to the many students who have spent countless hours developing MATLAB code and HTML code, as well as running the laboratories at Georgia Tech and coaching the sophomores who have been exposed to this new format for signal processing education. In many cases, they have developed complex demonstrations and have found great challenge and reward in outperforming their professors. In other cases, they have brought skills and insights to the project that would have eluded us. We would like to recognize each of the following major contributors: Dr. Jeff Schodorf, who served as a TA for the EE–2200 course at Georgia Tech for over two years and who did most of the reconstruction demos in Chapter 4, as well as overseeing the archiving of much of the lab material for the CD-ROM. David Anderson, who served as the primary TA for 2200 and has developed new labs and refined old ones for the CD-ROM, as well as providing much needed quality testing of the final CD-ROM. Craig Ulmer, who developed PeZ, a full-featured pole-zero editor, to demonstrate the interrelationships among the time, frequency, and transform domains; and Brad North, who made improvements in PeZ. Emily Loadholt, who provided the music recordings for the labs and the music spectrogram viewer. Amer Abufadel, who created the FIR image-filtering demos, and Joseph Stanley, who provided a real boost by creating the artwork that transformed the CD-ROM from its bland beginnings. Scott McClellan, who scanned most of the homework problems and solutions late at night. Robbie Griffin, who was inspired by the music synthesis labs, and immediately after taking the course became a developer of new labs for these topics.

We want to acknowledge the contributions of our editor, Tom Robbins at Prentice-Hall. Very early on, he bought into the concept of *DSP First* and supported and encouraged us at every step in the project. He also arranged for reviews of the book and the CD-ROM by some very thoughtful and careful reviewers including Filson Glanz, S. Hossein Mousavinezhad, Geoffrey Orsak, Rogelio Palomera-Garcia, Stan Reeves, and Mitch Wilkes. An extraordinary thorough review by Robert Strum had a significant impact on the final result.

We also want to thank our colleagues and key members of our respective Institutes for their support and for providing resources such as: MAY's sabbatical from Rose-Hulman, which led to this team of authors; the John and Mary Franklin Foundation, which has provided continuous support of RWS; and support from the Georgia Tech Foundation and EduTech that financed additional course development by MAY and the graduate students. JHMc and RWS would like to acknowledge the freedom and opportunities afforded them by other faculty in the Georgia Tech School of ECE to initiate this new course as a part of on-going curriculum revisions. Likewise, MAY

---

[2] K. Steiglitz, *An Introduction to Discrete Systems,* John Wiley & Sons, 1972.

[3] K. Steiglitz, *A Digital Signal Processing Primer: with Applications to Computer Music,* Addison-Wesley Publishing Company, 1996.

would like to thank his colleagues at Rose-Hulman for their support and encouragement while developing this new approach and for designing a curriculum that uses this new foundation.

Finally, we want to recognize the understanding and support of our wives (Carolyn McClellan, Dorothy Schafer, and Sarah Yoder) who watched in amazement as this project consumed most of our energy and time during the final few fanatical fortnights. Also, MAY would like to thank his eight children for understanding "Why Daddy has to go to work when it is Saturday." In closing, we make the observation that this project is not really finished. At the outset, we decided to create material for this course on a rapid time schedule and to deliver that material to others as soon as possible. As a result, many ideas are left unfulfilled. But that is the appeal of the Web-based approach. It can easily grow to incorporate the innovative visualizations and experiments that others will provide.

1

Introduction

This is a book about signals and systems. In this age of multimedia computers, audio and video entertainment systems, and digital communication systems, it is almost certain that you, the reader of this text, have formed some impression of the meaning of the terms *signal* and *system*, and you probably use the terms often in daily conversation.

It is likely that your usage and understanding of the terms are correct within some rather broad definitions. For example, you may think of a signal as "something" that carries information. Usually, that something is a pattern of variations of a physical quantity that can be manipulated, stored, or transmitted by physical processes. Examples include speech signals, audio signals, video or image signals, radar signals, and seismic signals, to name just a few. An important point is that signals can take many equivalent forms or representations. For example, a speech signal is produced as an acoustic signal, but it can be converted to an electrical signal by a microphone, or a pattern of magnetization on a magnetic tape, or even a string of numbers as in digital audio recording.

The term *system* may be somewhat more ambiguous and subject to interpretation. For example, we often use "system" to refer to a large organization that administers or implements some process, such as the "Social Security system" or the "airline transportation system." However, we will be interested in a much narrower definition that is very closely linked to signals. More specifically, a system, for our purposes, is something that can manipulate, change, record, or transmit signals. For example, an audio compact disk (CD) recording stores or represents a music signal as sequence of numbers. A CD player is a system for converting the numbers stored on the disk (i.e., the numerical representation of the signal) to an acoustic signal that we can hear. In general, systems *operate* on signals to produce new signals or new signal representations.

Our goal in this text is to develop a framework wherein it is possible to make precise statements about both signals and systems. Specifically, we want to show that mathematics is an appropriate language for describing and understanding signals and systems. We want to show that the representation of signals and systems by mathematical equations allows us to understand how signals and systems interact and how we can design and implement systems that achieve a prescribed purpose.

1.1 MATHEMATICAL REPRESENTATION OF SIGNALS

Signals are patterns of variations that represent or encode information. They have a central role in measurement, in probing other physical systems, in medical technology, and in telecommunication, to name just a few areas.

Many signals are naturally thought of as a pattern of variations in time. A good example is a speech signal, which initially arises as a pattern of changing air pressure in the vocal tract. This pattern, of course, evolves with time, creating what we often call a "time waveform." Figure 1.1 shows a plot of a speech waveform.

Waveform Plot of a Part of a Speech Signal $s(t)$

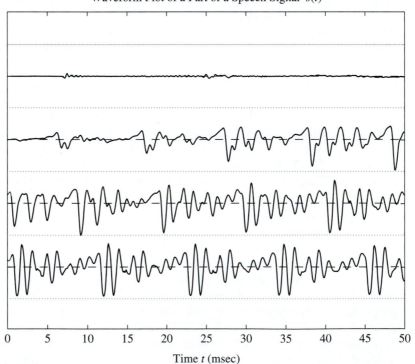

Time t (msec)

Figure 1.1   Example of a signal that can be represented as a function of a single (time) variable.

In this plot, the vertical axis represents air pressure and the horizontal axis represents time. Notice that there are four plots in the figure corresponding to four contiguous time segments of the speech waveform. The second plot is a continuation of the first, and so on, with each graph corresponding to a time interval of 50 milliseconds (msec).

The speech signal in Fig. 1.1 is an example of a one-dimensional continuous-time signal. Such signals can be represented mathematically as a function of a single independent variable, which is normally denoted t. Although in this particular case we cannot write a simple equation that describes the graph of Fig. 1.1 in terms of familiar mathematical functions, we can nevertheless associate a function $s(t)$ with the graph. Indeed, the graph itself can be taken as a definition of the function that assigns a number $s(t)$ to each instant of time (each value of t).

Many, if not most, signals originate as continuous-time signals. However, for a variety of reasons that will become increasingly obvious as we progress through this text, it is often desirable to obtain a discrete-time representation of a signal. This can be done by *sampling* a continuous-time signal at isolated, equally spaced points in time. The result is a sequence of numbers that can be represented as a function of an index variable that takes on only discrete values. This can be represented mathematically as $s[n] = s(nT_s)$, where n is an integer; i.e., $\{\ldots, -2, -1, 0, 1, 2, \ldots\}$, and T_s is the *sampling period*.[1] This is, of course, exactly what we do when we plot values of a function on graph paper or on a computer screen. We cannot evaluate the function at every possible value of a continuous variable, only at a set of discrete points. Intuitively, we know that the closer the spacing of the points, the more the sequence retains the shape of the original continuous-variable function. Figure 1.2 shows an example of a short segment of a discrete-time signal that was derived by

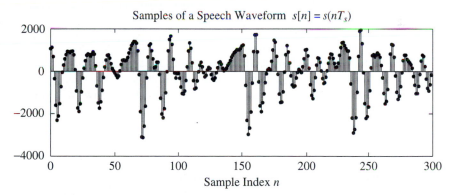

Figure 1.2   Example of a discrete-time signal that can be represented by a one-dimensional sequence or function of a discrete variable.

---

[1] Note that our convention will be to use parentheses ( ) to enclose the independent variable of a continuous-variable function, and square brackets [ ] to enclose the independent variable of a discrete-variable function (sequence).

sampling the speech waveform of Fig. 1.1 with a sampling period of $T_s = 1/8$ msec. In this case, a vertical line with a dot at the end shows the location and the size of each of the isolated sequence values.

While many signals can be thought of as evolving patterns in time, many other signals are not time-varying patterns. For example, an image formed by focusing light through a lens is a spatial pattern, and thus is appropriately represented mathematically as a function of two spatial variables. Such a signal would be considered, in general, as a function of two independent variables; i.e., a picture might be denoted $p(x, y)$. A photograph is another example, such as the "gray-scale image" shown in Fig. 1.3. In this case, the value $p(x_0, y_0)$ represents the shade of gray at position (x_0, y_0) in the image.

Figure 1.3   Example of a signal that can be represented by a function of two spatial variables.

Images such as that in Fig. 1.3 are generally considered to be two-dimensional continuous-variable signals, since we normally consider space to be a continuum. However, sampling can likewise be used to obtain a discrete-variable two-dimensional signal from a continuous-variable two-dimensional signal. Such a two-dimensional discrete-variable signal would be represented by a two-dimensional sequence or an array of numbers, and would be denoted $p[m, n] = p(m\Delta_x, n\Delta_y)$, where both m and n would take on only integer values, and Δ_x and Δ_y are the horizontal and vertical sampling periods, respectively.

Two-dimensional functions are appropriate mathematical representations of still images that do not change with time; on the other hand, videos are time-varying images that would require a third independent variable for time; i.e., $v(x, y, t)$. Video signals are intrinsically three-dimensional, and, depending on the type of video system, either two or all three variables may be discrete.

Our purpose in this section has been simply to introduce the idea that signals can be represented by mathematical functions. Although we will soon see that many familiar functions are quite valuable in the study of signals and systems, we have not even attempted to demonstrate that fact. Our sole concern is to make the connection between functions and signals, and, at this point, functions simply serve as abstract symbols for signals. Thus, for example, now we can refer to "the speech signal $s(t)$" or "the sampled image $p[m, n]$." Although this may not seem highly significant, we will see in the next section that it is indeed a very important step toward our goal of using mathematics to describe signals and systems in a systematic way.

1.2   MATHEMATICAL REPRESENTATION OF SYSTEMS

As we have already suggested, a system is something that transforms signals into new signals or different signal representations. This is a rather abstract definition, but it is useful as a starting point. To be more specific, we say that a one-dimensional continuous-time system takes an input signal $x(t)$ and produces a corresponding output signal $y(t)$. This can be represented mathematically by

$$y(t) = \mathcal{T}\{x(t)\} \tag{1.2.1}$$

which means that the input signal (waveform, image, etc.) is operated on by the system (symbolized by the operator $\mathcal{T}$) to produce the output $y(t)$. While this sounds very abstract at first, a simple example will show that this need not be mysterious. Consider a system such that the output signal is the square of the input signal. The mathematical description of this system is simply

$$y(t) = [x(t)]^2 \tag{1.2.2}$$

which says that at each time instant the value of the output is equal to the square of the input signal value at that same time. Such a system would logically be termed a "squarer system." Figure 1.4 shows the output of the squarer for the input of Fig. 1.1. As would be expected from the properties of the squaring operation, we see that the output signal is always nonnegative and the large signal values are emphasized relative to the small signal values.

The squarer system defined by (1.2.2) is an example of a *continuous-time system*, i.e., a system whose input and output are continuous-time signals. Can we build a physical system that acts like the squarer system? The answer is that the system of (1.2.2) can be approximated through appropriate connections of electronic circuits. On the other hand, if the input and output of the system are both discrete-time signals (sequences of numbers) related by

$$y[n] = (x[n])^2 \tag{1.2.3}$$

then the system would be a *discrete-time system*. The implementation of the discrete-time squarer system would be trivial; one simply multiplies each discrete signal value by itself.

Waveform Plot of the Output of a Squarer System $y(t) = [x(t)]^2$

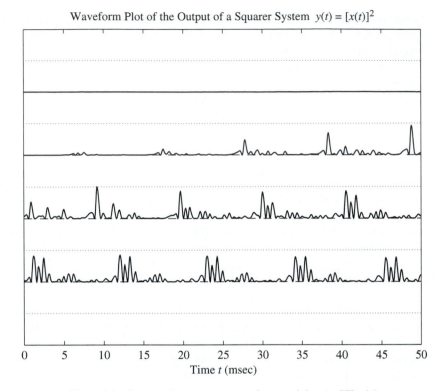

Time t (msec)

Figure 1.4   Output of a squarer system for speech input of Fig. 1.1

In thinking and writing about systems, it is often useful to have a visual representation of the system. For this purpose, engineers use "block diagrams" to represent operations performed in an implementation of a system and to show the interrelations among the many signals that may exist in an implementation of a complex system. An example of the general form of a block diagram is shown in Fig. 1.5. What this diagram shows is simply that the signal $y(t)$ is obtained from the signal $x(t)$ by the operation $\mathcal{T}\{\ \}$.

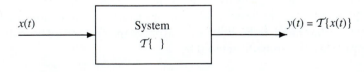

Figure 1.5   Block diagram representation of a continuous-time system.

Another example of a system was suggested earlier when we discussed the sampling relationship between continuous-time signals and discrete-time signals. Therefore, we would define a "sampler" as a system whose input is a continuous-time signal $x(t)$ and whose output is the corresponding sequence of samples, defined by the equation

$$x[n] = x(nT_s) \tag{1.2.4}$$

which simply states that the sampler "takes an instantaneous snapshot" of the continuous-time input signal once every T_s seconds. Thus, the operation of sampling fits our definition of a system, and it can be represented by the block diagram in Fig. 1.6. Often we will refer to the sampler system as an "ideal continuous-to-discrete converter" or "ideal C-to-D converter." In this case, as in the case of the squarer, the name that we give to the system is really just a description of what the system does.

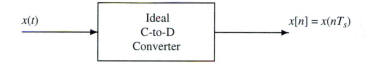

Figure 1.6   Block diagram representation of a sampler.

1.3   THINKING ABOUT SYSTEMS

Block diagrams are useful for representing complex systems in terms of simpler systems, which are more easily understood. For example, Fig. 1.7 shows a block diagram representation of the process of recording and playback of an audio CD. This block diagram breaks the operation down into four subsystems. The first operation is A-to-D (analog-to-digital) conversion, which is a physical approximation to the ideal C-to-D converter defined in (1.2.4). An A-to-D converter produces finite-precision numbers as samples of the input signal (quantized to a limited number of bits), while the ideal C-to-D converter produces samples with infinite precision. For the high-accuracy A-to-D converters used in precision audio systems, the difference between an A-to-D converter and our idealized C-to-D converter is slight, but the distinction is very important. Only finite-precision quantized sample values can be stored in digital memory of finite size!

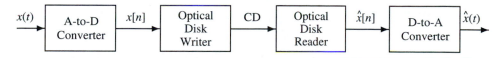

Figure 1.7   Simplified block diagram for recording and playback of an audio CD.

Figure 1.7 shows that the output of the A-to-D converter is the input to a system that writes the numbers $\hat{x}[n]$ onto the optical disc. This is a complex process, but for our purposes it is sufficient to simply show it as a single operation. Likewise, the complex mechanical/optical system for reading the numbers off the optical disk is shown as a single operation. Finally, the conversion of the signal from discrete-time form to continuous-time (acoustic) form is shown as a system called a D-to-A (digital-to-analog) converter. This system takes finite precision binary numbers in sequence and fills in a continuous-time function between the samples. The resulting continuous-time signal could then be fed to other systems, such as amplifiers, loudspeakers, and headphones, for conversion to sound.

Systems like the CD audio system are all around us. Most of the time we do not need to think about how such systems work, but this example illustrates the value of thinking about a complex system in a hierarchical form. In this way, we can first understand the individual parts, then the relationship among the parts, and finally the whole system. By looking at the CD audio system in this manner, we see that a very important issue is the conversion from continuous-time to discrete-time and back to continuous-time, and we see that it is possible to consider these operations separately from the other parts of the system. The effect of connecting the parts is then relatively easy to understand. Details of some parts can be left to experts in other fields who, for example, can develop more detailed breakdowns of the optical disk subsystems.

1.4   THE NEXT STEP

The CD audio system is a good example of a discrete-time system. Buried inside the blocks of Fig. 1.7 are many discrete-time subsystems and signals. While we do not promise to explain all the details of CD players or any other complex system, we do hope to establish the foundations for the understanding of discrete-time signals and systems so that this knowledge can be applied to understanding components of more complicated systems. In Chapter 2, we will start at a basic mathematical level and show how the well-known sine and cosine functions from trigonometry play a fundamental role in signal and system theory. Next, we will show how complex numbers can simplify the algebra of trigonometric functions. Subsequent chapters will introduce the concept of the frequency spectrum of a signal and the concept of filtering with a linear time-invariant system. By the end of the book, the diligent reader who has worked the problems, experienced the demonstrations, and done the laboratory exercises will be rewarded with a solid understanding of many of the key concepts underlying the digital multimedia information systems that are rapidly becoming commonplace.

2

Sinusoids

We begin our discussion by introducing a general class of signals that are commonly called *cosine signals* or, equivalently, *sine signals*.[1] Collectively, such signals are called *sinusoidal signals* or, more concisely, *sinusoids*. Although sinusoidal signals have simple mathematical representations, they are the most basic signals in the theory of signals and systems, and it is essential to become familiar with their properties. The most general mathematical formula for a cosine signal is

$$x(t) = A\cos(\omega_0 t + \phi) \qquad (2.0.1)$$

where $\cos(\cdot)$ denotes the cosine function that is familiar from the study of trigonometry. When defining a continuous-time signal, we typically use a function whose independent variable is t, a continuous variable that represents time. From (2.0.1) it follows that $x(t)$ is a mathematical function in which the angle of the cosine function is, in turn, a function of the variable t. The parameters A, ω_0, and ϕ are fixed numbers for a particular cosine signal. Specifically, A is called the *amplitude*, ω_0 the *radian frequency*, and ϕ the *phase-shift* of the cosine signal.

Figure 2.1 shows a plot of the continuous-time cosine signal

$$x(t) = 10\cos\left(2\pi(440)t - 0.4\pi\right)$$

i.e., $A = 10$, $\omega_0 = 2\pi(440)$, and $\phi = -0.4\pi$ in (2.0.1). Note that $x(t)$ oscillates between A and $-A$, and that it repeats the same pattern of oscillations every $1/440 = 0.00227$ sec (approximately). This time interval is called the *period* of the sinusoid. We will show later in this chapter that most features of the sinusoidal waveform are directly dependent on the choice of the parameters A, ω_0, and ϕ.

---

[1] It is also common to refer to cosine or sine signals as cosine or sine *waves*, particularly when referring to acoustic or electrical signals.

Sinusoidal Signal $x(t)$

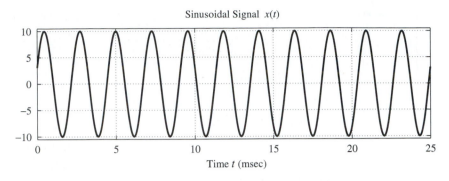

Time t (msec)

Figure 2.1   Sinusoidal signal $x(t) = 10\cos(2\pi(440)t - 0.4\pi)$.

2.1   AN EXPERIMENT WITH A TUNING FORK

One of the reasons that cosine waves are so important is that many physical systems generate signals that can be modeled (i.e., represented mathematically) as sine or cosine functions versus time. Among the most prominent of these are signals that are audible to humans. The tones or notes produced by musical instruments are perceived as different pitches. Although it is an oversimplification to equate notes to sinusoids and pitch to frequency, the mathematics of sinusoids is an essential first step to understanding complicated sound signals.

To provide some motivation for our study of sinusoids, we will begin by considering a very simple and familiar system for generating a sinusoidal signal. This system is a *tuning fork*, an example of which is shown in Fig. 2.2. When struck sharply, the tines of the tuning fork vibrate and emit a "pure" tone. This tone is at a single frequency, which is usually stamped on the tuning fork. It is common to find "A–440" tuning forks, because 440 hertz (Hz) is the frequency of A above middle C on a musical scale, and is often used as the reference note for tuning a piano and other musical instruments. If you can obtain a tuning fork, perform the following experiment:

Strike the tuning fork against your knee, and then hold it close to your ear. You should hear a distinct "hum" at the frequency designated for the tuning fork. The sound will persist for a rather long time if you have struck the tuning fork properly; however, it is easy to do this experiment incorrectly. If you hit the tuning fork sharply on a hard surface such as a table, you will hear a high pitched metallic "ting" sound. This is *not* the characteristic sound that you are seeking. If you hold the tuning fork close to your ear, you will hear two tones: The higher-frequency "ting" will die away rapidly, and then the lower-frequency "hum" will be heard.

TUNING
FORK DEMO   With a microphone and a computer equipped with an A-to-D converter, we can make a digital recording of the signal produced by the tuning fork. The microphone

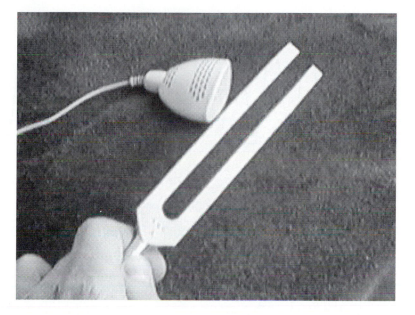

Figure 2.2   Picture of a tuning fork and a microphone.

converts the sound into an electrical signal, which in turn is converted to a sequence of numbers stored in the computer. Then these numbers can be plotted on the computer screen. A typical plot is shown in Fig. 2.3 for an A–440 tuning fork. In this case, the A-to-D converter sampled the output of the microphone at a rate of 5563.6 samples/sec.[2] The plot was constructed by connecting the sample values by straight lines. It appears that the signal generated by the tuning fork is very much like the cosine signal of Fig. 2.1. It oscillates between symmetric limits of amplitude and it also repeats periodically with a period of about 2.27 msec (0.00227 sec). As we will see in Section 2.3.1, this period is proportional to the reciprocal of ω_0; i.e., $2\pi / (2\pi (440)) \approx 0.00227$.

This experiment shows that common physical systems produce signals whose graphical representations look very much like cosine signals; i.e., they look very much like the graphical plots of the mathematical functions defined in (2.0.1). Later, in Section 2.7, we will add further credence to the sinusoidal model for the tuning fork sound by showing that cosine functions arise as solutions to the differential equation that (through the laws of physics) describes the motion of the tuning fork's tines. Before looking at the physics of the tuning fork, however, we should become more familiar with sinusoids and sinusoidal signals.

---

[2] This rate is one-quarter of the A-to-D conversion rate on a Macintosh computer.

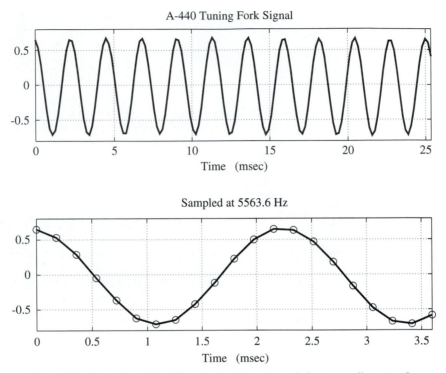

Figure 2.3   Recording of A–440 tuning fork signal sampled at a sampling rate of 5563.6 samples/sec. The bottom plot is approximately the first 3.5 msec taken from the top plot.

2.2   REVIEW OF SINE AND COSINE FUNCTIONS

Sinusoidal signals are defined in terms of the familiar sine and cosine functions of trigonometry. A brief review of the properties of these basic trigonometric functions is useful, since these properties determine the properties of sinusoidal signals.

The sine and cosine functions are often introduced and defined through a diagram like Fig. 2.4. The trigonometric functions sine and cosine take an angle as their argument. We often think of angles in degrees, but where sine and cosine functions are concerned, angles must be dimensionless. Angles are therefore specified in radians. If the angle θ is in the first quadrant ($0 \leq \theta < \pi/2$ rad), then the sine of θ is the length y of the side of the triangle opposite the angle θ divided by the length r of the hypotenuse of the right triangle. Similarly, the cosine of θ is the ratio of the length of the adjacent side x to the length of the hypotenuse.

Note that as θ increases from 0 to $\pi/2$, $\cos \theta$ decreases from 1 to 0 and $\sin \theta$ increases from 0 to 1. When the angle is greater than $\pi/2$ radians, the algebraic signs of x and y come into play, x being negative in the second and third quadrants and y being negative in the third and fourth quadrants. This is most easily shown

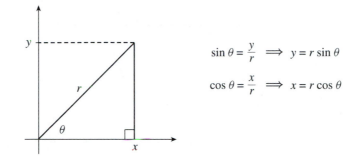

$$\sin\theta = \frac{y}{r} \implies y = r\sin\theta$$

$$\cos\theta = \frac{x}{r} \implies x = r\cos\theta$$

Figure 2.4   Definition of sine and cosine of an angle θ within a right triangle.

by plotting the values of $\sin\theta$ and $\cos\theta$ as a function of θ, as in Fig. 2.5.[3] Several features of these plots are worthy of comment. The two functions have exactly the same shape. Indeed, the sine function is just a cosine function that is shifted to the right by $\pi/2$; i.e., $\sin\theta = \cos(\theta - \pi/2)$. Both functions oscillate between $+1$ and -1, and they repeat the same pattern periodically with period 2π. Furthermore, the sine function is an odd function of its argument, and the cosine is an even function. A summary of these and other properties is presented in Table 2.1.

Property	Equation
Equivalence	$\sin\theta = \cos(\theta - \pi/2)$ or $\cos(\theta) = \sin(\theta + \pi/2)$
Periodicity	$\cos(\theta + 2\pi k) = \cos\theta$, when k is an integer
Evenness of cosine	$\cos(-\theta) = \cos\theta$
Oddness of sine	$\sin(-\theta) = -\sin\theta$
Zeros of sine	$\sin(\pi k) = 0$, when k is an integer
Ones of cosine	$\cos(2\pi k) = 1$ when k is an integer
Minus ones of cosine	$\cos\left[2\pi\left(k + \frac{1}{2}\right)\right] = -1$, when k is an integer

Table 2.1: Basic properties of the sine and cosine functions.

Clearly, the sine and cosine functions are very closely related. This often leads to opportunities for simplification of expressions involving both sine and cosine functions. In calculus, we have the interesting property that the sine and cosine functions are derivatives of each other:

$$\frac{d\sin\theta}{d\theta} = \cos\theta \quad \text{and} \quad \frac{d\cos\theta}{d\theta} = -\sin\theta$$

---

[3] It is a good idea to memorize the form of these plots, and to be able to sketch them accurately.

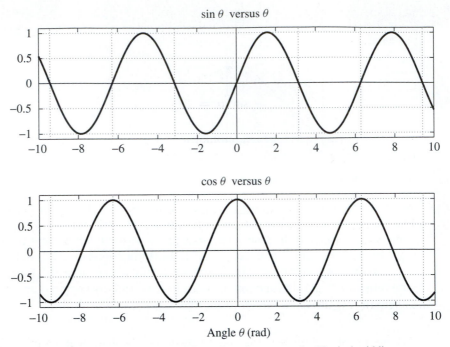

Figure 2.5 Sine and cosine functions plotted versus angle. Vertical grid lines (dotted) are drawn at multiples of π.

That is, the cosine function gives the slope of the sine function, and the sine function is the negative of the slope of the cosine function. In trigonometry, there are many identities that can be used in simplifying expressions involving combinations of sinusoidal signals. Table 2.2 gives a brief table of *trigonometric identities* that will be useful. Recall from your study of trigonometry that these identities are not independent; e.g., identity 3 can be obtained from identity 4 by substituting $\alpha = \beta = \theta$.

Identity Number	Equation
1	$\sin^2 \theta + \cos^2 \theta = 1$
2	$\cos 2\theta = \cos^2 \theta - \sin^2 \theta$
3	$\sin 2\theta = 2 \sin \theta \cos \theta$
4	$\sin(\alpha \pm \beta) = \sin \alpha \cos \beta \pm \cos \alpha \sin \beta$
5	$\cos(\alpha \pm \beta) = \cos \alpha \cos \beta \mp \sin \alpha \sin \beta$

Table 2.2: Some basic trigonometric identities.

Also, these identities can be combined to derive other identities. For example, combining identity 1 with identity 2 leads to the identities

$$\cos^2 \theta = \tfrac{1}{2}(1 + \cos 2\theta)$$

$$\sin^2 \theta = \tfrac{1}{2}(1 - \cos 2\theta)$$

A more extensive table of trigonometric identities may be found in any book on trigonometry or in a book of mathematical tables.

Exercise 2.1. Use trigonometric identity 5 to derive an expression for $\cos 8\theta$ in terms of $\cos 9\theta$, $\cos 7\theta$, and $\cos \theta$. (Solutions to all exercises can be found on the CD-ROM.)

Solutions on
CD

2.3   SINUSOIDAL SIGNALS

The most general mathematical formula for a sinusoidal time signal is obtained by making the argument (i.e., the angle) of the cosine function be a function of t. The following equation gives two equivalent forms:

$$x(t) = A \cos(\omega_0 t + \phi) = A \cos(2\pi f_0 t + \phi) \tag{2.3.1}$$

The two forms are related by defining $\omega_0 = 2\pi f_0$. In either form, given in (2.3.1), there are three independent parameters. The names and interpretations of these parameters are as follows:

1. A is called the *amplitude*. The amplitude is a scaling factor that determines how large the cosine signal will be. Since the function $\cos \theta$ oscillates between $+1$ and -1, the signal $x(t)$ in (2.3.1) oscillates between $+A$ and $-A$.

2. ϕ is called the *phase shift*. The units of phase shift must be radians, since the argument of the cosine must be in radians. We will generally prefer to use the cosine function when defining the phase shift. If we happen to have a formula containing sine, e.g., $x(t) = A \sin(\omega_0 t + \phi')$, then we can rewrite it in terms of cosine if we use the equivalence property in Table 2.1. The result is:

$$x(t) = A \sin(\omega_0 t + \phi') = A \cos(\omega_0 t + \phi' - \pi/2)$$

so we define the phase shift to be $\phi = \phi' - \pi/2$ in (2.3.1). For simplicity and to prevent confusion, we should avoid using the sine function.

3. ω_0 is called the *radian frequency*. Since the argument of the cosine function must be in radians, which is dimensionless, the quantity $\omega_0 t$ must likewise be dimensionless. Thus, ω_0 must have units of rad/sec if t has units of sec. Similarly, $f_0 = \omega_0/2\pi$ is called the *cyclic frequency*, and f_0 must have units of sec^{-1}.

As an example, Fig. 2.6 shows a plot of the signal

PLOTTING
SINUSOIDS

$$x(t) = 20\cos(2\pi(40)t - 0.4\pi) \qquad (2.3.2)$$

In terms of our definitions, $A = 20$, $\omega_0 = 2\pi(40)$, $f_0 = 40$, and $\phi = -0.4\pi$ for this signal. The dependence of the signal on the amplitude parameter A is obvious: Its maximum and minimum peaks are $+20$ and -20, respectively. The maxima occur at

$$t = \ldots, -0.02, 0.005, 0.03, \ldots$$

and the minima at

$$\ldots, -0.0325, -0.0075, 0.0175, \ldots$$

The time interval between successive maxima of the signal is $1/f_0 = 0.025$ sec. To understand why the signal has these properties, we will need to do a bit of analysis.

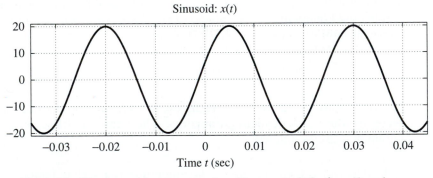

Figure 2.6   Sine wave with parameters $A = 20$, $\omega_0 = 2\pi(40)$, $f_0 = 40$, and $\phi = -0.4\pi$.

2.3.1 Relation of Frequency to Period

The sinusoid in Fig. 2.6 is clearly a periodic signal. The *period* of the sinusoid, denoted by T_0, is the length of one cycle of the sinusoid. In general, the frequency of the sinusoid determines its period, and the relationship can be found by examining the following equations:

$$x(t + T_0) = x(t)$$

$$A\cos(\omega_0(t + T_0) + \phi) = A\cos(\omega_0 t + \phi)$$

$$\cos(\omega_0 t + \omega_0 T_0 + \phi) = \cos(\omega_0 t + \phi)$$

Since the cosine function has a period of 2π, the equality above holds for all values of t if

$$\omega_0 T_0 = 2\pi \qquad \Longrightarrow \qquad T_0 = \frac{2\pi}{\omega_0}$$

$$(2\pi f_0)T_0 = 2\pi \qquad \Longrightarrow \qquad T_0 = \frac{1}{f_0} \qquad\qquad (2.3.3)$$

SINE DRILL

Since T_0 is the period of the signal, $f_0 = 1/T_0$ is the number of periods (cycles) per second. Therefore, cycles per second is an appropriate unit for f_0, and it was in general use until the 1960s.[4] When dealing with ω_0, the unit of radian frequency is rad/sec. The units of f_0 are more convenient when describing the sinusoid, because cycles per second naturally define the period.

It is exceedingly important to understand the effect of the frequency parameter. Figure 2.7 shows this effect for several choices of f_0 in the signal

$$x(t) = 5\cos(2\pi f_0 t)$$

Notice, first of all, that $f_0 = 0$ is a perfectly acceptable value, and when this value is used, the resulting signal is constant, since $5\cos(2\pi \cdot 0 \cdot t) = 5$ for all values of t. Thus, the constant signal, often called DC,[5] is, in fact, a sinusoid of zero frequency.

The lower two plots in Fig. 2.7 show the effect of increasing f_0. As we expect, the waveform shape is the same for both values of frequency. However, for the higher frequency, the signal varies more rapidly with time; i.e., the cycle length is a shorter time interval. We have already seen that this is true because the period of a cosine signal is the reciprocal of the frequency (2.3.3). Note that when the frequency doubles ($100 \rightarrow 200$), the period is halved. This is an illustration of the general principle that the higher the frequency, the more rapid the signal waveform changes with time. The DC case, $f_0 = 0$, is consistent with this principle. The frequency is so low that the signal does not change at all. Throughout this book we will see more examples of the inverse relationship between time and frequency.

2.3.2 Relation of Phase Shift to Time Shift

The phase shift parameter ϕ (together with the frequency) determines the time locations of the maxima and minima of a cosine wave. To be specific, notice that the sinusoid (2.3.1) with $\phi = 0$ has a positive peak at $t = 0$. When $\phi \neq 0$, the phase shift

---

[4] The unit hertz (abbreviated Hz) was adopted in 1933 by the Electrotechnical Commission in honor of Heinrich Hertz, who first demonstrated the existence of radio waves.

[5] Electrical engineers use the abbreviation DC standing for direct current, which is a constant current.

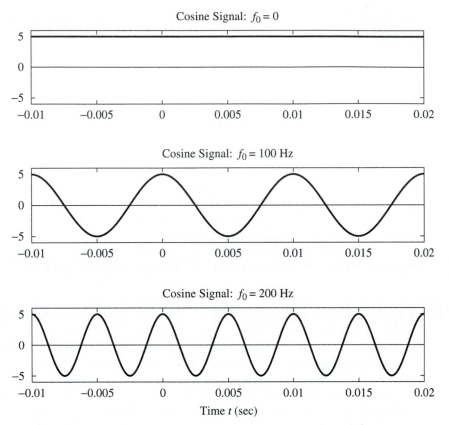

Figure 2.7   Cosine signals $x(t) = 5\cos(2\pi f_0 t)$ for several values of f_0: upper plot, $f_0 = 0$; middle plot, $f_0 = 100$ Hz; lower plot, $f_0 = 200$ Hz.

determines how much the maximum of the sinusoidal signal is shifted with respect to $t = 0$.

Before we examine this point in detail for sinusoids, it is useful to become familiar with the general concept of *time-shifting* a signal. Suppose that a signal $s(t)$ is defined by a known formula or graph. A simple example is the following triangularly shaped function:

$$s(t) = \begin{cases} t & 0 \leq t \leq 1 \\ \frac{1}{2}(3 - t) & 1 \leq t \leq 3 \\ 0 & \text{elsewhere} \end{cases} \tag{2.3.4}$$

This simple function has a slope of 1 for $0 \leq t < 1$ and a negative slope of $-\frac{1}{2}$ for $1 < t \leq 3$. Now consider the function $x_1(t) = s(t - 2)$. From the definition of $s(t)$, it is clear that $x_1(t)$ is nonzero for

$$0 \leq (t - 2) \leq 3 \quad \Longrightarrow \quad 2 \leq t \leq 5.$$

Within the time interval $[2, 5]$, the formula for the shifted signal is:

$$x_1(t) = \begin{cases} t - 2 & 2 \le t \le 3 \\ \frac{1}{2}(5 - t) & 3 \le t \le 5 \\ 0 & \text{elsewhere} \end{cases} \tag{2.3.5}$$

In other words, $x_1(t)$ is simply the $s(t)$ function with its origin shifted to the *right* by 2 seconds. Similarly, $x_2(t) = s(t+1)$ is the $s(t)$ function shifted to the *left* by 1 second; its nonzero portion is $-1 \le t \le 2$. The three signals $x(t) = s(t)$, $x_1(t) = s(t - 2)$, and $x_2(t) = s(t + 1)$ are all shown in Fig. 2.8.

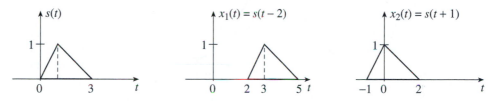

Figure 2.8   Illustration of time-shifting.

Exercise 2.2. Derive the equations for the shifted signal $x_2(t) = s(t + 1)$.

We will have many occasions to consider time-shifted signals. Whenever a signal can be expressed in the form $x_1(t) = s(t - t_1)$, we say that $x_1(t)$ is a time-shifted version of $s(t)$. If t_1 is a positive number, then the shift is to the right, and we say that the signal $s(t)$ has been *delayed* in time. When t_1 is a negative number, then the shift is to the left, and we say that the signal $s(t)$ was *advanced* in time. In summary, time shift is essentially a redefinition of the time origin of the signal. In general, any function of the form $s(t - t_1)$ has its origin moved to the location $t = t_1$.

One way to determine the time shift for a cosine signal would be to find the positive peak of the sinusoid that is closest to $t = 0$. In the plot of Fig. 2.6, the time where this positive peak occurs is $t_1 = 0.005$ sec. Since the peak in this case occurs at a positive time (to the right of $t = 0$), we say that the time shift is a delay of the zero-phase cosine signal. Let $x_0(t) = A\cos(\omega_0 t)$ denote a cosine signal with zero phase shift. A delay of $x_0(t)$ can be converted to a phase shift ϕ by making the following comparison:

$$x_0(t - t_1) = A\cos(\omega_0(t - t_1)) = A\cos(\omega_0 t + \phi)$$

$$\cos(\omega_0 t - \omega_0 t_1) = \cos(\omega_0 t + \phi)$$

Since this equation must hold for all t, we must have $-\omega_0 t_1 = \phi$, which leads to

$$t_1 = -\frac{\phi}{\omega_0} = -\frac{\phi}{2\pi f_0}$$

Notice that the phase shift is negative when the time shift is positive (a delay). In terms of the period $(T_0 = 1/f_0)$ we get the more intuitive formula

$$\phi = -2\pi f_0 t_1 = -2\pi \left(\frac{t_1}{T_0}\right) \tag{2.3.6}$$

SINE DRILL

which states that the phase shift is 2π times the fraction of a cycle given by the ratio of the time shift to the period.

Since the positive peak nearest to $t = 0$ must always lie within $|t_1| \leq T_0/2$, the phase shift can always be chosen to satisfy $-\pi < \phi \leq \pi$. However, the phase shift is also ambiguous because adding a multiple of 2π to the argument of a cosine function does not change the value of the cosine. This is a direct consequence of the fact that the cosine is periodic with period 2π. Each different multiple of 2π corresponds to picking a different peak of the periodic waveform. Thus, another way to compute the phase shift is to find any positive peak of the sinusoid and measure its corresponding time location. After the time location is converted to phase shift using (2.3.6), an integer multiple of 2π can be added to or subtracted from the phase shift to produce a final result between $-\pi$ and $+\pi$. This gives a final result identical to locating the peak that is within half a period of $t = 0$. The operation of adding or subtracting multiples of 2π is referred to as *reducing modulo* 2π, because it is similar to modulo reduction in mathematics, which amounts to dividing by 2π and taking the remainder. The value of phase shift that falls between $-\pi$ and $+\pi$ is called the *principal value* of the phase shift.

Exercise 2.3. In Fig. 2.6, it is possible to measure both a positive and a negative value of t_1 and then calculate the corresponding phase shifts. Which phase shift is within the range $-\pi < \phi \leq \pi$? Verify that the two phase shifts differ by 2π.

Exercise 2.4. Starting with the plot in Fig. 2.6, sketch a plot of $x(t - t_1)$ when $t_1 = 0.0075$. Repeat for $t_1 = -0.01$. Make sure that you shift in the correct direction. For each case, compute the phase shift of the shifted sinusoid.

Exercise 2.5. If $x(t) = 20\cos(2\pi(40)t - 0.4\pi)$ as in Fig. 2.6, find G and t_1 so that $y(t) = Gx(t - t_1) = 5\cos(2\pi(40)t)$; i.e., obtain an expression for $y(t) = 5\cos(2\pi(40)t)$ in terms of $x(t)$.

2.4   SAMPLING AND PLOTTING SINUSOIDS

All of the plots of sinusoids in this chapter were created using MATLAB. This had to be done with care, because MATLAB deals only with discrete signals represented by row or column matrices, but we are actually plotting a continuous function $x(t)$. For example, if we wish to plot a function such as

$$x(t) = 20\cos(2\pi(40)t - 0.4\pi)$$

which is shown in Fig. 2.6, we must evaluate $x(t)$ at a discrete set of times, $t_n = nT_s$, where n is an integer. If we do so, we obtain the sequence of samples

$$x(nT_s) = 20 \cos(80\pi nT_s - 0.4\pi)$$

where T_s is called the *sample spacing* or *sampling period,* and n is an integer. When plotting the function using the `plot( )` function in MATLAB, we must provide a pair of row or column vectors, one containing the time values and the other the computed function values to be plotted. For example, the MATLAB statements

```
n = -7:5;
Ts = 0.005;
tn = n*Ts;
xn = 20*cos(80*pi*tn - 0.4*pi);
plot(tn,xn)
```

would create a row vector `tn` of 13 numbers between -0.035 and 0.025 spaced by the sampling period 0.005, and a row vector `xn` of samples of $x(t)$. Then the `plot( )` function draws the corresponding points, connecting them with straight line segments. Constructing the curve between sample points in this way is called *linear interpolation.* The solid curve in the upper plot of Fig. 2.9 shows the result of linear interpolation when the sample spacing is $T_s = 0.005$. Intuitively, we realize that if the points are very close together, we will see a smooth curve. The important question is, "How small must we make the sample spacing, so that the cosine signal can be accurately reconstructed between samples by linear interpolation?" A qualitative answer to this question is provided by Fig. 2.9, which shows plots produced by three different sampling periods.

Obviously, the sample spacing of $T_s = 0.005$ is not sufficiently close to create an accurate plot when the sample points are connected by straight lines. Note that sample points are shown as dots in the upper two plots.[6] With a spacing of $T_s = 0.0025$, the plot starts to approximate a cosine, but it is still possible to see places where the points are connected by straight lines rather than the smooth cosine function. Only in the lower plot of Fig. 2.9, where the spacing is $T_s = 0.0001$, does the curve appear to our eye to be a faithful representation of the cosine function.[7] A precise answer to the question posed above would require a mathematical definition of accuracy; our subjective judgment would be too prone to variability among different observers. However, we learn from this example that as the sampling period decreases, more samples are taken across one cycle of the periodic cosine signal. When $T_s = 0.005$, there are 5 samples per cycle; when $T_s = 0.0025$ there are 10 samples per cycle; and when $T_s = 0.0001$, there are 250 samples per cycle. It seems that 10 samples per cycle is not quite enough, and 250 samples per cycle is probably more than necessary,

---

[6] This was achieved by adding a second plot using MATLAB's `hold` and `stem` functions.

[7] Here the points are so close together that we cannot show the discrete samples as individual large dots.

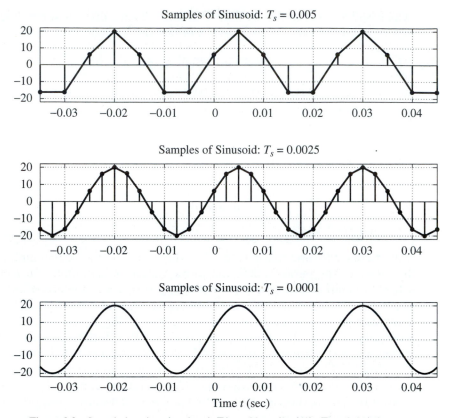

Figure 2.9   Sampled cosine signals $x(nT_s) = 20\cos(2\pi(40)nT_s - 0.4\pi)$ for several values of T_s: upper plot, $T_s = 0.005$ sec; middle plot, $T_s = 0.0025$ sec; lower plot, $T_s = 0.0001$ sec.

but, in general, the more samples per cycle, the smoother and more accurate is the linearly interpolated curve.

The choice of T_s also depends on the frequency of the cosine signal, because it is the number of samples per cycle that matters in plotting. For example, if the frequency of the cosine signal were 2000 Hz instead of 40 Hz, then a sample spacing of $T_s = 0.0001$ would yield only 5 samples per cycle. The key to accurate reconstruction is to sample frequently enough so that the cosine signal does not change very much between sample points. This will depend directly on the frequency of the cosine signal.

The problem of plotting a cosine signal from a set of discrete samples depends on the interpolation method used. With MATLAB's built-in plotting function, linear interpolation is used to connect points by straight-line segments. An insightful question would be: "If a more sophisticated interpolation method can be used, how *large* can the sample spacing be such that the cosine signal can be reconstructed accurately from the samples?" Surprisingly, the theoretical answer to this question is that the cosine signal can be reconstructed *exactly* from its samples if the sample spacing is less than half the period, i.e., the average number of samples per cycle need be only slightly more than two! Linear interpolation certainly cannot achieve this result, but, in Chapter 4, where we examine the sampling process in more detail, we will illustrate how this remarkable result can be achieved. For now, our observation that smooth and accurate sinusoidal curves can be reconstructed from samples if the sampling period is "small enough" will be adequate for our purposes.

2.5   COMPLEX EXPONENTIALS AND PHASORS

We have shown that cosine signals are useful mathematical representations for signals that arise in a practical setting, and that they are simple to define and interpret. However, it turns out that the analysis and manipulation of sinusoidal signals is often greatly simplified by dealing with related signals called *complex exponential signals*. Although the introduction of the unfamiliar and seemingly artificial concept of complex exponential signals may at first seem to be making the problem more difficult, we will soon see the value in this new representation. Before introducing the complex exponential signal, we will first review some basic concepts concerning complex numbers.[8]

2.5.1 Review of Complex Numbers

A complex number z is an ordered pair of real numbers. Complex numbers may be represented by the notation $z = (x, y)$, where $x = \Re e\{z\}$ is the *real part* and $y = \Im m\{z\}$ is the *imaginary part* of z. Electrical engineers use the symbol j for $\sqrt{-1}$ instead of i, so we can also represent a complex number as $z = x + jy$. These two representations are called the "Cartesian form" of the complex number. Complex numbers are often represented as points in a "complex plane," where the real and imaginary parts are the horizontal and vertical coordinates, respectively, as shown in Fig. 2.10(a). With the Cartesian notation and the understanding that any number multiplied by j is included in the imaginary part, the operations of complex addition, complex subtraction, complex multiplication, and complex division can be defined in terms of real operations on the real and imaginary parts. For example, the sum of two complex numbers is defined as the complex number whose real part is the sum of the real parts and whose imaginary part is the sum of the imaginary parts.

Since complex numbers can be represented as points in a plane, it follows that complex numbers are analogous to vectors in a two-dimensional space. This leads to a useful geometric interpretation of a complex number as a vector, shown in Fig. 2.10(b). Vectors have length and direction, so another way to represent a

---

[8] Appendix A provides a more detailed review of the fundamentals of complex numbers. Readers who know the basics of complex numbers and how to manipulate them can skim the Appendix and skip to Section 2.5.2.

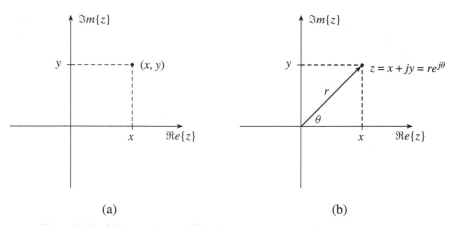

Figure 2.10   (a) Cartesian and (b) polar representations of complex numbers in the complex plane.

complex number is the "polar form" in which the complex number is represented by r, its vector length, together with θ, its angle with respect to the real axis. The length of the vector is also called the *magnitude* of z (denoted $|z|$), and the angle with the real axis is called the *argument* of z (denoted $\arg z$). This is indicated by the descriptive notation $z \longleftrightarrow r\angle\theta$, which is interpreted to mean that the vector representing z has length r and makes an angle θ with the real axis.

It is important to be able to convert between the Cartesian and polar forms of complex numbers. Figure 2.10(b) shows a complex number z and the quantities involved in both the Cartesian and polar representations. Using this figure, as well as some simple trigonometry and the Pythagorean theorem, we can derive a method for computing the Cartesian coordinates (x, y) from the polar variables $r\angle\theta$:

$$x = r \cos\theta \qquad \text{and} \qquad y = r \sin\theta \qquad (2.5.1)$$

and, likewise, for going from Cartesian to polar form:

$$r = \sqrt{x^2 + y^2} \qquad \text{and} \qquad \theta = \arctan\left(\frac{y}{x}\right) \qquad (2.5.2)$$

Many calculators and computer programs have these two sets of equations built in, making the conversion between polar and Cartesian forms simple and convenient.

The $r\angle\theta$ notation is clumsy, and does not lend itself to ordinary algebraic rules. A much better polar form is given by using Euler's famous formula for the complex exponential

$$e^{j\theta} = \cos\theta + j \sin\theta \qquad (2.5.3)$$

The Cartesian pair $(\cos\theta, \sin\theta)$ can represent any point on a circle of radius 1, so a slight generalization of (2.5.3) gives a representation valid for any complex number z

$$z = re^{j\theta} = r\cos\theta + jr\sin\theta \tag{2.5.4}$$

The complex exponential polar form of a complex number is most convenient when calculating a complex multiplication or division (see Appendix A for more details). It also serves as the basis for the complex exponential signal, which is introduced in the next section.

2.5.2 Complex Exponential Signals

The *complex exponential signal* is defined as[9]

$$\bar{x}(t) = A\,e^{j(\omega_0 t + \phi)} \tag{2.5.5}$$

Observe that the complex exponential signal is a complex-valued function of t, where the magnitude of $\bar{x}(t)$ is $|\bar{x}(t)| = A$ and the angle of $\bar{x}(t)$ is $\arg\bar{x}(t) = (\omega_0 t + \phi)$. Using Euler's formula (2.5.3), the complex exponential signal can be expressed in Cartesian form as

$$\bar{x}(t) = A\,e^{j(\omega_0 t + \phi)} = A\cos(\omega_0 t + \phi) + jA\sin(\omega_0 t + \phi) \tag{2.5.6}$$

As with the real sinusoid, A is the *amplitude*, and should be a positive real number; ϕ is the *phase shift*; and ω_0 is the *frequency* in rad/sec. In (2.5.6) it is clear that the real part of the complex exponential signal is a real cosine signal as defined in (2.3.1), and that its imaginary part is a real sine signal. Figure 2.11 shows a plot of the following complex exponential signal:

$$\bar{x}(t) = 20\,e^{j(2\pi(40)t - 0.4\pi)}$$

$$= 20\cos(2\pi(40)t - 0.4\pi) + j20\sin(2\pi(40)t - 0.4\pi)$$

$$= 20\cos(2\pi(40)t - 0.4\pi) + j20\cos(2\pi(40)t - 0.4\pi - \pi/2)$$

Plotting a complex signal as a function of time requires two graphs, one for the real part and another for the imaginary part. Observe that the real and the imaginary parts of the complex exponential signal are both real sinusoidal signals, and they differ by only a phase shift of $\pi/2$ rad.

The main reason that we are interested in the complex exponential signal is that it is an alternative *representation* for the real cosine signal. This is because we can always write

$$x(t) = \Re\left\{Ae^{j(\omega_0 t + \phi)}\right\} = A\cos(\omega_0 t + \phi) \tag{2.5.7}$$

---

[9] Note that we have placed a bar over x to call attention to the fact that $\bar{x}(t)$ is a complex function of t.

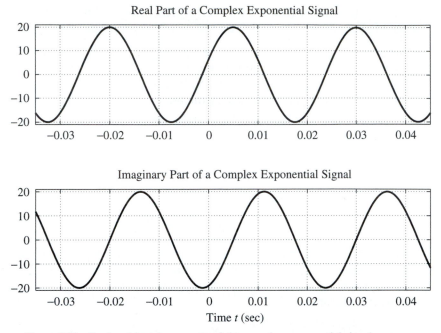

Figure 2.11   Real and imaginary parts of the complex exponential signal $\bar{x}(t) = 20e^{j(2\pi(40)t-0.4\pi)}$. The phase difference between the two waves is $90°$ or $\pi/2$ rad.

In fact, the real part of the complex exponential signal shown in Fig. 2.11 is identical to the cosine signal plotted in Fig. 2.6. Although it may seem that we have complicated things by first introducing the imaginary part to obtain the complex exponential signal and then throwing it away by taking only the real part, we will see that many calculations are simplified by using properties of the exponents. It is possible, for example, to replace all trigonometric manipulations with algebraic operations on the exponents.

Exercise 2.6.   Demonstrate that expanding the real part of $e^{j(\alpha+\beta)} = e^{j\alpha}e^{j\beta}$ will lead to identity 5 in Table 2.2. It should also be apparent that identity 4 is obtained from the imaginary part.

2.5.3 The Rotating Phasor Interpretation

When two complex numbers are multiplied, it is best to use the polar form for both numbers. To illustrate this, consider $z_3 = z_1 z_2$, where $z_1 = r_1 e^{j\theta_1}$ and $z_2 = r_2 e^{j\theta_2}$ and

$$z_3 = r_1 e^{j\theta_1} r_2 e^{j\theta_2} = r_1 r_2 e^{j\theta_1} e^{j\theta_2} = r_1 r_2 e^{j(\theta_1+\theta_2)}.$$

We have used the law of exponents to combine the two complex exponentials. From this result we conclude that to multiply two complex numbers, we multiply the magnitudes and add the angles. If we consider one of the complex numbers to be represented by a fixed vector in the complex plane, then multiplication by a second complex number scales the length of the first vector by the magnitude of the second complex number and rotates it by the angle of the second complex number. This is illustrated in Fig. 2.12 where it is assumed that $r_1 > 1$ so that $r_1 r_2 > r_2$.

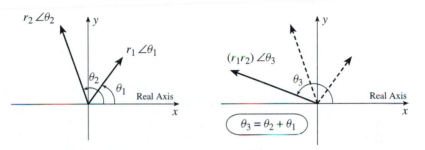

Figure 2.12   Geometric view of complex multiplication $z_3 = z_1 z_2$.

This geometric view of complex multiplication leads to a useful interpretation of the complex exponential signal as a complex vector that rotates as time increases. If we define the complex number

$$X = A e^{j\phi} \qquad (2.5.8)$$

then (2.5.5) can be expressed as

$$\bar{x}(t) = X e^{j\omega_0 t} \qquad (2.5.9)$$

i.e., $\bar{x}(t)$ is the product of the complex number X and the complex-valued function $e^{j\omega_0 t}$. The complex number X, which is aptly called the *complex amplitude*, is a polar representation created from the amplitude and the phase shift of the complex exponential signal. Taken together, the complex amplitude $X = A e^{j\phi}$ and the frequency ω_0 are sufficient to represent $\bar{x}(t)$, as well as the real cosine signal, $x(t) = A \cos(\omega_0 t + \phi)$, using (2.5.7). The complex amplitude is also called a *phasor*. Use of this terminology is common in electrical circuit theory, where complex exponential signals are used to greatly simplify the analysis and design of circuits. Since it is a complex number, X can be represented graphically as a vector in the complex plane, where the vector's magnitude ($|X| = A$) is the amplitude, and vector's angle ($\angle X = \phi$) is the phase shift. In the remainder of the text, the terms *phasor* and *complex amplitude* will be used interchangeably, because they refer to the same quantity defined in (2.5.8).

The complex exponential signal defined in (2.5.9) can also be written as

$$\bar{x}(t) = X e^{j\omega_0 t} = A e^{j\phi} e^{j\omega_0 t} = A e^{j\theta(t)}$$

where

$$\theta(t) = \omega_0 t + \phi \qquad \text{(Radians)}$$

At a given instant in time, t, the value of the complex exponential signal, $\bar{x}(t)$, is a complex number whose magnitude is A and whose argument is $\theta(t)$. Like any complex number, $\bar{x}(t)$ can be represented as a vector in the complex plane. In this case, the tip of the vector always lies on the perimeter of a circle of radius A. Now, if t increases, the complex vector $\bar{x}(t)$ will simply rotate at a constant rate, determined by the radian frequency ω_0. In other words, multiplying the phasor X by $e^{j\omega_0 t}$ as in (2.5.9) causes the fixed phasor X to rotate. (Since $|e^{j\omega_0 t}| = 1$, no scaling occurs.) Thus, another name for the complex exponential signal is *rotating phasor*.

If the frequency ω_0 is positive, the direction of rotation is counterclockwise, because $\theta(t)$ will increase with increasing time. Similarly, when ω_0 is negative, the angle $\theta(t)$ changes in the negative direction as time increases, so the complex phasor rotates clockwise. Thus, rotating phasors are said to have *positive frequency* if they rotate counterclockwise, and *negative frequency* if they rotate clockwise.

A rotating phasor makes one complete revolution every time the angle $\theta(t)$ changes by 2π radians. The time it takes to make one revolution is also equal to the period, T_0, of the complex exponential signal, so

$$\omega_0 T_0 = (2\pi f_0) T_0 = 2\pi \qquad \Longrightarrow \qquad T_0 = \frac{1}{f_0}$$

Notice that the phase shift ϕ defines where the phasor is pointing when $t = 0$. For example, if $\phi = \pi/2$, then the phasor is pointing straight up when $t = 0$, whereas if $\phi = 0$, the phasor is pointing to the right when $t = 0$.

The plots in Fig. 2.13(a) illustrate the relationship between a single complex rotating phasor and the cosine signal waveform. The upper left plot shows the complex plane with two vectors. The vector at an angle in the third quadrant represents the signal

$$\bar{x}(t) = e^{j(t - \pi/4)}$$

at the specific time $t = 1.5\pi$. The horizontal vector pointing to the left represents the real part of the vector $\bar{x}(t)$ at the particular time $t = 1.5\pi$; i.e.,

$$x(1.5\pi) = \Re e\{\bar{x}(1.5\pi)\} = \cos(1.5\pi - \pi/4) = \cos(5\pi/4)$$

ROTATING
PHASOR
DEMO

As t increases, the rotating phasor $\bar{x}(t)$ rotates in the counterclockwise direction, and its real part $x(t)$ oscillates left and right along the real axis. This is shown in the lower left plot, which shows how the real part of the rotating phasor has varied for $0 \le t \le 1.5\pi$. (An animated version of this figure is available on the CD-ROM.)

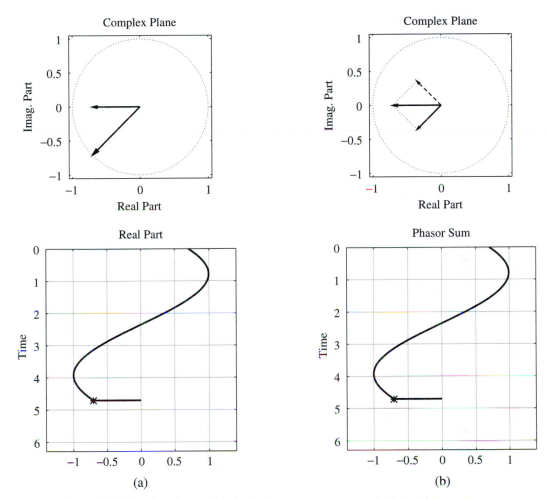

Figure 2.13   Rotating phasors: (a) single phasor rotating counter-clockwise; (b) complex conjugate rotating phasors.

2.5.4 Inverse Euler Formulas

The inverse Euler formulas allow us to write the cosine function in terms of complex exponentials:

$$\cos\theta = \frac{e^{j\theta} + e^{-j\theta}}{2} \qquad (2.5.10)$$

and also for the sine function:

$$\sin\theta = \frac{e^{j\theta} - e^{-j\theta}}{2j} \qquad (2.5.11)$$

(See Appendix A for more details.)

Equation (2.5.10) can be used to express $\cos(\omega_0 t + \phi)$ in terms of a positive and a negative frequency complex exponential as follows:

$$A \cos(\omega_0 t + \phi) = A \left(\frac{e^{j(\omega_0 t + \phi)} + e^{-j(\omega_0 t + \phi)}}{2} \right)$$

$$= \tfrac{1}{2} X e^{j\omega_0 t} + \tfrac{1}{2} X^* e^{-j\omega_0 t}$$

$$= \tfrac{1}{2} \bar{x}(t) + \tfrac{1}{2} \bar{x}^*(t)$$

$$= \Re e\{\bar{x}(t)\}$$

where $*$ denotes complex conjugation.

This formula has an interesting interpretation. The real cosine signal with frequency ω_0 is actually composed of two complex exponential signals; one with positive frequency (ω_0) and the other with negative frequency $(-\omega_0)$. The complex amplitude of the positive frequency complex exponential signal is $\tfrac{1}{2} X = \tfrac{1}{2} A e^{j\phi}$, and the complex amplitude of the negative frequency complex exponential is $\tfrac{1}{2} X^* = \tfrac{1}{2} A e^{-j\phi}$. In other words, the real cosine signal can be represented as the sum of two complex rotating phasors that are complex conjugates of each other.

Figure 2.13(b) illustrates how the sum of the two half-amplitude complex conjugate rotating phasors becomes the real cosine signal. In this case, the vector at an angle in the third quadrant is the complex rotating phasor $\tfrac{1}{2} \bar{x}(t)$ at time $t = 1.5\pi$. As t increases after that time, the angle would increase in the counterclockwise direction. Similarly, the vector in the second quadrant (plotted with a dotted line) is the complex rotating phasor $\tfrac{1}{2} \bar{x}^*(t)$ at time $t = 1.5\pi$. As t increases after that time, the angle of $\tfrac{1}{2} \bar{x}^*(t)$ will increase in the clockwise direction. The horizontal vector pointing to the right is the sum of these two complex conjugate rotating phasors. Clearly, the result is the same as the real vector in the plot on the left, and therefore the real cosine wave traced out as a function of time is the same in both cases. The lower right shows the variation of the real values of $\cos(t - \pi/4)$ for $0 \le t \le 1.5\pi$.

This representation of real sinusoidal signals in terms of their positive and negative frequency components is a remarkably useful concept. The negative frequencies, which arise due to the complex exponential representation, turn out to lead to many simplifications in the analysis of signal and systems problems. We will develop this representation further in Chapter 3, where we introduce the idea of the spectrum of a signal.

Exercise 2.7. Show that the following representation can be derived for the real sine signal:

$$A \sin(\omega_0 t + \phi) = \tfrac{1}{2} X e^{-j\pi/2} e^{j\omega_0 t} + \tfrac{1}{2} X^* e^{j\pi/2} e^{-j\omega_0 t}$$

CDROM

ROTATING
PHASOR
DEMO

where $X = Ae^{j\phi}$. In this case, the interpretation is that the sine signal is also composed of two complex exponentials with the same positive and negative frequencies, but the complex coefficients multiplying the terms are different from those of the cosine signal. Specifically, the sine signal requires additional phase shifts of $\mp\pi/2$ applied to the complex amplitude X and X^*, respectively.

2.6   PHASOR ADDITION

There are many situations in which it is necessary to add two or more sinusoidal signals. When all signals have the same frequency, the problem simplifies. This problem arises in electrical circuit analysis, and it will arise again in Chapter 5, where we introduce the concept of discrete-time filtering. Thus it is useful to develop a mechanism for adding several sinusoids having the same frequency, *but with different amplitudes and phases*. Our goal is to prove that the following statement is true:

$$\sum_{k=1}^{N} A_k \cos(\omega_0 t + \phi_k) = A \cos(\omega_0 t + \phi) \tag{2.6.1}$$

Equation (2.6.1) states that a sum of N cosine signals of differing amplitudes and phase shifts, but with the same frequency, can always be reduced to a single cosine signal of the same frequency. A proof of (2.6.1) can be accomplished by using trigonometric identities such as

$$A_k \cos(\omega_0 t + \phi_k) = A_k \cos\phi_k \cos(\omega_0 t) - A_k \sin\phi_k \sin(\omega_0 t) \tag{2.6.2}$$

We can expand the sum (2.6.1) into sines and cosines, collect terms involving $\cos(\omega_0 t)$ and those involving $\sin(\omega_0 t)$, and then finally use the same identity (2.6.2) in the reverse direction. However, this is exceedingly tedious to do as a numerical computation (see Exercise 2.8), and it leads to some very messy expressions if we wish to obtain a general formula. As we will see, a much simpler approach can be based on the complex exponential representation of the cosine signals.

Exercise 2.8. Use (2.6.2) to show that the sum

$$1.7 \cos(2\pi(10)t + 70\pi/180) + 1.9 \cos(2\pi(10)t + 200\pi/180)$$

reduces to $A \cos(2\pi(10)t + \phi)$, where

$$A = \{[1.7\cos(70\pi/180) + 1.9\cos(200\pi/180)]^2$$
$$+ [1.7\sin(70\pi/180) + 1.9\sin(200\pi/180)]^2\}^{1/2} = 1.532$$

and

$$\phi = \arctan\left\{\frac{1.7\sin(70\pi/180) + 1.9\sin(200\pi/180)}{1.7\cos(70\pi/180) + 1.9\cos(200\pi/180)}\right\} = 141.79\pi/180$$

The value of ϕ, given in radians, corresponds to $141.79°$.

2.6.1 Addition of Complex Numbers

When two complex numbers are added, it is necessary to use the Cartesian form. If $z_1 = x_1 + jy_1$ and $z_2 = x_2 + jy_2$, then $z_3 = z_1 + z_2 = (x_1 + x_2) + j(y_1 + y_2)$; i.e., the real and imaginary parts of the sum are the sum of the real and imaginary parts, respectively. Using the vector interpretation of complex numbers, where both z_1 and z_2 are viewed as vectors with their tails at the origin, the sum z_3 is the result of vector addition, and is constructed as follows:

1. Draw a copy of z_1 with its tail at the head of z_2. Call this displaced vector $\hat{z}_1$.

2. Draw the vector from the origin to the head of $\hat{z}_1$. This is the sum z_3.

This process is depicted in Fig. 2.14 for the case $z_1 = 4 - j3$ and $z_2 = 2 + j5$.

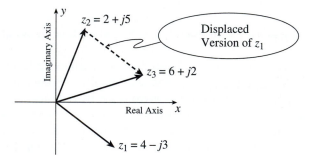

Figure 2.14   Graphical construction of complex number addition $z_3 = z_1 + z_2$.

2.6.2 Phasor Addition Rule

The phasor representation of cosine signals can be used to show the following result:

$$x(t) = \sum_{k=1}^{N} A_k \cos(\omega_0 t + \phi_k) = A\cos(\omega_0 t + \phi) \tag{2.6.3}$$

where N is any integer. That is, the sum of two or more cosine signals each having the same frequency, but having different amplitudes and phase shifts, can be expressed as a single equivalent cosine signal. The resulting amplitude (A) and phase (ϕ) of the term on the right-hand side of (2.6.3) can be computed from the individual amplitudes (A_k) and phases (ϕ_k) on the left-hand side by doing the following addition of complex numbers:

$$\sum_{k=1}^{N} A_k e^{j\phi_k} = A e^{j\phi} \tag{2.6.4}$$

Equation (2.6.4) is the essence of the phasor addition rule. Proof of the phasor rule requires the following two pieces of information:

1. Any sinusoid can be written in the form:

$$A \cos(\omega_0 t + \phi) = \Re e\{A e^{j(\omega_0 t + \phi)}\} = \Re e\{A e^{j\phi} e^{j\omega_0 t}\}$$

2. For any set of complex numbers $\{X_k\}$ the sum of the real parts is equal to the real part of the sum, so we have

$$\Re e\left\{\sum_{k=1}^{N} X_k\right\} = \sum_{k=1}^{N} \Re e\{X_k\}$$

Proof of the phasor addition rule involves the following algebraic manipulations:

$$\sum_{k=1}^{N} A_k \cos(\omega_0 t + \phi_k) = \sum_{k=1}^{N} \Re e\left\{A_k e^{j(\omega_0 t + \phi_k)}\right\}$$

$$= \Re e\left\{\sum_{k=1}^{N} A_k e^{j\phi_k} e^{j\omega_0 t}\right\}$$

$$= \Re e\left\{\left(\sum_{k=1}^{N} A_k e^{j\phi_k}\right) e^{j\omega_0 t}\right\}$$

$$= \Re e\left\{\left(A e^{j\phi}\right) e^{j\omega_0 t}\right\}$$

$$= \Re e\left\{A e^{j(\omega_0 t + \phi)}\right\} = A \cos(\omega_0 t + \phi)$$

This completes the proof. Note that the crucial step in the proof (fourth line in proof) is replacing the summation term in parentheses with $A e^{j\phi}$, as defined in (2.6.4).

2.6.3 Phasor Addition Rule: Example

We now return to the example of Exercise 2.8, where

$$x_1(t) = 1.7\cos(2\pi(10)t + 70\pi/180)$$

$$x_2(t) = 1.9\cos(2\pi(10)t + 200\pi/180)$$

and the sum was found to be

$$x_3(t) = x_1(t) + x_2(t) = 1.532\cos(2\pi(10)t + 141.79\pi/180)$$

The frequency of both sinusoids is 10 Hz, so the period is $T_0 = 0.1$ sec. The waveforms of the three signals are shown Fig. 2.15(b) and the phasors used to solve the problem are shown on the left in Fig. 2.15(a). Notice that the times where the maximum of each cosine signal occurs can be derived from the phase through the formula

$$t_m = -\frac{\phi T_0}{2\pi}$$

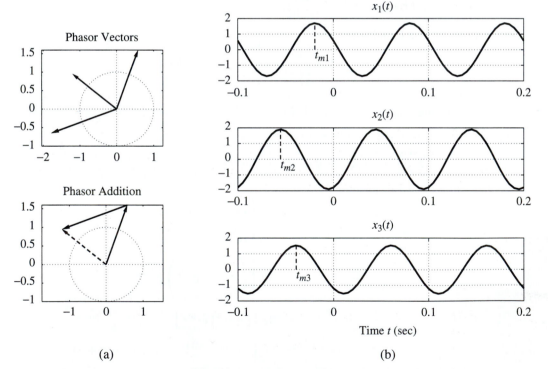

(a)   (b)

Figure 2.15 Adding sinusoids by (a) doing a phasor addition which is actually a graphical vector sum. ($X_3 = X_1 + X_2$ is dashed vector) (b) The time of the signal maximum is marked on each $x_i(t)$ plot.

which gives

$$t_{m1} = -0.0194 \qquad t_{m2} = -0.0556 \qquad t_{m3} = -0.0394 \text{ sec.}$$

These times are marked with vertical dotted lines in the corresponding waveform plots in Fig. 2.15(b). The phasor addition of the two signals is computed in four steps.

1. Represent $x_1(t)$ and $x_2(t)$ by the phasors:

$$X_1 = A_1 e^{j\phi_1} = 1.7 e^{j70\pi/180} \text{ and } X_2 = A_2 e^{j\phi_2} = 1.9 e^{j200\pi/180}$$

2. Convert both phasors to rectangular form:

$$X_1 = 0.5814 + j1.597 \text{ and } X_2 = -1.785 - j0.6498$$

3. Add the real parts and the imaginary parts:

$$X_3 = X_1 + X_2 = (0.5814 + j1.597) + (-1.785 - j0.6498) = -1.204 + j0.9476$$

4. Convert back to polar form, obtaining $X_3 = 1.532 e^{j141.79\pi/180}$

Therefore, the final formula for $x_3(t)$ is

$$x_3(t) = 1.532 \cos(20\pi t + 141.79\pi/180)$$

$$\text{or} \quad x_3(t) = 1.532 \cos(20\pi(t + 0.0394))$$

2.6.4 MATLAB Demo of Phasors

The process of phasor addition can be accomplished easily using MATLAB. The answer generated by MATLAB and printed with the special function zprint (provided on the CD-ROM) for this particular phasor addition is:

```
Z   =     X    +     jY     Magnitude   Phase    Ph/pi   Ph(deg)
Z1      0.5814      1.597        1.7     1.222    0.389    70.00
Z2     -1.785     -0.6498        1.9    -2.793   -0.889  -160.00
Z3     -1.204      0.9476      1.532     2.475    0.788   141.79
```

Help on zprint gives:

```
ZPRINT    printout complex # in rect and polar form
    usage:    zprint(z)
       z = vector of complex numbers
```

DSP FIRST
TOOLBOX

The MATLAB code that generates Fig. 2.15 is given in the first lab (Appendix C). It uses the special MATLAB functions (provided on the CD-ROM) zprint, zvect, zcat, ucplot, and zcoords to make the vector plots.

2.6.5 Summary of the Phasor Addition Rule

In this section, we have shown how a real cosine signal can be represented as the real part of a complex exponential signal (complex rotating phasor), and we have applied this representation to show how to simplify the process of adding several cosine signals of the same frequency.

$$x(t) = \sum_{k=1}^{N} A_k \cos(\omega_0 t + \phi_k) = A \cos(\omega_0 t + \phi)$$

In summary, all we have to do to get the cosine signal representation of the sum is:

1. Obtain the phasor representation $X_k = A_k e^{j\phi_k}$ of each of the individual signals.

2. Add the phasors of the individual signals to get $X = X_1 + X_2 + \ldots = A e^{j\phi}$. This requires polar-to-Cartesian-to-polar format conversions.

3. Multiply the result by $e^{j\omega_0 t}$ to get $\bar{x}(t) = A e^{j\phi} e^{j\omega_0 t}$.

4. Take the real part to get $x(t) = \Re\{A e^{j\phi} e^{j\omega_0 t}\} = A \cos(\omega_0 t + \phi) = x_1(t) + x_2(t) + \ldots$.

In other words, A and ϕ must be calculated by doing a vector sum of all the X_k phasors.

Exercise 2.9. Consider the two sinusoids,

$$x_1(t) = 5 \cos(2\pi (100)t + \pi/3)$$

$$x_2(t) = 4 \cos(2\pi (100)t - \pi/4)$$

Obtain the phasor representations of these two signals, add the phasors, plot the two phasors and their sum in the complex plane, and show that the sum of the two signals is

$$x_3(t) = 5.536 \cos(2\pi (100)t + 0.2747)$$

In degrees the phase should be $15.74°$. Examine the plots in Fig. 2.16 to see whether you can identify the cosine waves $x_1(t)$, $x_2(t)$, and $x_3(t) = x_1(t) + x_2(t)$.

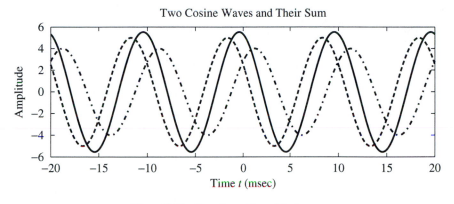

Figure 2.16   Two sinusoids and their sum.

2.7   PHYSICS OF THE TUNING FORK

In Section 2.1, we described a simple experiment in which a tuning fork was seen to generate a signal whose waveform looked very much like that of a sinusoidal signal. Now that we know a lot more about sinusoidal signals, it is worthwhile to take up this issue again. Is it a coincidence that the tuning-fork signal looks like a sinusoid, or is there a deeper connection between vibrations and sinusoids? In this section, we present a simple analysis of the tuning-fork system that shows that the tuning fork does indeed vibrate sinusoidally when given a displacement from its equilibrium position. The sinusoidal motion of the tuning-fork tines is transferred to the surrounding air particles, thereby producing the acoustic signal that we hear. This simple example illustrates how mathematical models of physical systems derived from fundamental physical principles can lead to concise mathematical descriptions of physical phenomena and of signals that result.

2.7.1 Equations from Laws of Physics

A schematic diagram of the tuning fork is shown in Fig. 2.17. As we have seen experimentally, when struck against a firm surface, the tines of the tuning fork vibrate and produce a "pure" tone. We are interested in deriving the equations that describe the physical behavior of the tuning fork so that we can understand the basic mechanism by which the sound is produced.[10] Newton's second law, $F = ma$, will lead to a differential equation whose solution is a sine or cosine function, or a complex exponential.

---

[10] The generation and propagation of sound is treated in any general college physics text.

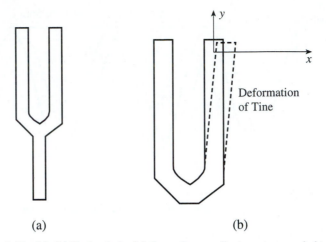

Figure 2.17   (a), (b) Tuning fork. (b) shows the coordinate system needed to write equations for the vibration of the tine.

When the tuning fork is struck, one of the tines is deformed slightly from its rest position, as depicted in Fig. 2.17(b). We know from experience that unless the deformation was so large as to break or bend the metal, there would be a tendency for the tine to return to its original rest position. The physical law that governs this movement is Hooke's law. Although the tuning fork is made of a very stiff metal, we can think of it as an elastic material when the deformation is tiny. Hooke's law states that the restoring force is directly proportional to the amount of deformation. If we set up a coordinate system as in Fig. 2.17(b), the deformation is along the x-axis, and we can write

$$F = -kx$$

where the parameter k is the elastic constant of the material (i.e., its stiffness). The minus sign indicates that this restoring force acts in the negative direction when the displacement of the tine is in the positive x direction; i.e., it acts to pull the tine back toward the neutral position.

Now this restoring force due to stiffness produces an acceleration as dictated by Newton's second law, i.e.,

$$F = ma = m \frac{d^2x}{dt^2}$$

where m is the mass of the tine, and the second derivative with respect to time of position x is the acceleration of the mass along the x-axis. Since these two forces must balance each other (i.e., the sum of the forces is zero), we get a second-order differential equation that describes the motion $x(t)$ of the tine for all values of time t

$$m \frac{d^2x(t)}{dt^2} = -k\, x(t) \qquad (2.7.1)$$

This particular differential equation is rather easy to solve, because we can, in fact, guess the solution. From the derivative properties of sine and cosine functions, we are motivated to try as a solution the function

$$x(t) = \cos \omega_0 t$$

where the parameter ω_0 is a constant that must be determined. The second derivative of $x(t)$ is

$$\frac{d^2x(t)}{dt^2} = \frac{d^2}{dt^2}(\cos \omega_0 t)$$

$$= \frac{d}{dt}(-\omega_0 \sin \omega_0 t)$$

$$= -\omega_0^2 \cos \omega_0 t$$

Notice that the second derivative of the cosine function is the same cosine function multiplied by a constant $(-\omega_0^2)$. Therefore, when we substitute $x(t)$ into (2.7.1), we get

$$m \frac{d^2x(t)}{dt^2} = -k\, x(t)$$

$$-m\, \omega_0^2 \cos \omega_0 t = -k \cos \omega_0 t$$

Since this equation must be satisfied for all t, it follows that the coefficients of $\cos \omega_0 t$ must be equal, which leads to the following algebraic equation

$$-m\, \omega_0^2 = -k$$

This equation can be solved for ω_0, obtaining

$$\omega_0 = \pm \sqrt{\frac{k}{m}} \qquad (2.7.2)$$

Therefore, we conclude that one solution of the differential equation is

$$x(t) = \cos\left(\sqrt{\frac{k}{m}}\, t\right)$$

From our model, $x(t)$ describes the motion of the tuning-fork tine. Therefore we conclude that the tines oscillate sinusoidally. This motion is, in turn, transferred to the particles of air in the locality of the tines producing the tiny variations in air pressure that make up an acoustic wave. The formula for the frequency lets us draw two conclusions:

TUNING
FORK DEMO

1. Of two tuning forks having the same mass, the stiffer one will produce a higher frequency. This is because the frequency is proportional to $\sqrt{k}$, which is in the numerator of (2.7.2).

2. Of two tuning forks having the same stiffness, the heavier one will produce a lower frequency. This is because the frequency is inversely proportional to the square root of the mass, $\sqrt{m}$, which is in the denominator of (2.7.2).

2.7.2 General Solution to the Differential Equation

There are many possible solutions to the tuning-fork differential equation (2.7.1). We can prove that the following function

$$x(t) = A\cos(\omega_0 t + \phi)$$

will satisfy the differential equation (2.7.1) by substituting back into (2.7.1), and taking derivatives. Once again the frequency must be $\omega_0 = \sqrt{k/m}$. Only the frequency ω_0 is constrained by our simple model; the specific values of the parameters A and ϕ are not important. From this we can conclude that any scaled or time-shifted sinusoid with the correct frequency will satisfy the differential equation that describes the motion of the tuning fork's tine. This implies that an infinite number of different sinusoidal waveforms can be produced in the tuning fork experiment. For any particular experiment, A and ϕ would be determined by the exact strength and timing of the sharp force that gave the tine its initial displacement. However, the frequency of all these sinusoids will be determined only by the mass and stiffness of the tuning-fork metal.

Exercise 2.10. Demonstrate that a complex exponential signal can also be a solution to the tuning-fork differential equation:

$$\frac{d^2x}{dt^2} = -\frac{k}{m}x(t)$$

By substituting $\bar{x}(t)$ and $\bar{x}^*(t)$ into both sides of the differential equation, show that the equation is satisfied for all t by both of the signals

$$\bar{x}(t) = X\,e^{j\omega_0 t} \quad \text{and} \quad \bar{x}^*(t) = X^*\,e^{-j\omega_0 t}$$

Determine the value of ω_0 for which the differential equation is satisfied.

2.7.3 Listening to Tones

The observer is an important part of any physical experiment. This is particularly true when the experiment involves listening to the sound produced. In the tuning-fork experiment, we perceive a tone with a certain pitch (related to the frequency) and loudness (related to the amplitude). The human ear and neural processing system respond to the frequency and amplitude of a sustained sound like that produced by the tuning fork, but the phase is not perceptible. This is because phase is really due to an arbitrary definition of the starting time of the sinusoid; i.e., a sustained tone sounds the same now as it did 5 minutes ago. On the other hand, we could pick up the sound with a microphone and sample or display the signal on an oscilloscope. In this case, it would be possible to make precise measurements of frequency and amplitude, but phase would be measured accurately only with respect to the time base of the sampler or the oscilloscope.

2.8 TIME SIGNALS: MORE THAN FORMULAS

The purpose of this chapter has been to introduce the concept of a sinusoidal signal and to illustrate how sinusoidal signals can arise in real situations. Signals, as we have defined them, are varying patterns that convey or represent information, usually about the state or behavior of a physical system. We have seen by both theory and observation that a tuning fork generates a signal that can be represented mathematically as a sinusoidal signal. In the context of the tuning fork, the cosine wave conveys and represents information about the state of the tuning fork. Encoded in the sinusoidal waveform is information such as whether the tuning fork is vibrating or at rest, and, if it is vibrating, its frequency and the amplitude of its vibrations. This information can be extracted from the signal by human listeners, or it can be recorded for later processing by either humans or computers.

Although the solution to the differential equation of the tuning fork (2.7.1) is a cosine function, the resulting mathematical formula is simply a model that results from an idealization of the tuning fork. It is important to recall that the signal is a separate entity from the formula. The actual waveform produced by a tuning fork is probably not exactly sinusoidal. The signal is *represented* by the mathematical formula $x(t) = A\cos(\omega_0 t + \phi)$, which can be derived from an idealized model based on physical principles. This model is a good approximation of reality, but an approximation nevertheless. Even so, this model is extremely useful since it leads directly to a useful mathematical representation of the signal produced by the tuning fork.

In the case of a complicated signal generated by a musical instrument, the signal cannot be so easily reduced to a mathematical formula. Figure 2.18 shows a short segment of a recording of orchestra music. Just the appearance of the waveform suggests a much more complex situation. Although it oscillates like the cosine wave, it clearly is not periodic (at least, as far as we can see from the given segment). Orchestra music consists of many instruments sounding different notes together.

If each instrument produced a pure sinusoidal tone at the frequency of the note that is assigned to it, then the composite orchestra signal would be simply a sum of sinusoids with different frequencies, amplitudes, and phase shifts. While this is far from being a correct model for most instruments, it is actually a highly appropriate way to think about the orchestra signal. In fact, we will see very soon that sums of sinusoids of different frequencies, amplitudes, and phase shifts can result in an infinite variety of waveforms. Indeed, it is true that almost any signal can be represented as a sum of sinusoidal signals. When this concept was first introduced by Jean-Baptiste Joseph Fourier in 1807, it was received with great skepticism by the famous mathematicians of the world. Nowadays, this notion is commonplace (although no less remarkable). The mathematical and computational techniques of Fourier analysis underlie the frequency–domain analysis tools used extensively in electrical engineering and other areas of science and engineering.

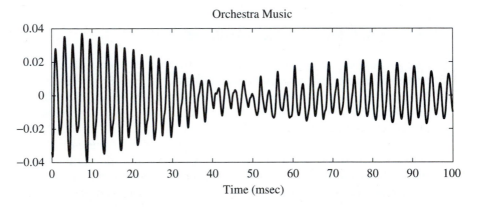

Figure 2.18   A short segment of an orchestra music signal.

2.9 SUMMARY AND LINKS

We have introduced sinusoidal signals in this chapter. We have attempted to show that they arise naturally as a result of simple physical processes and that they can be represented by familiar mathematical functions, and also by complex exponentials. The value of the mathematical representation of a signal is twofold. First, the mathematical representation provides a convenient formula to consistently describe the signal. For example, the cosine signal is completely described in terms of just three parameters. Second, by representing both signals and systems through mathematical expressions, we can make precise statements about the interaction between signals and systems.

In connection with this chapter, two laboratories are found in Appendix C. Lab C.1 involves some introductory exercises on the basic elements of the MATLAB programming environment, and its use for manipulating complex numbers and plotting sinusoids. Appendix B is also available for a quick overview of essential ideas about MATLAB. Lab C.2 (a) deals with sinusoids and phasor addition. In this lab, students must re-create a phasor addition demonstration similar to Fig. 2.15. Copies of the lab write-ups are also found on the CD-ROM.

On the CD-ROM, one can find the following resources:

LINKS

1. A tuning-fork movie that shows the experiment of striking the tuning fork and recording its sound. Several recorded sounds from different tuning forks are available, as well as sounds from clay whistles.

2. A drill program written in MATLAB for working with the amplitude, phase, and frequency of sinusoidal plots.

3. A set of movies showing rotating phasor(s) and how they generate sinusoids through the real part operator.

Finally, the CD-ROM also contains a wealth of solved homework problems that may be used for practice and self-study.

PROBLEMS

2.1 Define $x(t)$ as

$$x(t) = 3\cos(\omega_0 t - \pi/4)$$

For $\omega_0 = \pi/5$, make a plot of $x(t)$ over the range $-10 \le t \le 20$.

2.2 Make a carefully labeled sketch for each of the following functions.

(a) Sketch $\cos\theta$ for values of θ in the range $0 \le \theta \le 6\pi$.

(b) Sketch $\cos(0.2\pi t)$ for values of t such that three periods of the function are shown.

(c) Sketch $\cos(2\pi t/T_0)$ for values of t such that three periods of the function are shown. Label the horizontal axis in terms of the parameter T_0.

(d) Sketch $\cos(2\pi t/T_0 + \pi/2)$ for values of t such that three periods of the function are shown.

2.3 The figure shown at the top of the next page is a plot of a sinusoidal wave. From the plot, determine values for the amplitude (A), phase (ϕ), and frequency (ω_0) needed in the representation:

$$x(t) = A\cos(\omega_0 t + \phi)$$

Give the answer as numerical values, *including the units* where applicable.

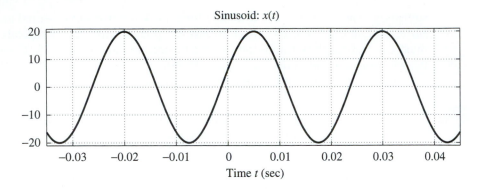

2.4 Use the series expansions for e^x, $\cos(\theta)$, and $\sin(\theta)$ given here to show Euler's formula.

$$e^x = 1 + x + \frac{x^2}{2!} + \frac{x^3}{3!} + \cdots$$

$$\cos(\theta) = 1 - \frac{\theta^2}{2!} + \frac{\theta^4}{4!} + \cdots$$

$$\sin(\theta) = \theta - \frac{\theta^3}{3!} + \frac{\theta^5}{5!} + \cdots$$

2.5 Use complex exponentials (i.e., phasors) to show the following trigonometric identities:

(a) $\cos(\theta_1 + \theta_2) = \cos(\theta_1)\cos(\theta_2) - \sin(\theta_1)\sin(\theta_2)$

(b) $\cos(\theta_1 - \theta_2) = \cos(\theta_1)\cos(\theta_2) + \sin(\theta_1)\sin(\theta_2)$

2.6 Use Euler's formula for the complex exponential to prove the formula:

$$(\cos\theta + j\sin\theta)^n = \cos n\theta + j\sin n\theta$$

This formula is called DeMoivre's formula (see p. 463, Appendix A). Use it to evaluate $(3/5 + j4/5)^{100}$.

2.7 Simplify the following complex-valued expressions:

(a) $3e^{j\pi/3} + 4e^{-j\pi/6}$

(b) $(\sqrt{3} - j3)^{10}$

(c) $(\sqrt{3} - j3)^{-1}$

(d) $(\sqrt{3} - j3)^{1/3}$

(e) $\Re\{je^{-j\pi/3}\}$

Give the answers in *both* Cartesian form $(x + jy)$ and polar form $(re^{j\theta})$.

2.8 Suppose that MATLAB is used to plot a sinusoidal signal. The following MATLAB code generates the signal and makes the plot. Derive a formula for the signal; then draw a sketch of the plot that will be done by MATLAB.

```
dt = 1/100;
tt = -1 : dt : 1;
Fo = 2;
xx = 300*real( exp( j*(2*pi*Fo*(tt - 0.75) ) ) );
%
plot( tt, xx ),    grid
title( 'SECTION of a SINUSOID' ), xlabel('TIME    (sec)')
```

2.9 Define $x(t)$ as

$$x(t) = 2\sin(\omega_0 t + 45°) + \cos(\omega_0 t)$$

(a) Express $x(t)$ in the form $x(t) = A\cos(\omega_0 t + \phi)$

(b) Assume that $\omega_0 = 5\pi$. Make a plot of $x(t)$ over the range $-1 \le t \le 2$. How many periods are included in the plot?

(c) Find a complex-valued signal $\bar{x}(t)$ such that $x(t) = \Re\{\bar{x}(t)\}$.

2.10 Define $x(t)$ as

$$x(t) = 5\cos(\omega_0 t + 90°) + 5\cos(\omega_0 t - 30°) + 5\cos(\omega_0 t - 120°)$$

Simplify $x(t)$ into the standard form: $x(t) = A\cos(\omega_0 t + \phi)$. Use phasors to do the algebra, but also provide a plot of the vectors representing each of the three phasors.

2.11 Solve the following equation for θ:

$$\Re\{(1 + j)e^{j\theta}\} = -1$$

Give the answers in radians. Make sure that you find *all* possible answers.

2.12 Give two possible complex-valued solutions to the following differential equation:

$$\frac{d^2 x(t)}{dt^2} = -100\, x(t)$$

ROTATING
PHASOR
DEMO

2.13 Define the following complex exponential signal:

$$s(t) = 5e^{j\pi/3}e^{j10\pi t}$$

(a) Make a plot of $s_i(t) = \Im\{s(t)\}$. Pick a range of values that will include exactly three periods of the signal.

(b) Make a plot of $q(t) = \Im\{\dot{s}(t)\}$, where the dot means differentiation with respect to time t. Again plot three cycles of the signal.

2.14 For the sinusoidal waveform shown in the following figure, determine the complex phasor representation

$$X = A e^{j\phi}$$

i.e., find ω_0, ϕ, and A such that the waveform is represented by

$$x(t) = \Re e\{X\, e^{j\omega_0 t}\}$$

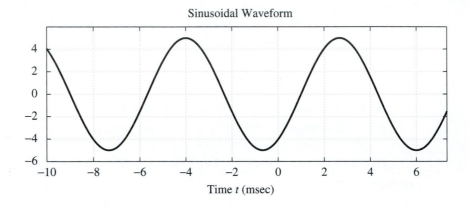

Sinusoidal Waveform

2.15 Define $x(t)$ as

$$x(t) = 5\cos(\omega_0 t + \pi/3) + 7\cos(\omega_0 t - 5\pi/4) + 3\cos(\omega_0 t + 3\pi/2)$$

Express $x(t)$ in the form $x(t) = A\cos(\omega_0 t + \phi)$. Use complex phasor manipulations to obtain the answer. Explain your answer by giving a phasor diagram.

2.16 The phase of a sinusoid can be related to time shift as follows:

$$x(t) = A\cos(2\pi f_0 t + \phi) = A\cos(2\pi f_0 (t - t_1))$$

In the following parts, assume that the period of the sinusoidal wave is $T_0 = 8$ sec.

(a) "When $t_1 = -2$ sec, the value of the phase is $\phi = \pi/2$." Explain whether this is True or False.

(b) "When $t_1 = 3$ sec, the value of the phase is $\phi = 3\pi/4$." Explain whether this is True or False.

(c) "When $t_1 = 7$ sec, the value of the phase is $\phi = \pi/4$." Explain whether this is True or False.

2.17 Define $x(t)$ as

$$x(t) = 5\cos(\omega_0 t + 3\pi/2) + 4\cos(\omega_0 t + 2\pi/3) + 4\cos(\omega_0 t + \pi/3)$$

(a) Express $x(t)$ in the form $x(t) = A\cos(\omega_0 t + \phi)$ by finding the numerical values of A and ϕ.

(b) Plot all the phasors used to solve the problem in (a) in the complex plane.

2.18 Solve the following simultaneous equations by using the phasor method. Is the answer for A_1, A_2, ϕ_1, ϕ_2 unique? Provide a geometrical diagram to explain the answer.

$$\cos(\omega_0 t) = A_1 \cos(\omega_0 t + \phi_1) + A_2 \cos(\omega_0 t + \phi_2)$$

$$\sin(\omega_0 t) = 2A_1 \cos(\omega_0 t + \phi_1) + A_2 \cos(\omega_0 t + \phi_2)$$

2.19 Solve the following equation for M and ψ. Obtain *all* possible answers. Use the phasor method, and provide a geometrical diagram to explain the answer.

$$5\cos(\omega_0 t) = M \cos(\omega_0 t - \pi/6) + 5\cos(\omega_0 t + \psi)$$

Hint: Describe the figure in the z-plane given by the set $\{z : z = 5e^{j\psi} - 5\}$ where $0 \le \psi \le 2\pi$.

3

Spectrum Representation

This chapter introduces the concept of the *spectrum*, a graphical representation of the frequency content of a signal. In Chapter 2, we learned about the properties of sinusoidal waveforms of the form

$$x(t) = A\cos(2\pi f_0 t + \phi) = \Re e\left\{ X e^{j2\pi f_0 t} \right\}$$

where $X = Ae^{j\phi}$ is a phasor, and we showed how phasors can simplify the addition of sinusoids of the *same* frequency. In this chapter, we will show how more complicated sinusoidal waveforms can be constructed out of sums of sinusoidal signals of the form

$$x(t) = A_0 + \sum_{k=1}^{N} A_k \cos(2\pi f_k t + \phi_k) = X_0 + \Re e\left\{ \sum_{k=1}^{N} X_k e^{j2\pi f_k t} \right\}$$

where $X_0 = A_0$ is a real constant, and $X_k = A_k e^{j\phi_k}$ is the complex amplitude (i.e., phasor) for the complex exponential of frequency f_k. The spectrum is a graphical presentation of the individual sinusoidal components that make up the signal. This visual form allows us to see interrelationships among the different frequency components and their relative amplitudes quickly and easily.

3.1 THE SPECTRUM OF A SUM OF SINUSOIDS

One of the reasons that sinusoids are so important for our study is that they are the basic building blocks for making more complicated signals. Later in this chapter, we will show some extraordinarily complicated waveforms that can be constructed from rather simple combinations of the basic cosine wave. The most general and powerful method for producing new signals from sinusoids is the *additive linear combination*, where a signal is created by adding together a constant and N sinusoids, each with a different frequency, amplitude, and phase. Mathematically, this signal may be represented by the equation

$$x(t) = A_0 + \sum_{k=1}^{N} A_k \cos(2\pi f_k t + \phi_k) \tag{3.1.1}$$

where each amplitude, phase, and frequency[1] may be chosen independently. Such a signal may also be represented in terms of the phasor representations of the individual sinusoidal components, i.e.,

$$x(t) = X_0 + \sum_{k=1}^{N} \Re e \left\{ X_k e^{j2\pi f_k t} \right\} \tag{3.1.2}$$

where $X_0 = A_0$ represents a real constant component, and each phasor

$$X_k = A_k e^{j\phi_k}$$

represents the magnitude and phase of a rotating phasor whose frequency is f_k.

The *inverse Euler formula* gives a way to represent $x(t)$ in the alternative form

$$x(t) = X_0 + \sum_{k=1}^{N} \left\{ \frac{X_k}{2} e^{j2\pi f_k t} + \frac{X_k^*}{2} e^{-j2\pi f_k t} \right\} \tag{3.1.3}$$

As in the case of individual sinusoids, this form follows from the fact that the real part of a complex number is equal to one-half the sum of that number and its complex conjugate. Equation (3.1.3) also has the interesting feature that each sinusoid in the sum decomposes into two rotating phasors, one with positive frequency, f_k, and the other with negative frequency, $-f_k$.

We define the *two-sided spectrum* of a signal composed of sinusoids as in (3.1.3) to be the set of $2N + 1$ complex phasors and the $2N + 1$ frequencies that specify the signal in the representation of (3.1.3). Although it is a somewhat awkward mathematical representation, our definition of the spectrum is just the set of pairs

$$\left\{ (X_0, 0), \ \left(\tfrac{1}{2}X_1, \ f_1\right), \ \left(\tfrac{1}{2}X_1^*, \ -f_1\right), \ \left(\tfrac{1}{2}X_2, \ f_2\right), \ \left(\tfrac{1}{2}X_2^*, \ -f_2\right), \ldots \right\} \tag{3.1.4}$$

---

[1] For this chapter, we prefer cyclic frequency f_k to radian frequency $\omega_k = 2\pi f_k$, because it is easier to describe physical quantities such as musical notes in Hz.

Each pair $(\frac{1}{2}X_k,\ f_k)$ indicates the size and relative phase of the sinusoidal component contributing at frequency f_k. It is common to refer to the spectrum as the *frequency-domain representation* of the signal. Instead of giving the time waveform itself (i.e., the time-domain representation), the frequency-domain representation simply gives the information required to synthesize it with (3.1.3)

Example 3.1   For example, consider the sum of a constant and two sinusoids:

$$x(t) = 10 + 14\cos(200\pi t - \pi/3) + 8\cos(500\pi t + \pi/2)$$

When we apply the inverse Euler formula (3.1.3), we get the following five terms:

$$x(t) = 10 + 7e^{-j\pi/3}e^{j2\pi(100)t} + 7e^{j\pi/3}e^{-j2\pi(100)t}$$
$$+ 4e^{j\pi/2}e^{j2\pi(250)t} + 4e^{-j\pi/2}e^{-j2\pi(250)t} \qquad (3.1.5)$$

In the list form suggested in (3.1.4), the spectrum of this signal is the set of five rotating phasors represented by

$$\{(10,\, 0),\ (7e^{-j\pi/3},\, 100),\ (7e^{j\pi/3},\, -100),\ (4e^{j\pi/2},\, 250),\ (4e^{-j\pi/2},\, -250)\}$$

The constant component of the signal, often called the "DC component", can be expressed as a complex exponential signal with zero frequency, i.e., $10e^{j0t} = 10$. $\diamond$

3.1.1 Graphical Plot of the Spectrum

A plot of the spectrum is much more revealing than the list of $(\frac{1}{2}X_k,\ f_k)$ pairs. Each frequency component can be represented by a vertical line at the appropriate frequency, and the length of the line can be drawn proportional to the magnitude, $|\frac{1}{2}X_k|$. This is shown in Fig. 3.1 for the signal in (3.1.5). Each *spectral line* is labeled with the value of $\frac{1}{2}X_k$ to complete the information needed to define the spectrum. This simple but effective plot makes it easy to see two things: the relative location of the frequencies, and the relative amplitudes of the sinusoidal components. This is why the spectrum is widely used as a graphical representation of the signal. As we will see in Chapters 4 and 6, another reason that the frequency-domain representation is so useful is that it is often very easy to see how systems affect a signal by determining what happens to the signal spectrum as it is transmitted through the system. This is why the spectrum is the key to understanding most complex processing systems such as radios, televisions, CD players, and the like.

Notice that, for the example in Fig. 3.1, the complex amplitude of each negative frequency component is the complex conjugate of the complex amplitude at the corresponding positive-frequency component. This is a general property of the spectrum whenever $x(t)$ is a real signal, because the complex rotating phasors with positive and negative frequency must combine to form a real signal (see Fig. 2.13(b) and the movie found on the CD).

Chapter 2
ROTATING
PHASOR
DEMO

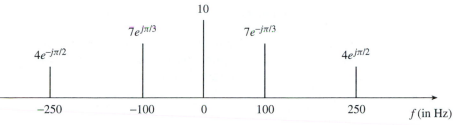

Figure 3.1   Spectrum of the signal $x(t) = 10 + 14\cos(200\pi t - \pi/3) + 8\cos(500\pi t + \pi/2)$. Positive and negative frequency components must be included even though the negative-frequency ones are the conjugate of the positive-frequency components.

A general procedure for computing and plotting the spectrum for any signal requires the study of Fourier analysis. We will find that focusing on signals consisting of constants, sinusoids, or sums of sinusoids is not a severe limitation. For such signals, the procedure is straightforward: It is necessary only to express the cosines and sines as complex exponentials (by using the inverse Euler relation) and then to plot the complex amplitude of each of the positive- and negative-frequency components at the corresponding frequency. In other words, the process of analyzing the signal to find its *spectral components* involves simply looking at an equation for the signal in the form of (3.1.3), and picking off the amplitude, phase, and frequency of each of its rotating phasor components.

In many other cases, spectrum analysis is not so simple, but it is nevertheless possible. For example, it is possible to represent any periodic waveform (even discontinuous signals) as a sum of complex exponential signals where the frequencies are all integer multiples of a common frequency, called the *fundamental frequency*. Likewise, most (nonperiodic) signals can also be represented as a superposition of complex exponential signals. The mathematical tools for doing this analysis are called Fourier series and Fourier transforms. We will introduce these techniques in Chapter 9, but the rigorous mathematics of Fourier analysis is generally the subject of later courses. Until then, we will treat spectrum analysis as a useful concept that can be implemented easily for finite sums of cosine waves, as found in (3.1.1).

3.2   BEAT NOTES

BEAT NOTES

When we multiply two sinusoids having different frequencies, we can create an interesting audio effect called a *beat note*. The phenomenon, which may sound like a warble, is best heard by picking one of the frequencies to be very small (e.g., 10 Hz), and the other around 1 kHz. Some musical instruments naturally produce these beating tones. Another use for multiplying sinusoids is modulation for radio broadcasting. AM radio stations use this method, which is called *amplitude modulation*.

3.2.1 Multiplication of Sinusoids

Our spectrum representation demands that the signal be expressed as an additive linear combination of complex exponential signals. Other combinations of sinusoids must be rewritten in the additive form in order to display their spectrum representation. For example, if we define a beat signal as the product of two sinusoids

$$x(t) = \sin(10\pi t)\cos(\pi t) \qquad (3.2.1)$$

it is essential to rewrite $x(t)$ as a sum before its spectrum can be defined. The following technique for doing this relies on the inverse Euler formula:

$$x(t) = \left(\frac{e^{j10\pi t} - e^{-j10\pi t}}{2j}\right)\left(\frac{e^{j\pi t} + e^{-j\pi t}}{2}\right)$$

$$= \tfrac{1}{4}e^{-j\pi/2}e^{j11\pi t} + \tfrac{1}{4}e^{-j\pi/2}e^{j9\pi t} - \tfrac{1}{4}e^{-j\pi/2}e^{-j9\pi t} - \tfrac{1}{4}e^{-j\pi/2}e^{-j11\pi t}$$

$$= \tfrac{1}{2}\cos(11\pi t - \pi/2) + \tfrac{1}{2}\cos(9\pi t - \pi/2)$$

Now it is obvious that there are four terms in the additive combination, and the four spectrum components are at frequencies 5.5, 4.5, −4.5, and −5.5 Hz. It is worth noting that neither of the original frequencies used to define $x(t)$ are in the spectrum.

Exercise 3.1. Let $x(t) = \sin^2(10\pi t)$. Find an additive combination in the form of (3.1.3) for $x(t)$, and then plot the spectrum. Count the number of frequency components in the spectrum. What is the highest frequency contained in $x(t)$? Use the inverse Euler formula rather than a trigonometric identity.

3.2.2 Beat Note Waveform

Beat notes are produced by adding two sinusoids with nearly identical frequencies, e.g., by playing two neighboring piano keys. As the example (3.2.1) suggests, the sum of two sinusoids can also be written as a product. We can derive a general relationship between the beat signal, its spectrum, and the product form if we start with an additive combination of two closely spaced sinusoids:

$$x(t) = \cos(2\pi f_1 t) + \cos(2\pi f_2 t) \qquad (3.2.2)$$

The two frequencies can be expressed as $f_1 = f_c - f_\Delta$ and $f_2 = f_c + f_\Delta$, if we define a *center frequency* $f_c = \frac{1}{2}(f_1 + f_2)$ and a *deviation frequency* $f_\Delta = \frac{1}{2}(f_2 - f_1)$, which is usually much smaller than f_c. The spectrum of this beat signal is plotted in Fig. 3.2.

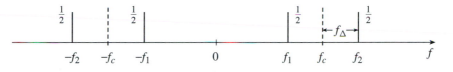

Figure 3.2   Spectrum of the beat signal in (3.2.2).

Using the complex exponential representation of the two cosines, we can rewrite $x(t)$ as a product of two cosines, and thereby have a form that is easier to plot in the time domain.

$$
\begin{aligned}
x(t) &= \Re\left\{e^{j2\pi f_1 t}\right\} + \Re\left\{e^{j2\pi f_2 t}\right\} \\
&= \Re\left\{e^{j2\pi(f_c - f_\Delta)t} + e^{j2\pi(f_c + f_\Delta)t}\right\} \\
&= \Re\left\{e^{j2\pi f_c t}\left(e^{-j2\pi f_\Delta t} + e^{j2\pi f_\Delta t}\right)\right\} \\
&= \Re\left\{e^{j2\pi f_c t}\left(2\cos(2\pi f_\Delta t)\right)\right\} \\
&= 2\cos(2\pi f_\Delta t)\cos(2\pi f_c t) \qquad (3.2.3)
\end{aligned}
$$

For a numerical example, we take $f_c = 200$ and $f_\Delta = 20$ Hz.

$$
x(t) = 2\cos(2\pi(20)t)\cos(2\pi(200)t) \qquad (3.2.4)
$$

A plot of this signal is given in Fig. 3.3. The top panel of Fig. 3.3 shows the two sinusoidal components, $2\cos(2\pi(20)t)$ and $\cos(2\pi(200)t)$, that make up the product in (3.2.4). The plot of the beat note is constructed by first drawing $2\cos(2\pi(20)t)$ and its negative version $-2\cos(2\pi(20)t)$ to define boundaries inside of which we then draw the higher frequency signal. The resulting beat note is plotted in the bottom panel of Fig. 3.3, where it can be seen that the effect of multiplying the higher-frequency sinusoid (200 Hz) by the lower-frequency sinusoid (at 20 Hz) is to change the amplitude envelope of the higher-frequency waveform. If we listen to such an $x(t)$ we can hear that the f_Δ variation causes the signal to fade in and out because the signal envelope is rising and falling, as in the lower panel of Fig. 3.3. This is the phenomenon called "beating" of tones in music.

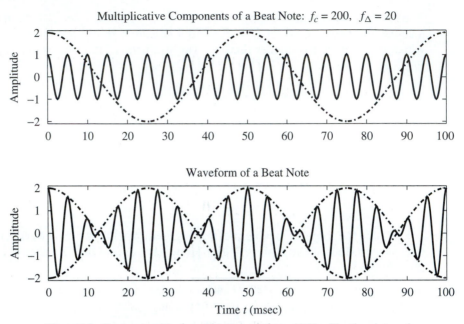

Figure 3.3   Beat note with $f_c = 200$ Hz and $f_\Delta = 20$ Hz. The time interval between nulls is $\frac{1}{2}(1/f_\Delta) = 25$ msec, which is dictated by the frequency difference.

SPECTRUM
DEMO:
SIMPLE
SOUNDS

If f_Δ is decreased to 9 Hz, we see in Fig. 3.4 (bottom) that the envelope of the 200 Hz tone changes much more slowly. The time interval between nulls of the envelope is $\frac{1}{2}(1/f_\Delta)$, so the more closely spaced the sinusoids, the slower the envelope variation. These figures are simplified somewhat by using cosines for both terms in (3.2.2), but other phase relationships would give similar patterns. Finally, remember that the spectrum for $x(t)$ in Fig. 3.3 contains frequency components at ± 220 Hz and ± 180 Hz, while the spectrum for Fig. 3.4 has frequencies ± 209 Hz and ± 191 Hz.

Musicians use this phenomenon as an aid in tuning two instruments to the same pitch. When two notes are close but not identical in frequency, the beating phenomenon is heard. As one pitch is changed to become closer to the other, the effect disappears, and the two instruments are then "in tune."

3.2.3 Amplitude Modulation

Multiplying sinusoids is also useful in modulation for communication systems. *Amplitude modulation* is the process of multiplying a low-frequency signal by a high-frequency sinusoid. It is the technique used to broadcast AM radio: In fact "AM" is just the abbreviation for amplitude modulation. The AM signal is a product of the form

$$x(t) = v(t)\cos(2\pi f_c t) \qquad (3.2.5)$$

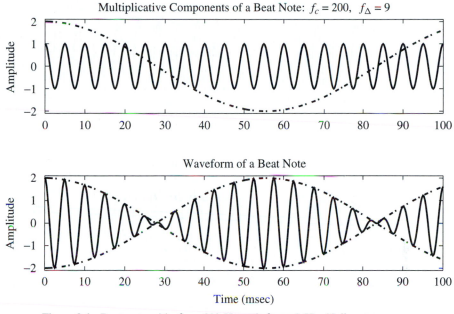

Figure 3.4   Beat note with $f_c = 200$ Hz and $f_\Delta = 9$ Hz. Nulls are now $\frac{1}{2}(1/f_\Delta) = 55.6$ msec. apart.

where it is assumed that the frequency of the cosine term (f_c Hz) is much higher than any frequencies contained in the spectrum of $v(t)$, which represents the voice or music signal to be transmitted. The cosine wave in (3.2.5) is called the *carrier signal*, and its frequency is called the *carrier frequency*.

With our limited knowledge, the form of $v(t)$ in (3.2.5) must be restricted to be a sum of sinusoids, but that is sufficient to understand how the modulation process works. If we let $v(t) = 5 + 2\cos(40\pi t)$ and $f_c = 200$ Hz, then the AM signal is a multiplication similar to the beat signal:

$$x(t) \;=\; (5 + 2\cos(40\pi t))\,\cos(400\pi t) \tag{3.2.6}$$

A plot of this signal is given in Fig. 3.5, where it can be seen that the effect of multiplying the higher-frequency sinusoid (200 Hz) by the lower-frequency sinusoid (at 20 Hz) is to "modulate" (or change) the amplitude envelope of the carrier waveform—hence the name amplitude modulation for a signal like $x(t)$. The primary difference between this AM signal and the beat signal is that the envelope never goes to zero. When the carrier frequency becomes very high compared to the frequencies in $v(t)$ as in Fig. 3.6, it is possible to see the outline of the modulating cosine without drawing the envelope signal explicitly. When detecting the modulation signal $v(t)$ in AM radio, this characteristic simplifies the implementation.

In the frequency domain, the AM signal spectrum is nearly the same as the beat signal, the only difference being a large term at $f = f_c$. The spectrum can be

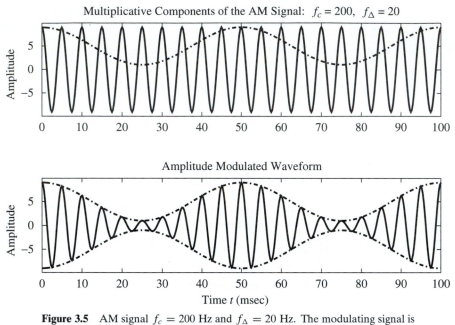

Figure 3.5 AM signal $f_c = 200$ Hz and $f_\Delta = 20$ Hz. The modulating signal is clearly visible.

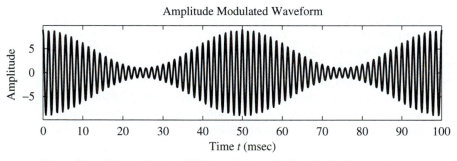

Figure 3.6 AM signal $f_c = 700$ Hz and $f_\Delta = 20$ Hz. The higher carrier frequency makes it possible to see the outline of the modulating cosine without drawing the envelope.

derived by first breaking the time–domain signal into two terms

$$x(t) = 5\cos(400\pi t) + 2\cos(40\pi t)\cos(400\pi t) \qquad (3.2.7)$$

and then using our previous knowledge about the beat signal to get the following additive combination for the spectrum:

$$x(t) = \tfrac{5}{2}e^{j400\pi t} + \tfrac{1}{2}e^{j440\pi t} + \tfrac{1}{2}e^{j360\pi t} + \tfrac{5}{2}e^{-j400\pi t} + \tfrac{1}{2}e^{-j440\pi t} + \tfrac{1}{2}e^{-j360\pi t}$$
$$(3.2.8)$$

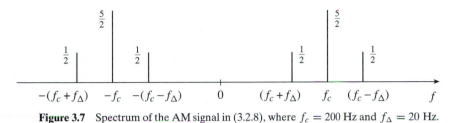

Figure 3.7   Spectrum of the AM signal in (3.2.8), where $f_c = 200$ Hz and $f_\Delta = 20$ Hz.

Thus there are six spectral components for $x(t)$ at the frequencies ± 220 Hz and ± 180 Hz, and also at the carrier frequency ± 200 Hz (Fig. 3.7). It is interesting to note that the spectrum for $x(t)$ contains two identical subsets, one centered at $f = f_c$ and the other at $f = -f_c$. These subsets each contain three spectral lines, and it is easy to show that they are just a frequency shifted version of the two-sided spectrum of $v(t)$.

Exercise 3.2.   Derive the spectrum of $v(t) = 5 + 2\cos(2\pi(20)t)$ and make a plot versus frequency. Compare this result to the spectral plot for the AM signal in Fig. 3.7.

3.3   PERIODIC WAVEFORMS

We have shown that when two cosine waves of the same frequency are added, the result is a cosine wave of the same frequency. Furthermore, the signal addition can be accomplished by adding the corresponding complex phasors, and then converting to the cosine form. We have also shown that multiplying sinusoids is equivalent to adding components at different frequencies. Another interesting situation occurs when we add two or more cosine waves having *harmonically* related frequencies, as in

$$x(t) = A_0 + \sum_{k=1}^{N} A_k \cos(2\pi k f_0 t + \phi_k) \qquad (3.3.1)$$

i.e., $x(t)$ is the sum of $N+1$ cosine waves whose frequencies are all integer multiples of f_0.[2] The frequency, f_k, of the kth cosine component in (3.3.1) is actually

$$f_k = k f_0 \qquad \text{(harmonic frequencies)}$$

This frequency is called a *harmonic* of f_0 because it is an integer multiple of the basic frequency f_0, which is called the *fundamental frequency*.

---

[2] The DC component is a cosine signal with zero frequency.

Using the phasor representation of the cosines, we can write

$$x(t) = X_0 + \Re\left\{\sum_{k=1}^{N} X_k e^{j2\pi k f_0 t}\right\} \tag{3.3.2}$$

where $X_0 = A_0$ and

$$X_k = A_k e^{j\phi_k} \tag{3.3.3}$$

What is the period of $x(t)$? By substituting into either (3.3.1) or (3.3.2), it is easily shown that $x(t + T_0) = x(t)$ for all t if $T_0 = 1/f_0$. Since T_0 is the reciprocal of the fundamental frequency, it is called the *fundamental period*.

3.3.1 Synthetic Vowel

As an example, consider the signal

VOWEL
DEMO

$$x(t) = \Re\left\{X_2 e^{j2\pi 2 f_0 t} + X_4 e^{j2\pi 4 f_0 t} + X_5 e^{j2\pi 5 f_0 t} + X_{16} e^{j2\pi 16 f_0 t} + X_{17} e^{j2\pi 17 f_0 t}\right\} \tag{3.3.4}$$

where the fundamental frequency is $f_0 = 100$ Hz and the complex amplitudes are listed in Table 3.1.[3] This signal approximates the waveform produced by a man

k	f_k **(Hz)**	X_k	**Mag**	**Phase (rad)**
1	100	0	0	0
2	200	$(771 + j12202)$	12,226	1.508
3	300	0	0	0
4	400	$(-8865 + j28048)$	29,416	1.876
5	500	$(48001 - j8995)$	48,836	−0.185
6	600	0	0	0
⋮	⋮	⋮	⋮	
15	1500	0	0	0
16	1600	$(1657 - j13520)$	13,621	−1.449
17	1700	$4723 + j0$	4723	0

Table 3.1: Complex amplitudes for harmonic signal that approximates the vowel sound "ah".

---

[3] Note that to simplify our expressions we have used a "one-sided" spectrum representation in (3.3.4) and in Table 3.1.

speaking the vowel sound "ah." The two-sided spectrum of this signal is plotted in Fig. 3.8. Note that all the frequencies are multiples of 100 Hz, even though there is no spectral component at 100 Hz itself. Also note that the negative frequency components have phase angles that are the negative of the phase angles of the corresponding positive frequency components; i.e., the complex phasors for the negative frequencies are the complex conjugates of the phasors for the corresponding positive frequencies.

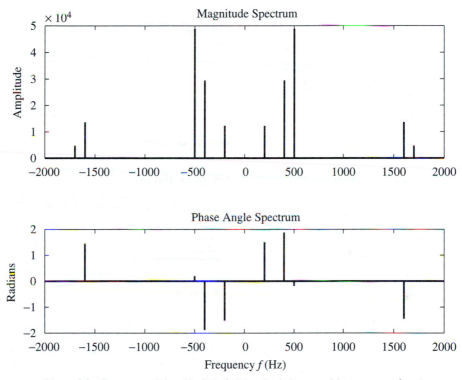

Figure 3.8   Spectrum of signal in (3.3.4). Magnitude is even with respect to $f = 0$; phase is odd.

The synthetic vowel signal has ten spectral components, but only five when the real part is taken as in (3.3.4). It is interesting to examine the contribution of each real component separately. We can do this by successively plotting the waveforms corresponding to only one term, then two terms, etc. Figure 3.9 (top) shows a plot of the first term in (3.3.4) alone. Note that since the frequency of this component is $2f_0 = 200$ Hz, the waveform is periodic with period $1/200 = 5$ msec. Figure 3.9 (bottom) shows a plot of the sum of the first two terms.

Now notice that the two frequencies are multiples of 200 Hz, so the period of the waveform is still 5 msec. Figure 3.10 (top) shows a plot of the sum of the first three terms. Now we see that the period of the waveform is increased to 10 msecs. This is because the three frequencies, 200, 400, and 500 Hz are integer multiples of 100 Hz; i.e., the fundamental frequency is now 100 Hz. Figure 3.11 shows the sum of all the terms in (3.3.4).

Note that even though there is no component with frequency f_0, the waveform is nevertheless periodic with period $T_0 = 10$ msec. Furthermore, the waveform becomes increasingly complicated and more rapidly varying as higher-frequency components such as the 16th and 17th harmonics are added. The waveform in Fig. 3.11 is typical of waveforms for vowel sounds in speech.

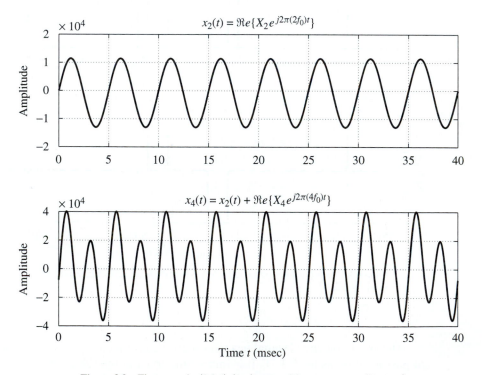

Figure 3.9   First term in (3.3.4) (top); sum of first two terms (bottom).

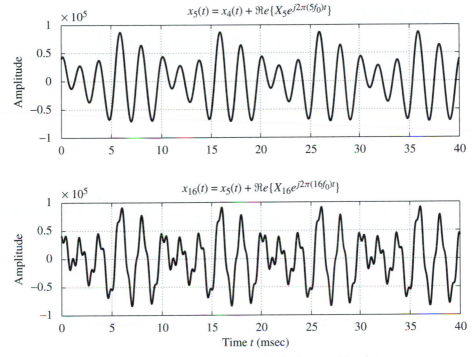

$$x_5(t) = x_4(t) + \Re e\{X_5 e^{j2\pi(5f_0)t}\}$$

$$x_{16}(t) = x_5(t) + \Re e\{X_{16} e^{j2\pi(16f_0)t}\}$$

Time t (msec)

Figure 3.10   Sum of first three terms in (3.3.4) (top); sum of first four terms (bottom).

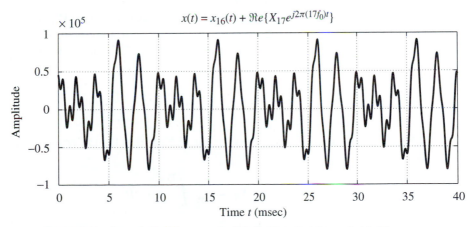

$$x(t) = x_{16}(t) + \Re e\{X_{17} e^{j2\pi(17f_0)t}\}$$

Time t (msec)

Figure 3.11   Sum of all of the terms in (3.3.4). Note that the period is 10 msec, which equals $1/f_0$.

3.4 MORE PERIODIC SIGNALS

We can synthesize *periodic* waveforms by using a sum of sinusoids, as long as we constrain the frequencies to be harmonically related. In fact, the theory underlying Fourier series tells us that any periodic signal can be approximated with a sum of harmonically related sinusoids, although the sum may need an infinite number of terms. Thus we consider the general synthesis formula:

$$x(t) = A_0 + \sum_{k=1}^{\infty} A_k \cos(2\pi k f_0 t + \phi_k) = X_0 + \Re e \left\{ \sum_{k=1}^{\infty} X_k e^{j2\pi k f_0 t} \right\} \quad (3.4.1)$$

where the frequencies kf_0 are all multiples of a fundamental frequency f_0. By clever choice of the complex amplitudes $X_k = A_k e^{j\phi_k}$ in (3.4.1), we can approximate a number of interesting waveforms, such as square waves, triangle waves, and so on. The fact that a discontinuous square wave can be approximated with a large number of sinusoids was one of the amazing parts of Fourier's famous thesis of 1807.

To demonstrate harmonic synthesis of waveshapes that do not look sinusoidal, we work out the following two cases, which can be synthesized from a simple formula for the X_k. If you have access to MATLAB, it is straightforward to write a function that takes a list of frequencies and a list of complex amplitudes and then produces a signal as the sum of several cosines or complex exponentials as in (3.4.1). This is an exercise in Lab C.2.

3.4.1 Fourier Series: Analysis

How do we derive the coefficients for the harmonic sum (3.4.1), i.e., how do we go from $x(t)$ to X_k? The answer is that we use *Fourier series*. Starting from $x(t)$ and calculating X_k is called *Fourier analysis*.

The reverse process of starting from X_k and generating $x(t)$ is called *Fourier synthesis*. We have already done several synthesis examples. The analysis problem is a bit harder, so we are not going to derive the procedure; instead, we will just state the answer. The complex amplitudes for any periodic signal can be calculated with the Fourier integral

$$X_k = \frac{2}{T_0} \int_0^{T_0} x(t) e^{-j2\pi k t / T_0} \, dt \quad (3.4.2)$$

where T_0 is the fundamental period of $x(t)$.[4] Note that the DC component is obtained by

$$X_0 = \frac{1}{T_0} \int_0^{T_0} x(t) \, dt \quad (3.4.3)$$

---

[4] The factor of 2 in (3.4.2) is due to our representation of the Fourier series as a sum of cosines. In other texts, this factor may not appear due to a different definition of the coefficients.

which is simply the average value of the signal over one period. Many issues are involved with a careful definition of the Fourier series, issues that occupied great mathematicians for much of the nineteenth century. We will not present a detailed exploration of the Fourier series representation. We have introduced it here simply to show that it is possible to analyze a signal into its sinusoidal components. Two examples will be presented to illustrate this point.

The integral (3.4.2) is convenient if we have a formula for $x(t)$. On the other hand, if $x(t)$ is known only as a recording, then numerical methods are needed. Some of these will be discussed in Chapter 9.

3.4.2 The Square Wave

The simplest example to consider is the periodic square wave, which is defined for one cycle by

$$x(t) = \begin{cases} 1 & 0 \le t < \tfrac{1}{2}T_0 \\ -1 & \tfrac{1}{2}T_0 \le t < T_0 \end{cases} \tag{3.4.4}$$

Exercise 3.3. Draw a plot of the square wave defined in (3.4.4) for $T_0 = 0.04$ sec.

We will use (3.4.2) to calculate a formula for the complex amplitudes X_k. First observe that the average value of this signal is zero, so $X_0 = 0$. Then, substituting the definition of $x(t)$ into the integral (3.4.2) and splitting the integral into two terms gives

$$X_k = \frac{2}{T_0} \int_0^{\frac{1}{2}T_0} (1)e^{-j2\pi kt/T_0}\, dt + \frac{2}{T_0} \int_{\frac{1}{2}T_0}^{T_0} (-1)e^{-j2\pi kt/T_0}\, dt$$

which can be manipulated as follows:[5]

$$
\begin{aligned}
X_k &= \frac{2}{T_0} \frac{e^{-j2\pi k(\frac{1}{2}T_0)/T_0} - e^{-j2\pi k(0)/T_0}}{-j2\pi k/T_0} + \frac{(-2)}{T_0} \frac{(e^{-j2\pi kT_0/T_0} - e^{-j2\pi k(\frac{1}{2}T_0)/T_0})}{-j2\pi k/T_0} \\[2mm]
&= \frac{e^{-j\pi k} - 1}{-j\pi k} + \frac{e^{-j\pi k} - e^{-j2\pi k}}{-j\pi k} \\[2mm]
&= \frac{2 - 2e^{-j\pi k}}{j\pi k} = \frac{2(1 - (-1)^k)}{j\pi k}
\end{aligned}
$$

The last form for X_k has a numerator that is either 0 (for k even) or 4 (for k odd), so we get the final answer for the Fourier series coefficients of the square wave:

---

[5] We use the fact that $e^{-j2\pi k} = 1$ when k is an integer.

$$X_k = \begin{cases} \dfrac{4}{j\pi k} & k = \pm 1, \pm 3, \pm 5, \ldots \\ 0 & k = 0, \pm 2, \pm 4, \pm 6, \ldots \end{cases} \qquad (3.4.5)$$

The magnitude of these coefficients is shown in Fig. 3.12. The phase angles are $-\pi/2$ for $k > 0$, and $\pi/2$ for $k < 0$. Note that if $f_0 = 1/T_0 = 25$ Hz, only the frequencies at $\pm 25, \pm 75, \pm 125$, etc. are in the spectrum.

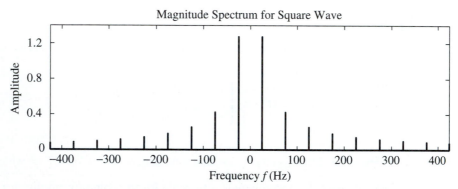

Figure 3.12   Spectrum of the square-wave signal whose Fourier series coefficients are given in (3.4.5) with $f_0 = 1/T_0 = 25$ Hz.

Using a simple MATLAB M-file, a synthesis was done with the fundamental frequency $f_0 = 25$ Hz and $f_k = kf_0$. In Fig. 3.13, the plots are shown for three different cases where the number of terms in the sum is $N = 3, 7$, and 17. Notice how the period of the synthesized waveform is always the same, because it is related to the fundamental frequency.

Obviously, a square-wave signal waveshape is approximated with this sum of cosines as more terms are added. However, notice what happens as N increases: The sum of cosines appears to converge to the constant values +1 and −1, but the convergence is not uniformly good—the "ears" at the discontinuous step never go away completely. This behavior, which will occur at any discontinuity of the waveform, is called the *Gibbs phenomenon*, and it is one of the interesting subtleties of Fourier theory that is extensively studied in advanced treatments.

3.4.3 Triangle Wave

Another set of interesting coefficients is obtained by squaring and scaling the complex amplitudes given in (3.4.5):

$$X_k = \begin{cases} \dfrac{-8}{\pi^2 k^2} & k \text{ an odd integer} \\ 0 & k \text{ an even integer} \end{cases} \qquad (3.4.6)$$

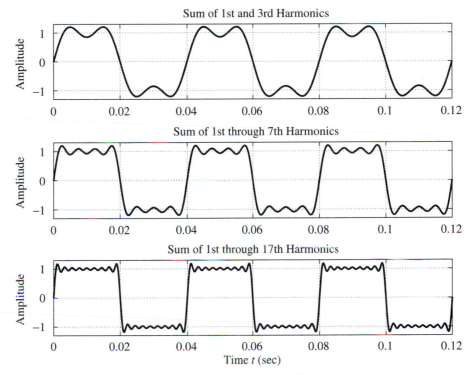

Figure 3.13   Summing harmonic components: three (top); seven (middle); and seventeen (bottom).

With the fundamental frequency equal to $f_0 = 25$ Hz, two cases ($N = 3$ and 11) are shown in Fig. 3.14. This set of coefficients seems to approximate a triangularly shaped waveform. Note that the periodic triangular wave is not discontinuous and the approximation in this case is much smoother. Indeed, the triangular wave is the integral of the square wave. The theory of Fourier series would confirm that squaring the coefficients of the square wave would result in the triangular wave.

3.4.4 Example of a Non-periodic Signal

When we add harmonically related complex exponentials, we get a periodic result. What happens when the frequencies have no simple relation to one another? The sinusoidal synthesis formula

$$x(t) = A_0 + \sum_{k=1}^{N} A_k \cos(2\pi f_k t + \phi_k)$$

$$= A_0 + \sum_{k=1}^{N} \left(\tfrac{1}{2} A_k e^{j\phi_k} e^{j2\pi f_k t} + \tfrac{1}{2} A_k e^{-j\phi_k} e^{-j2\pi f_k t} \right)$$

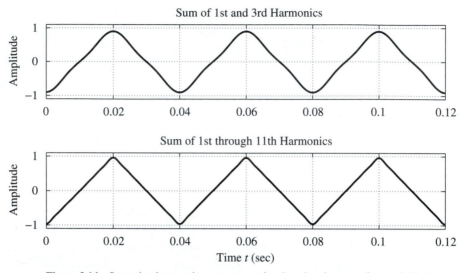

Figure 3.14 Summing harmonic components for the triangle wave: first and third harmonics (top); up to and including the eleventh harmonic (bottom).

is still valid, but now we will make no assumptions about the individual frequencies f_k.

We want to demonstrate only that periodicity is tied to harmonic frequencies. We can do this by taking a specific example. Consider the harmonic signal $x_h(t)$ made up from the first, third, and fifth harmonics of a square wave with fundamental frequency $f_0 = 10$ Hz:

$$x_h(t) = 2\cos(20\pi t) - \tfrac{2}{3}\cos(20\pi(3)t) + \tfrac{2}{5}\cos(20\pi(5)t)$$

A plot of $x_h(t)$ is shown in Fig. 3.15 using a "strip chart" format. The plot consists of three lines, each one containing 2 sec of the signal. The first line starts at $t = 0$, the second at $t = 2$, and the third at $t = 4$. This lets us view a long section of the signal, which in this case is clearly periodic, with period equal to 1/10 sec.

Sound of
Square Wave

Now we create a second signal that is just a slight perturbation from the first. Define $x_2(t)$ to be the sum of three sinusoids:

$$x_2(t) = 2\cos(20\pi t) - \tfrac{2}{3}\cos(20\pi\sqrt{8}t) + \tfrac{2}{5}\cos(20\pi\sqrt{27}t)$$

The amplitudes are the same, but the frequencies have been changed slightly. The plot of in Fig. 3.16 shows that $x_2(t)$ is not periodic.

Sound:
Non-periodic

Sum of Cosine Waves with Harmonic Frequencies

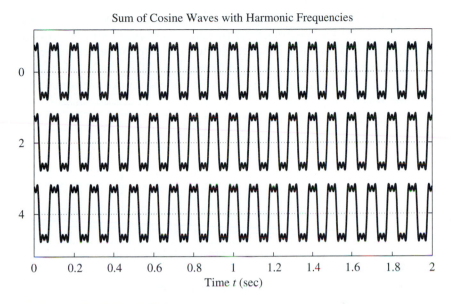

Figure 3.15   Sum of cosine waves of harmonic frequencies.

Sum of Cosine Waves with Nonharmonic Frequencies

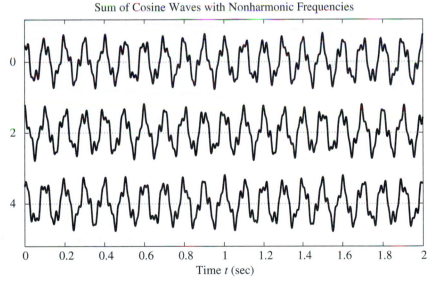

Figure 3.16   Sum of cosine waves of different frequencies. No matter how hard you try, you cannot find a repetition in this signal.

Figure 3.17 will help explain the difference between the signals in Figs. 3.16 and 3.15. In Fig. 3.15, the frequencies are integer multiples of a common frequency, $f_0 = 10$, so the waveform of Fig. 3.15 is periodic with period $T_0 = 1/10$ sec. The waveform of Fig. 3.16 is *nonperiodic*. We can see why this is so by examining Fig. 3.17, which shows the spectra of the two signals in Figs. 3.16 and 3.15. These *spectrum* plots show "how much" of each cosine wave is in the sum, and they are very similar. However, the frequencies are different: $10\sqrt{8} = 28.28\ldots \approx 30$ and $10\sqrt{27} = 51.96\ldots \approx 50$. These slight shifts of frequency make a dramatic difference in the time waveform.

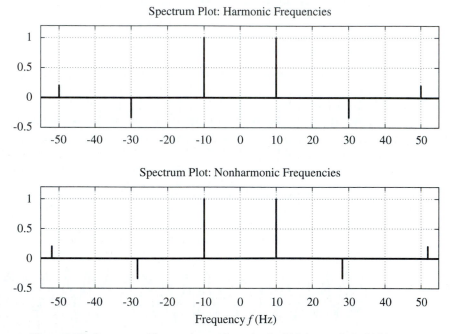

Figure 3.17   Spectrum of harmonic waveform (top) which has period of 1/10 sec, and a nonharmonic waveform (bottom) that is not periodic.

3.5 TIME–FREQUENCY SPECTRUM

We have seen that a wide range of interesting waveforms can be synthesized by the equation

$$x(t) = A_0 + \sum_{k=1}^{N} A_k \cos(2\pi f_k t + \phi_k) \tag{3.5.1}$$

These waveforms range from constants, to cosine signals, to general periodic signals, to complicated-looking signals that are not periodic. One assumption we have made so far is that the amplitudes, phases, and frequencies in (3.5.1) do not change with time. However, most real-world signals exhibit frequency changes over time. Music is the best example. For very short time intervals, the music may have a "constant" spectrum, but over the long term, the frequency content of the music changes dramatically. Indeed, the changing frequency spectrum is the very essence of music. Human speech is another good example. Vowel sounds, if held for a long time, exhibit a "constant" nature because the vocal tract resonates with its characteristic frequency components. However, as we speak different words, the frequency content is continually changing. In any event, most interesting signals can be modeled as a sum of sinusoids if we let the frequencies, amplitudes, and phases vary with time. Therefore, we need a way to describe such time–frequency variations. This leads us to the concept of a time–frequency spectrum or *spectrogram*.

The mathematical concept of a time–frequency spectrum is a sophisticated idea, but the intuitive notion of such a spectrum is supported by common, everyday examples. The best example to cite is musical notation (Fig. 3.18). A musical score specifies how a piece is to be played by giving the notes to be played, the time duration of each note, and the starting time of each. The notation itself is not completely obvious, but the horizontal "axis" in Fig. 3.18 is time, while the vertical axis is frequency. The time duration for each note varies depending on whether it is a whole note, half note, quarter note, eighth, sixteenth, etc. In Fig. 3.18 most of the notes are sixteenth notes, indicating that the piece should be played briskly. If we assign a time duration to a sixteenth note, then all sixteenth notes should have the same duration. An eighth note would have twice the duration of a sixteenth note, and a quarter note would have twice the duration of an eighth note, etc.

Figure 3.18   Sheet-music notation is a time–frequency diagram.

The vertical axis has a much more complicated notation to define frequency. If you look carefully at Fig. 3.18, you will see that the black dots that mark the notes lie either on one of the horizontal lines or in the space betweeen two lines. Each of these denotes a white key on the piano keyboard depicted in Fig 3.19. The black keys on the piano are denoted by "sharps" (♯) or flats" (♭). Figure 3.18 has a few notes sharped. The musical score is divided into a treble section (the top five lines) and a bass section (the bottom five lines). The vertical reference point for the notes is "middle C," which lies on an invisible horizontal line between the treble and bass sections (key number 40 in Fig. 3.19). Thus the bottom horizontal line in the treble section represents the white key (E) that is two above middle C; i.e. key number 44 in Fig. 3.19.

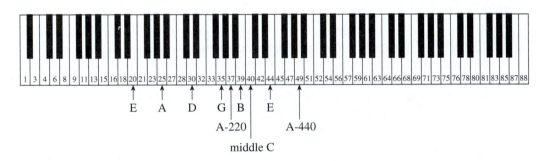

Figure 3.19   Piano keys can be numbered from 1 to 88. Middle C is key 40. A–440 is key 49.

Once the mapping from the musical score to the piano keys has been made, we can write a mathematical formula for the frequency. A piano, which has 88 keys, is divided into octaves containing twelve keys each. The meaning of the word *octave* is a doubling of the frequency. Within an octave, the neighboring keys maintain a constant frequency ratio. Since there are twelve keys per octave, the ratio (r) is

$$r^{12} = 2 \qquad \Longrightarrow \qquad r = 2^{1/12} = 1.0595$$

With this ratio, we can compute the frequencies of all keys if we have one reference. The convention is that the A key above middle C, called A–440, is at 440 Hz. Since A–440 is key number 49 and middle C is key number 40, the frequency of middle C is

$$f_{\text{middle C}} = 440 \times 2^{(40-49)/12} \approx 262 \text{ Hz}$$

LAB: CHIRP
SYNTHESIS

It is not our objective to explain how to read sheet music, although two of the lab projects will investigate methods for synthesizing waveforms to create songs and musical sounds. What is interesting about musical notation is that it uses a two-dimensional display to indicate frequency content that changes with time. If we adopt a similar notation, we can specify how to synthesize sinusoids with time-varying frequency content. Our notation is illustrated in Fig. 3.20.

3.5.1 Stepped Frequency

The simplest example of time-varying frequency content is to make a waveform whose frequency holds for a short duration and then steps to a higher (or lower) frequency. An example from music would be to play a "scale" that would be a succession of notes progressing over one octave. For example, the C-major scale consists of playing the notes {C, D, E, F, G, A, B, C} one after another, starting at middle C. This scale is played completely on the white keys. The frequencies of these notes are:

Middle C	D	E	F	G	A	B	C
262 Hz	294	330	349	392	440	494	523

Figure 3.20 should be interpreted as follows: Synthesize the frequency 262 Hz for 200 msec, then the frequency 294 Hz during the next 200 msec, and so on. The total waveform duration will be 1.6 sec. In music notation, the notes would be written as in Fig. 3.21(top), where each note is a quarter note.

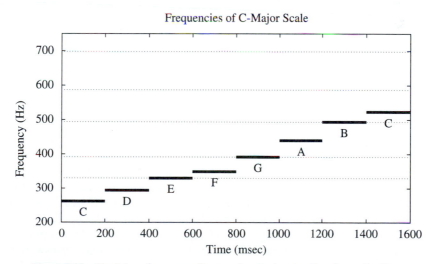

Figure 3.20   Ideal time-frequency diagram for playing the C-major scale. The horizontal dotted lines correspond to the five lines in the treble staff.

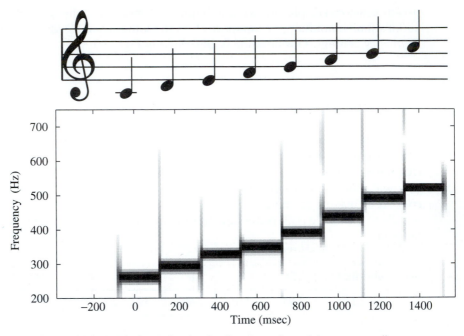

Figure 3.21   Musical notation for the C-major scale, and the corresponding spectrogram computed using MATLAB's `specgram` function.

3.5.2 Spectrogram Analysis

The frequency content of a signal can be considered from two points of view: analysis or synthesis. We have been dealing almost exclusively with synthesis. For example, the ideal time–frequency diagram in Figure 3.20 specifies a rule for synthesizing the C-major scale. Analysis is a more challenging problem, as we saw in Section 3.4.1, where the Fourier series analysis formulas were given.

Analysis for time-varying frequencies is usually considered a subject reserved for advanced graduate courses. One reason is that we cannot write a simple mathematical formula like the Fourier series integral to do the analysis. On the other hand, excellent numerical routines are now available for time–frequency analysis. Specifically, we can compute a *spectrogram*, which is a two-dimensional function of time and frequency that displays the time variation of the spectral content of a signal.

In MATLAB, the function `specgram` will compute the spectrogram, and its default values work well for most signals.[6] Therefore, it is reasonable to see what sort of output can be produced by the `specgram` function. Figure 3.21 shows the results of applying `specgram` to the stepped sinusoids that make up the C-major scale. The calculation is performed by doing a frequency analysis on short segments of the signal and plotting the results at the specific time at which the analysis is done. By repeat-

---

[6] The *DSP First* toolbox contains an equivalent function called `spectgr`.

ing this process with slight displacements in time, a two-dimensional array is created whose magnitude can be displayed as a gray-scale image whose horizontal axis is time and whose vertical axis is frequency. The time axis must be interpreted as the "time of analysis" because the frequency calculation is not instantaneous; rather, it is based on a finite segment of the signal—in this case, 25.6 msec.

It is quite easy to identify the frequency content due to each note, but there are also some interfering artifacts that make the spectrogram in Fig. 3.21 less than ideal. In Chapter 9, we will undertake a discussion of frequency *analysis* in order to explain how the spectrogram is calculated and how one should choose the analysis parameters to get a good result. Even though the spectrogram is a highly advanced idea in signal analysis, its application is relatively easy and intuitive, especially for music signals which are described symbolically by a notation that is very much like a spectrogram.

3.6   FREQUENCY MODULATION: CHIRP SIGNALS

LAB: CHIRP
SYNTHESIS

Section 3.5 revealed the possibility that interesting sounds can be created when the frequency varies as a function of time. In this section, we use a different mathematical formula to create signals whose frequency is time-varying. We will also pursue this idea in Lab C.4.

3.6.1 Chirp, or Linearly Swept Frequency

A "chirp" signal is a swept-frequency signal whose frequency changes linearly from some low value to a high one. For example, in the audible region, we might begin at 220 Hz and go up to 2320 Hz. One method for producing such a signal would be to concatenate a large number of short constant-frequency sinusoids, whose frequencies step from low to high. This approach has one notable disadvantage: The boundary between the short sinusoids will be discontinuous unless we are careful to adjust the initial phase of each small sinusoid. Figure 3.22 shows a time waveform where the frequency is being stepped. Notice the jumps at $t = 1, 2, 3, 4, 5$ secs, which, in this case, are due to using $\phi = 0$ for each small sinusoidal segment.

A better approach is to modify the formula for the sinusoid so that we get a time-varying frequency. Such a formula can be derived from the complex-exponential point of view. If we regard a constant-frequency sinusoid as the real part of a complex (rotating) phasor:

$$x(t) = \Re\{Ae^{j(\omega_0 t + \phi)}\} = A\cos(\omega_0 t + \phi) \tag{3.6.1}$$

then the *phase*[7] of this signal is the exponent $(\omega_0 t + \phi)$ which obviously changes *linearly* with time. The time derivative of the phase is ω_0, which equals the constant frequency.

---

[7] Here we use the term phase to mean the *angle* of the cosine wave. Recall that the constant ϕ is the *phase-shift*.

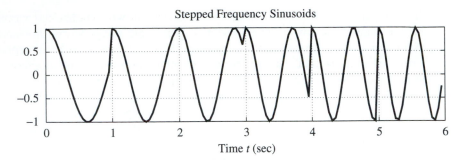

Figure 3.22   Stepped-frequency sinusoid, with frequencies 1 Hz, 1.1 Hz, 1.2 Hz, 1.3 Hz and so on. The frequency changes once per second.

Therefore, we adopt the following general notation for the class of signals with time-varying phase:

$$x(t) = \Re\{Ae^{j\psi(t)}\} = A\cos(\psi(t)) \tag{3.6.2}$$

where $\psi(t)$ denotes the phase as a function of time. For example, we can create a signal with quadratic phase by defining

$$\psi(t) = 2\pi\mu t^2 + 2\pi f_0 t + \phi \tag{3.6.3}$$

Now we can define the *instantaneous frequency* for these signals as the slope of the phase (i.e., its derivative)

$$\omega_i(t) = \frac{d}{dt}\psi(t) \quad \text{(rad/sec)} \tag{3.6.4}$$

where the units of $\omega_i(t)$ are rad/sec, or, if we divide by 2π

$$f_i(t) = \frac{1}{2\pi}\frac{d}{dt}\psi(t) \quad \text{(Hz)} \tag{3.6.5}$$

SPECTRO-
GRAM
DEMO:
CHIRP
SOUNDS

we obtain Hz. If the phase of $x(t)$ is quadratic, then its frequency changes *linearly* with time, i.e.,

$$f_i(t) = 2\mu t + f_0$$

The frequency variation produced by the time-varying phase is called frequency modulation, and signals of this class are called "FM signals." Finally, since the linear variation of the frequency can produce an audible sound similar to a siren or a chirp, the linear-FM signals are also called "chirp" signals, or simply "chirps."

The process can be reversed: if a certain linear frequency sweep is desired, the actual phase needed in (3.6.2) is obtained from the integral of $\omega_i(t)$. Returning to the example above, we can synthesize a frequency sweep from $f_1 = 220$ Hz to $f_2 = 2320$ Hz over the time interval $t = 0$ to $t = T_2 = 3$ sec if we first create a formula for the instantaneous frequency

$$f_i(t) = \frac{f_2 - f_1}{T_2}t + f_1 = \frac{2320 - 220}{3}t + 220,$$

and then integrate to get the following phase function:

$$\psi(t) = \int_0^t \omega_i(u)\, du$$

$$= \int_0^t 2\pi \left(\frac{2320 - 220}{3}u + 220\right) du$$

$$= \int_0^t 2\pi (700u + 220)\, du$$

$$= 700\pi t^2 + 440\pi t + \phi$$

where the phase-shift ϕ is an arbitrary constant.

3.6.2 A Closer Look at Instantaneous Frequency

It may be difficult to see why the derivative of the phase would be the instantaneous frequency. The following experiment provides a clue.

1. Use the following parameters to define a "chirp" signal:

$$f_1 = 100 \text{ Hz}$$
$$f_2 = 500 \text{ Hz}$$
$$T_0 = 0.04 \text{ sec}$$

In other words, determine μ and f_0 in (3.6.3) to define $x(t)$ so that it sweeps the specified frequency range.

2. Now make a plot of the signal synthesized in (1). In Fig. 3.23, this plot is the middle panel.

3. It is difficult to verify whether or not this chirp signal will have the correct frequency content. However, the rest of this experiment will demonstrate that the derivative of the phase is the "correct" definition of instantaneous

frequency. First of all, plot a 300-Hz sinusoid, $x_1(t)$ which is shown in the upper panel of Fig. 3.23.

4. Finally, generate and plot a 500-Hz sinusoid, $x_1(t)$ as in the bottom panel of Fig. 3.23.

5. Now compare the three signals in Fig. 3.23 with respect to the frequency content of the chirp. Concentrate on the frequency of the chirp in the time range $0.019 \leq t \leq 0.021$ sec. Notice which sinusoid matches the chirp in this time region. Evaluate the theoretical $f_i(t)$ in this region. Look for the region where the chirp frequency is equal (locally) to 500 Hz.

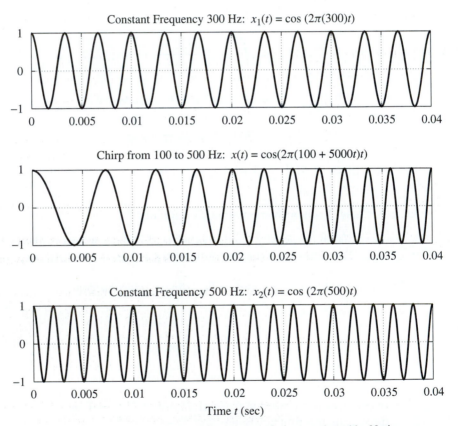

Figure 3.23 Comparing chirp signal to constant-frequency sinusoids. Notice where the local frequency of the chirp is equal to one of the sinusoids.

We have seen that for signals of the form $x(t) = A \cos(\psi(t))$, the instantaneous frequency of the signal is the derivative of the phase $\psi(t)$. If $\psi(t)$ is constant, the frequency is zero. If $\psi(t)$ is linear, $x(t)$ is a sinusoid at some fixed frequency. If

FM
SYNTHESIS

$\psi(t)$ is quadratic, $x(t)$ is a chirp signal whose frequency changes linearly versus time. More complicated variations of $\psi(t)$ can produce a wide variety of signals. One application of FM signals is in music synthesis. This application is illustrated with demos on the CD-ROM and in Lab C.4.

3.7   SUMMARY AND LINKS

This chapter introduced the concept of the *spectrum*, in which we represent a signal by its sinusoidal components. The spectrum is a graphical presentation of the complex amplitude for each frequency component in the signal. We showed how complicated signals can be formed from relatively simple spectra, and we ended with a discussion of how the spectrum can vary with time.

LINKS

At this point, so many different demonstrations and projects can be done that we must limit our list somewhat. Among the laboratory projects in Appendix C, we have provided three on the CD-ROM: Lab C.2 (b) contains exercises on the Fourier series representation of the square wave and a sawtooth wave. Lab C.3 requires students to develop a music synthesis program to play a piece such as Bach's "Jesu, Joy of Man's Desiring." This synthesis must be done with sinusoids, but can be refined with extras such as a tapered amplitude envelope. Lab C.4 deals with beat notes, chirp signals, and spectrograms. The second part of this lab involves a music synthesis method based on frequency modulation. The FM-synthesis algorithm can produce realistic sounds for instruments such as a clarinet or a drum. In addition, Lab C.7 involves some practical systems that work with sinusoidal signals, such as a touch-tone phone. This later lab, however, requires some knowledge of filtering. Write-ups of the labs are also found on the CD-ROM.

The CD-ROM also contains many demonstrations of sounds and their spectrograms:

1. Spectrograms of simple sounds such as sine waves, square waves, and other harmonics.

2. Spectrograms of realistic sounds, including a piano recording, a synthetic scale, and a synthesized music passage done by one of the students who took an early version of this course.

3. Spectrograms of chirp signals that show how the rate of change of the frequency affects the sound you hear.

4. An explanation of the FM-synthesis method for emulating musical instruments. Several example sounds are included for listening.

Finally, the reader is reminded of the large number of solved homework problems that are available for review and practice on the CD-ROM.

PROBLEMS

3.1 In Section 3.2.2, we discussed a simple example of the "beating" of one cosine wave against another. In this problem, you will consider a more general case. Let

$$x(t) = A \cos[2\pi(f_c - f_\Delta)t] + B \cos[2\pi(f_c + f_\Delta)t]$$

For the case discussed in Section 3.2.2, $A = B = 1$.

(a) Use phasors to obtain a complex signal $\bar{x}(t)$ such that

$$x(t) = \Re\{\bar{x}(t)\}$$

(b) By manipulating the expression for $\bar{x}(t)$ and then taking the real part, show that in the more general case above, $x(t)$ can be expressed in the form

$$x(t) = C \cos(2\pi f_\Delta t) \cos(2\pi f_c t) + D \sin(2\pi f_\Delta t) \sin(2\pi f_c t)$$

and find expressions for C and D in terms of A and B. Check your answer by substituting $A = B = 1$.

(c) Find values for A and B so that

$$x(t) = 2 \sin(2\pi f_\Delta t) \sin(2\pi f_c t)$$

Plot the spectrum of this signal.

3.2 A signal composed of sinusoids is given by the equation

$$x(t) = 10 \cos(800\pi t + \pi/4) + 7 \cos(1200\pi t - \pi/3) - 3 \cos(1600\pi t)$$

(a) Sketch the spectrum of this signal, indicating the complex size of each frequency component. Make separate plots for real/imaginary or magnitude/phase of the complex amplitudes at each frequency.

(b) Is $x(t)$ periodic? If so, what is the period?

(c) Now consider the signal $y(t) = x(t) + 5 \cos(1000\pi t + \pi/2)$. How is the spectrum changed? Is $y(t)$ periodic? If so, what is the period?

3.3 A signal $x(t)$ has the two-sided spectrum representation shown here.

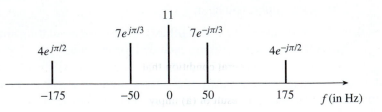

(a) Write an equation for $x(t)$ as a sum of cosines.

(b) Is $x(t)$ a periodic signal? If so, what is its period?

(c) Explain why "negative" frequencies are needed in the spectrum.

3.4 Let $x(t) = \sin^3(27\pi t)$.

(a) Determine a formula for $x(t)$ as the real part of a sum of complex exponentials.

(b) What is the fundamental period for $x(t)$?

(c) Plot the *spectrum* for $x(t)$.

3.5 Consider the signal

$$x(t) = 10 + 20\cos(2\pi(100)t + \pi/4) + 10\cos(2\pi(250)t) \qquad (3.8.1)$$

(a) Using Euler's relation, the signal $x(t)$ defined by (3.8.1) can be expressed as a sum of complex exponential signals in the form

$$x(t) = X_0 + \Re e\left\{\sum_{k=1}^{N} X_k e^{jk2\pi f_0 t}\right\} \qquad (3.8.2)$$

What is ω_0, and what are the values of X_k? *It is not necessary to evaluate any integrals to obtain X_k.*

(b) Is the signal $x(t)$ periodic? If so, what is the period?

(c) Plot the spectrum of this signal.

3.6 An amplitude-modulated (AM) cosine wave is represented by the formula

$$x(t) = [12 + 7\sin(\pi t - \pi/3)]\cos(13\pi t)$$

(a) Use phasors to show that $x(t)$ can be expressed in the form:

$$x(t) = A_1\cos(\omega_1 t + \phi_1) + A_2\cos(\omega_2 t + \phi_2) + A_3\cos(\omega_3 t + \phi_3)$$

where $\omega_1 < \omega_2 < \omega_3$; i.e., find A_1, A_2, A_3, ϕ_1, ϕ_2, ϕ_3, ω_1, ω_2, ω_3.

(b) Sketch the two-sided spectrum of this signal on a frequency axis. Be sure to label important features of the plot. Label your plot in terms of the numerical values of A_i, ϕ_i, and ω_i.

3.7 Consider a signal $x(t)$ such that

$$x(t) = 2\cos(\omega_1 t)\cos(\omega_2 t) = \cos[(\omega_2 + \omega_1)t] + \cos[(\omega_2 - \omega_1)t]$$

where $0 < \omega_1 < \omega_2$.

(a) What is the general condition that must be satisfied by $\omega_2 - \omega_1$ and $\omega_2 + \omega_1$ so that $x(t) = x(t + T_0)$, i.e., so that $x(t)$ is periodic with period T_0?

(b) What does the result of (a) imply about ω_1 and ω_2?

3.8   We have seen that musical tones can be modeled mathematically by sinusoidal signals. If you read music or play the piano, you know that the piano keyboard is divided into octaves, with the tones in each octave being twice the frequency of the corresponding tones in the next lower octave. To calibrate the frequency scale, the reference tone is the A above middle C, which is usually called A–440, since its frequency is 440 Hz. Each octave contains 12 tones, and the ratio between the frequencies of successive tones is constant. Thus, the ratio must be $2^{1/12}$. Since middle C is 9 tones below A–440, its frequency is approximately $(440)2^{-9/12} \approx 261$ Hz. The names of the tones (notes) of the octave starting with middle C and ending with high C are:

Note name	C	C#	D	E♭	E	F	F#	G	G#	A	B♭	B	C
Note number	40	41	42	43	44	45	46	47	48	49	50	51	52
Frequency													

(a) Make a table of the frequencies of the tones of the octave beginning with middle C, assuming that the A above middle C is tuned to 440 Hz.

(b) The above notes on a piano are numbered 40 through 52. If n denotes the note number and f denotes the frequency of the corresponding tone, give a formula for the frequency of the tone as a function of the note number.

(c) A *chord* is a combination of musical notes sounded simultaneously. A *triad* is a three-note chord. The D-major chord is composed of the tones of D, $F^{\#}$, and A sounded simultaneously. From the set of corresponding frequencies determined in (a), make a sketch of the essential features of the spectrum of the D-major chord assuming that each note is realized by a pure sinusoidal tone. (Do not specify the complex phasors precisely.)

3.9   A periodic signal is given by the equation

$$x(t) = 2 + 4\cos(40\pi t - \pi/5) + 3\sin(60\pi t) + 4\cos(120\pi t - \pi/3)$$

(a) Determine the fundamental frequency ω_0, the period T_0, and the coefficients in the representation

$$x(t) = X_0 + \Re e\left\{ \sum_{k=1}^{N} X_k e^{jk\omega_0 t} \right\}$$

for the above input. *Remember, you can do this problem **without** evaluating any integrals.*

(b) Sketch the spectrum of this signal indicating the complex size of each frequency component. You do not have to make separate plots for real/imaginary parts or magnitude/phase. Just indicate the complex phasor value at the appropriate frequency.

(c) Now consider a new signal $y(t) = x(t) + 10\cos(50\pi t - \pi/6)$. How is the spectrum changed? Is $y(t)$ still periodic? If so, what is the period?

3.10 A periodic signal $x(t) = x(t + T_0)$ is described over one period $-T_0/2 \le t \le T_0/2$ by the equation

$$x(t) = \begin{cases} 1 & |t| < t_c \\ 0 & t_c < |t| \le T_0/2 \end{cases}$$

where $t_c < T_0/2$.

    (a) Sketch the periodic function $x(t)$ for $-2T_0 < t < 2T_0$ for the case $t_c = T_0/4$.

    (b) Determine the DC coefficient X_0.

    (c) Determine a formula for the Fourier coefficients X_k in the representation

$$x(t) = X_0 + \Re e \left\{ \sum_{k=1}^{\infty} X_k e^{jk\omega_0 t} \right\} \quad \text{where } X_k = \frac{2}{T_0} \int_{-T_0/2}^{T_0/2} x(t) e^{-jk\omega_0 t} dt$$

    Your final result should depend on t_c and T_0.

    (d) Sketch the spectrum of $x(t)$ for the case $\omega_0 = 2\pi(100)$ and $t_c = T_0/4$ for frequencies between $-10\omega_0$ and $+10\omega_0$.

    (e) Sketch the spectrum of $x(t)$ for the case $\omega_0 = 2\pi(100)$ and $t_c = T_0/10$ for frequencies between $-10\omega_0$ and $+10\omega_0$.

    (f) From your results in (d) and (e), what do you conclude about the relationship between t_c and the relative size of the high-frequency components of $x(t)$?

3.11 A chirp signal is one that sweeps in frequency from $\omega_1 = 2\pi f_1$ to $\omega_2 = 2\pi f_2$ as time goes from $t = 0$ to $t = T_2$. The general formula for a chirp is

$$x(t) = A\cos(\alpha t^2 + \beta t + \phi) = \cos(\psi(t)) \tag{3.8.3}$$

where

$$\psi(t) = \alpha t^2 + \beta t + \phi$$

The derivative of $\psi(t)$ is the *instantaneous frequency* which is also the frequency heard if the frequencies are in the audible range.

$$\omega_i(t) = \frac{d}{dt}\psi(t) \quad \text{radians/sec} \tag{3.8.4}$$

    (a) For the chirp in (3.8.3), determine formulas for the beginning frequency (ω_1) and the ending frequency (ω_2) in terms of α, β, and T_2.

    (b) For the chirp signal

$$x(t) = \Re e\{e^{j(40t^2 + 27t + 13)}\}$$

    derive a formula for the *instantaneous* frequency versus time.

    (c) Make a plot of the *instantaneous* frequency (in Hz) versus time over the range $0 \le t \le 1$ sec.

3.12 It may be difficult to see why the derivative of the phase would be the instantaneous frequency. The following experiment provides a clue.

    (a) Use the following parameters to define a chirp signal:

$$f_1 = 1 \text{ Hz}$$

$$f_2 = 9 \text{ Hz}$$

$$T_2 = 2 \text{ sec}$$

In other words, determine α and β in (3.8.3) to define $x(t)$ so that it sweeps the specified frequency range.

    (b) It will be difficult to verify whether or not this chirp signal has the correct frequency content. However, the rest of this problem is devoted to an experiment that will demonstrate that the derivative of the phase is the "correct" definition of instantaneous frequency. First, make a plot of the instantaneous frequency $f_i(t)$ (in Hz) versus time.

    (c) Now make a plot of the signal synthesized in (a). Pick a time-sampling interval that is small enough so that the plot is very smooth. Put this plot in the middle panel of a 3 × 1 subplot, i.e., `subplot(3,1,2)`.

    (d) Now generate and plot a 4-Hz sinusoid. Put this plot in the upper panel of a 3 × 1 subplot, i.e., `subplot(3,1,1)`.

    (e) Finally, generate and plot an 8-Hz sinusoid. Put this plot in the lower panel of a 3 × 1 subplot, i.e., `subplot(3,1,3)`.

    (f) Compare the three signals and comment on the frequency content of the chirp. Concentrate on the frequency of the chirp in the time range $1.6 \leq t \leq 2$ sec. Which sinusoid matches the chirp in this time region? Compare the expected $f_i(t)$ in this region to 4 Hz and 8 Hz.

4

Sampling and Aliasing

This chapter is concerned with the conversion of signals between the analog (continuous-time) and digital (discrete-time) domains. The primary objective of our presentation is an understanding of the *sampling theorem,* which states that when the sampling rate is greater than twice the highest frequency contained in the spectrum of the analog signal, the original signal can be reconstructed exactly from the samples.

The process of converting from digital back to analog is called *reconstruction.* A good example is given by audio CDs. The music is stored in a digital form from which a CD player reconstructs the continuous (analog) waveform that we listen to. The reconstruction process is basically one of interpolation. In other words, we must "connect the dots" by drawing a smooth curve through the discrete-time sample values $x(t_n)$ at the appropriate sample times t_n. Although this process may be studied as time-domain interpolation, we will see that a frequency-domain view is very helpful in understanding the important issues in sampling.

4.1 SAMPLING

Sinusoidal waveforms of the form

$$x(t) = A \cos(\omega t + \phi) \tag{4.1.1}$$

are examples of *continuous-time* signals. We will also use the term *analog* signal to refer to such signals. Continuous-time signals are represented mathematically by functions of time, $x(t)$, where t is a continuous variable. In earlier chapters, we have "plotted" analog waveforms using MATLAB, but actually we did not plot the waveform. Instead, we really plotted the waveform only at isolated (discrete) points in time and then connected those points with straight lines. Indeed, digital computers cannot deal with continuous-time signals directly; they can represent and manipulate them symbolically (as with *Mathematica*) or numerically (as with MATLAB), but the key point is that any computer representation is *discrete*. (Recall the discussion in Section 2.4.)

A *discrete-time signal* is represented mathematically by an indexed sequence of numbers. When stored in a digital computer, the numbers are held in memory locations, so they would be indexed by memory address. We denote the values of such a sequence as $x[n]$, where n is the integer index indicating the order of the values in the sequence. The square brackets [ ] enclosing the argument n allow us to differentiate between the continuous-time signal $x(t)$ and the discrete-time signal $x[n]$.[1]

We can obtain discrete-time signals in either of the following ways:

1. We can *sample* a continuous-time signal $x(t)$, where $x(t)$ could represent any continuously varying signal such as speech or audio. In this case, the sequence $x[n]$ is obtained by recording the values of $x(t)$ at equally spaced time instants; i.e.,

$$x[n] = x(nT_s) \qquad -\infty < n < \infty$$

The result is a sequence of numbers whose individual values are *samples* of the analog signal. The sampling operation, which can be viewed as a transformation that acts on $x(t)$ to get the output $x[n]$, is an example of a system where the input is a continuous-time signal and the output is a discrete-time signal. The operation of sampling is represented in block diagrams as in Fig. 4.1. The system of Fig. 4.1 is a mathematical idealization of a system that we will call a continuous-to-discrete (C-to-D) converter. The actual hardware system for doing this operation is an analog-to-digital (A-to-D) converter, which approximates the perfect sampling of the C-to-D converter, but is degraded somewhat by real-world problems such as word-length quantization to 12 or 16 bits, jitter in the sampling times, and other factors.

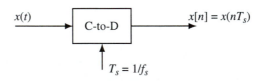

Figure 4.1   Block diagram representation of the ideal continuous-to-discrete (C-to-D) converter. The parameter T_s specifies uniform sampling of the input every T_s seconds.

---

[1] The terminology for discrete-time signals is not universal, so we may also use word *sequence* in place of signal, or the adjective *digital* in place of discrete-time, to refer to $x[n]$.

2. We can also *compute* the values of a discrete-time signal directly from a formula. A simple example is

$$w[n] = n^2 - 5n + 3$$

which determines the sequence $\{3, -1, -3, -3, -1, 3, 9, \ldots\}$ corresponding to the indices $n = 0, 1, 2, 3, 4, 5, 6, \ldots$. Although in cases such as this, there is no explicit underlying continuous function that is being sampled, we will nevertheless often refer to the individual values of the sequence as *samples*. Discrete-time signals described by formulas will be very common in our study of discrete-time signals and systems.

We will often find it useful to plot discrete-time signals as in Fig. 4.2, which shows eight values (samples) of the sequence $w[n]$. Such plots, which are sometimes called "lolly pop" or "tinker-toy" plots, show clearly that the signal has values only for integer indices; in between, the discrete-time signal is undefined.

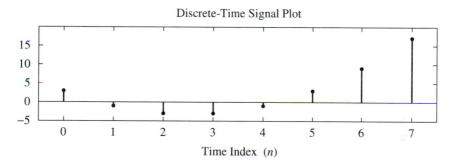

Figure 4.2   Plotting format for discrete-time signals.

4.1.1 Sampling Sinusoidal Signals

Sinusoidal signals are examples of continuous-time signals for which we can write a simple mathematical formula, and we have seen that they also are good models for real signals. Because more general continuous-time signals can be represented as sums of sinusoids, and because the effects of sampling are easily understood for sinusoids, we will use them as the basis for our study of sampling.

If we sample a signal of the form of (4.1.1), we obtain

$$x[n] = x(nT_s) = A\cos(\omega n T_s + \phi) = A\cos(\hat{\omega} n + \phi) \qquad (4.1.2)$$

where we have defined $\hat{\omega}$ to be

Normalized Radian Frequency

$$\hat{\omega} = \omega T_s \qquad (4.1.3)$$

The signal $x[n]$ in (4.1.2) is a *discrete-time cosine signal*, and $\hat{\omega}$ is its *discrete-time radian frequency*. We use a "hat" over the frequency variable to denote that the continuous-time radian frequency has been normalized by the sampling period. Note that since ω has units of rad/sec, $\hat{\omega} = \omega T_s$ has units of rad; i.e., $\hat{\omega}$ is a dimensionless quantity. This is entirely consistent with the fact that the index n in $x[n]$ is dimensionless. Once the samples are taken from $x(t)$, the time scale information is lost. The discrete-time signal is just a sequence of numbers, and these numbers carry no information about the sampling period, which is the information required to reconstruct the time scale. An immediate implication of this observation is that an infinite number of continuous-time sinusoidal signals can be transformed into the same discrete-time sinusoid by sampling. All we need to do is change the sampling period with changes in the frequency of the continuous-time sinusoid. For example, if $\omega = 200\pi$ rad/sec and $T_s = 1/2000$ sec, then $\hat{\omega} = 0.1\pi$ rad. On the other hand, if $\omega = 2000\pi$ rad/sec and $T_s = 1/20000$ sec, $\hat{\omega}$ is still equal to 0.1π rad.

The top panel of Fig. 4.3 shows $x(t) = \cos(200\pi t)$, a continuous-time sinusoid with frequency $f_0 = 100$ Hz. The middle panel of Fig. 4.3 shows the samples taken with sampling period $T_s = 0.5$ msec. The sequence is given by the formula $x[n] = x(nT_s) = \cos(0.1\pi n)$, so the discrete-time radian frequency is $\hat{\omega}_0 = 0.1\pi$. (Since $\hat{\omega}_0$ is dimensionless, we need not continue to specify units of rad.) Note that the rate of taking samples (sampling rate) is just $f_s = 1/T_s$, or, in this example, $f_s = 2000$ samples/sec. The sample values are plotted as discrete points as in Fig. 4.2. The points are not connected by a continuous curve because we do not have any *direct* information about the function between the sample values. In this case, there are 20 sample values per period of the signal, because the sampling frequency (2000 samples/sec) is 20 times higher than the frequency of the continuous signal (100 Hz). From the discrete-time plot, it is obvious that the sample values alone are sufficient to visually reconstruct a continuous-time cosine wave, but without knowledge of the sampling rate, we cannot tell what the frequency should be.

Another example of sampling is shown in the bottom panel of Fig. 4.3. In this case, the 100 Hz sinusoid is sampled at a lower rate ($f_s = 500$ samples/sec) resulting in the sequence of samples $x[n] = \cos(0.4\pi n)$. In this case, the discrete-time radian frequency is $\hat{\omega} = 0.4\pi$ rad. The time between samples is $T_s = 1/f_s = 2$ msec, so there are only 5 samples per period of the continuous-time signal. We see that without the original waveform superimposed, it would be difficult to discern the waveshape of the original continuous-time sinusoid.

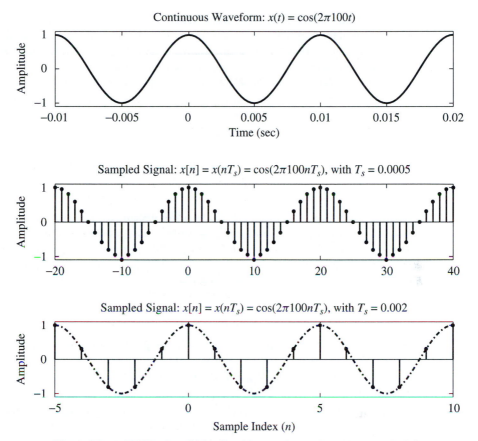

Figure 4.3   A 100 Hz sinusoid (top) and its samples at $f_s = 2000$ samples/sec (middle) and at $f_s = 500$ samples/sec (bottom).

4.1.2 The Sampling Theorem

The plots shown in Fig. 4.3 naturally raise the question of how frequently we must sample in order to retain enough information to reconstruct the original continuous-time signal from its samples. The amazingly simple answer is given by the following statement:

Shannon Sampling Theorem

A continuous-time signal $x(t)$ with frequencies no higher than f_{max} can be reconstructed exactly from its samples $x[n] = x(nT_s)$, if the samples are taken at a rate $f_s = 1/T_s$ that is greater than $2f_{max}$.

This is a statement of the *Shannon sampling theorem*, one of the theoretical pillars of modern digital communications, digital control, and digital signal processing.

Notice that the sampling theorem involves two issues. First, it talks about reconstruction of the signal from its samples, although it never specifies the algorithm for reconstruction. Second, it gives a minimum sampling rate that is dependent on the frequency content of $x(t)$, the continuous-time signal. This minimum sampling rate is called the *Nyquist rate*.[2] We can see examples of the sampling theorem in many commercial products. For example, audio CDs use a sampling rate of 44.1 kHz for storage of the digital audio signal. This number is slightly more than two times 20 kHz, which is the generally accepted upper limit for human hearing and perception of musical sounds.

4.1.3 Aliasing

The Shannon theorem states that reconstruction of a sinusoid is possible if we have at least two samples per period. What happens when we don't sample fast enough? We can derive a formula that gives the interrelationship between the signal frequency f_0 and the sampling rate f_s by considering in more detail the case of the continuous-time sinusoid

$$x(t) = A \cos(2\pi f_0 t + \phi) \qquad (4.1.4)$$

If we sample $x(t)$ with a sampling period of T_s, we get the sequence $x[n]$ with values

$$x[n] = x(nT_s) = A \cos(2\pi f_0 n T_s + \phi) \qquad (4.1.5)$$

Now consider another sinusoid with the same amplitude and phase, but with frequency $f_0 + \ell f_s$, where ℓ is an integer and $f_s = 1/T_s$.

$$y(t) = A \cos(2\pi (f_0 + \ell f_s)t + \phi)$$

If this second waveform, $y(t)$, is sampled with period T_s, we get

$$
\begin{aligned}
y[n] = y(nT_s) &= A \cos(2\pi (f_0 + \ell f_s)n T_s + \phi) \\
&= A \cos(2\pi f_0 n T_s + 2\pi \ell f_s T_s + \phi) \\
&= A \cos(2\pi f_0 n T_s + 2\pi \ell + \phi) \\
&= A \cos(2\pi f_0 n T_s + \phi) \\
&= x[n]
\end{aligned}
$$

---

[2] Harry Nyquist and Claude Shannon were researchers at Bell Telephone Laboratories who each made fundamental contributions to the theory of sampling and digital communication during the period from 1920–1950.

In other words, $y[n]$ has the same sample values as $x[n]$, so it is indistinguishable from $x[n]$. Since ℓ was specified only to be an integer (either positive or negative), this means that there are an infinite number of sinusoids that will give the same sequence of samples as $x[n]$. The frequencies $f_0 + \ell f_s$ are called *aliases* of the frequency f_0 with respect to the sampling frequency f_s, because all of them appear to be the same when sampled at the rate f_s.

4.1.4 Folding

A second source of aliased signals actually comes from the negative frequency component of the cosine wave. These frequencies are $-f_0 + \ell f_s$, where ℓ is a positive or negative integer. Consider a third signal

$$w(t) = A\cos(2\pi(-f_0 + \ell f_s)t - \phi)$$

whose initial phase is the negative of that in (4.1.4). If we sample $w(t)$ with a sampling period of T_s, we now get

$$
\begin{aligned}
w[n] = w(nT_s) &= A\cos(2\pi(-f_0 + \ell f_s)nT_s - \phi) \\
&= A\cos(-2\pi f_0 nT_s + 2\pi \ell f_s T_s - \phi) \\
&= A\cos(-2\pi f_0 nT_s + 2\pi \ell - \phi) \\
&= A\cos(2\pi f_0 nT_s + \phi) \\
&= x[n]
\end{aligned}
$$

The fourth line in this equation is true because the cosine function is an even function; i.e., $\cos(-\theta) = \cos\theta$.

Once again, the samples $w[n]$ and $x[n]$ are identical. The terminology *folded* frequency comes from the following experiment. Take a sinusoid whose frequency f_0 lies between $\frac{1}{2}f_s$ and f_s. When sampled at rate f_s, this sinusoid appears to be the same as one at frequency $f_1 = f_s - f_0$. If we plot the apparent frequency f_1 versus the input frequency f_0, as f_0 goes from $\frac{1}{2}f_s$ to f_s, then f_1 decreases from $\frac{1}{2}f_s$ to 0, as shown in Fig. 4.4. In other words, the two frequencies are mirror images with respect to $\frac{1}{2}f_s$, and would lie on top of one another if the graph were folded about the $\frac{1}{2}f_s$ line in Fig. 4.4.

In summary, we have demonstrated that there are two classes of alias signals, $y(t)$ and $w(t)$ given above, that are indistinguishable from $x(t)$ when they are sampled at a rate $f_s = 1/T_s$.

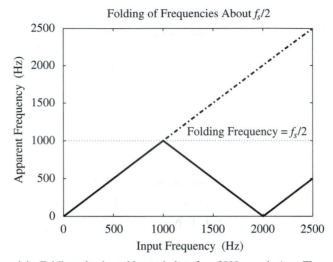

Figure 4.4   Folding of a sinusoid sampled at $f_s = 2000$ samples/sec. The apparent frequency is the lowest frequency of a sinusoid that has exactly the same samples as the input sinusoid.

4.2   SPECTRUM VIEW OF SAMPLING

DEMO:
ALIASING

We will now present a frequency spectrum explanation of the sampling process. The sampling process changes the location of the various spectrum components that make up a signal. The folding and aliasing of frequencies can be tracked via a spectrum diagram that includes *all* of the spectrum lines for a sinusoid and its aliases.

The following figures (4.5–4.8) all were constructed with a MATLAB M-file that plots the location of spectrum components involved in the sampling process. The test signal will be a continuous-time 100 Hz sinusoid of the form $x(t) = \cos(2\pi(100)t + \phi)$. The sampling period will be varied to show what happens at different sampling rates.

4.2.1   Over-Sampling

In general, we try to avoid the problems of aliasing and folding by sampling at a rate much higher than twice the highest frequency. This is called *over-sampling*. For example, when we sample the 100 Hz sinusoid, $x(t) = \cos(2\pi(100)t + \phi)$ at a sampling rate of $f_s = 1000$ samples/sec, we obtain the time- and frequency-domain plots shown in Fig. 4.5.

The upper panel of Fig. 4.5 gives the frequency-domain spectrum representation of the 100 Hz sinusoid. Note that the frequency axis has been *normalized* by division by the sampling frequency (or multiplication by the sampling period). Recall that earlier we introduced the normalized radian frequency variable $\hat{\omega} = \omega T_s$. Since we find it more convenient in our discussion in this chapter to use cyclic frequencies instead of radian frequencies, we define the normalized discrete-time cyclic frequency to be

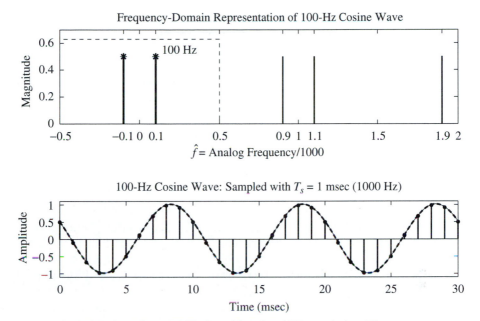

Figure 4.5 Sampling a 100 Hz sinusoid at $f_s = 1000$ samples/sec. The spectrum plot (top) shows the aliased spectrum components as well as the positive and negative frequency components of the original sinusoid. The time-domain plot (bottom) shows the samples as dots, the original signal as a thin solid line, and the reconstructed signal as a dashed line.

Normalized Cyclic Frequency

$$\hat{f} = \hat{\omega}/(2\pi) = fT_s = f/f_s \qquad (4.2.1)$$

As in the case of discrete-time radian frequency, the "hat" over the variable indicates that it is normalized by the sampling period. Along the discrete-time cyclic frequency axis ($\hat{f}$), all frequencies are plotted as a fraction of the sampling frequency, and therefore $\hat{f}$ is a dimensionless quantity. This is convenient because aliased frequencies are then displaced by ± 1, ± 2, ± 3, etc., along the $\hat{f}$ axis. Note that the spectrum of the original signal is represented by the vertical lines at $\hat{f} = \pm 0.1 = \pm (100/1000)$ with the marker $*$ at the top. The other vertical lines at $\hat{f} = 0.9, 1.1, 1.9$ show some aliases of 100 Hz with respect to a sampling frequency of 1000 Hz.

The sampling theorem suggests that a process exists for reconstructing a signal from its samples. As we will demonstrate in Section 4.4, this process (which we call D-to-C conversion) always reconstructs only the frequencies that fall inside the dotted box shown in the upper panel of Fig. 4.5. These are the frequencies between $-f_s/2$ and $+f_s/2$. Therefore, if the original frequency f_0 is such that $f_0 < f_s/2$ (or $\hat{f}_0 < 1/2$), the original waveform will be reconstructed. In the example of Fig. 4.5,

$f_0 = 100$ Hz and $f_s = 1000$, so the condition of the sampling theorem is satisfied, and the original signal will be reconstructed by the ideal D-to-C conversion process. The lower panel of Fig. 4.5 shows the time-domain reconstruction, with the samples superimposed on the waveform of $x(t)$, which is plotted as a thin solid line. The heavy dashed curve represents the signal that would be reconstructed by an ideal reconstruction system with the given samples as input. Since the sampling theorem is satisfied, the reconstructed signal is identical to the original input to the sampler. In this case, the signal is clearly oversampled because there are 10 samples per period.

4.2.2 Aliasing Due to Under-Sampling

When $f_s < f_0$, the signal is *under-sampled*. For example, if $f_s = 80$ Hz, aliasing distortion occurs. In the top portion of Fig. 4.6, only the positive-frequency component of the signal is shown at $\hat{f} = 1.25 = (100/80)$; the negative-frequency component is off the scale to the left. In the bottom part of Fig. 4.6, the 100 Hz sinusoid (thin solid line) is sampled too infrequently to be recognized. In fact, the same samples would have been obtained from a 20 Hz sinusoid, because $f_0 - f_s = 100 - 80 = 20$ Hz. This is the alias component at normalized frequency $\hat{f} = (f_0 - f_s)/f_s = (100 - 80)/80 = 0.25$. Since the reconstruction process always uses the frequencies in the band from $-f_s/2$ to $+f_s/2$, the restructured signal is the 20 Hz sinusoid shown as the heavy dashed line in Fig. 4.6 (bottom).

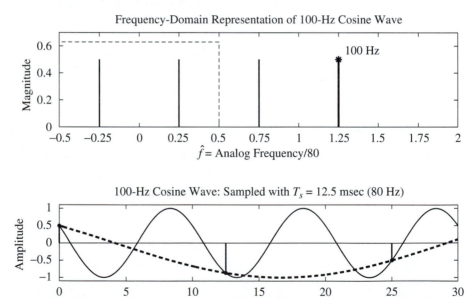

Figure 4.6   Sampling a 100 Hz sinusoid at $f_s = 80$ samples/sec. The time-domain plot (bottom) shows the samples as dots, the original sinusoid as a thin solid line, and the reconstructed signal as a dashed line. The dashed-line curve is a 20 Hz sinusoid that passes through the same sample points.

This aliasing of sinusoidal components can have some dramatic effects. Fig. 4.7 shows the case where the sampling rate and the frequency of the sinusoid are the same. Clearly, what happens is that the samples always hit at the same place on the waveform, so we get the equivalent of sampling a constant (DC), which is the same as a sinusoid with zero frequency.

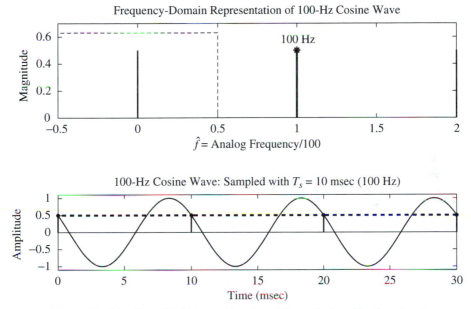

Figure 4.7   Sampling a 100 Hz sinusoid at $f_s = 100$ samples/sec. The time-domain plot (bottom) shows the samples as dots, the original sinusoid as a thin solid line, and the reconstructed signal as a dashed line.

4.2.3 Folding Due to Under-Sampling

Figure 4.8 shows the case where under-sampling leads to folding; here the sampling rate is $f_s = 125$ samples/sec. Once again, only the positive-frequency component of the signal is shown at $\hat{f} = 0.8 = (100/125)$. The negative-frequency component is offscale to the left. Now an interesting thing happens. The original frequency is outside the range $-\frac{1}{2}f_s < f < \frac{1}{2}f_s$, but two of the aliases are in this region. They are at $-f_0 + f_s = -100 + 125 = 25$ Hz and at $-f_s + f_0 = -125 + 100 = -25$ Hz ($\hat{f} = \pm 0.2$ on the normalized scale). What this means is that when we sample a 100 Hz sinusoid at a sampling rate of 125 samples/sec, we get the same samples that we would have gotten by sampling a 25 Hz sinusoid. To illustrate this, the lower part of Fig. 4.8 shows both the 25 Hz and the 100 Hz cosines together with the samples of both waveforms. This is the *folding* case, because 100 Hz minus 125 Hz is -25 Hz. In Fig. 4.8 (bottom) we can see the phase reversal that arises because the alias component at $25 = 125 - 100$ corresponds to the negative frequency -100 in the original signal.

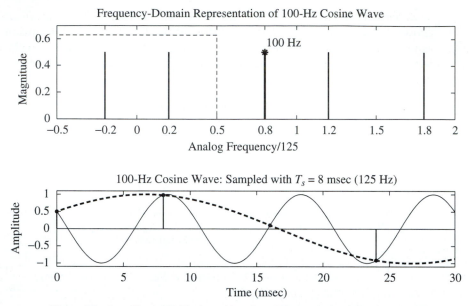

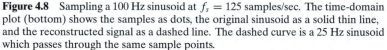

Figure 4.8   Sampling a 100 Hz sinusoid at $f_s = 125$ samples/sec. The time-domain plot (bottom) shows the samples as dots, the original sinusoid as a solid thin line, and the reconstructed signal as a dashed line. The dashed curve is a 25 Hz sinusoid which passes through the same sample points.

4.2.4 Maximum Reconstructed Frequency

Finally, we can say something definitive about the reconstruction process, i.e., going from the samples of a waveform back to the continuous-time signal. If we postulate a device for doing this, it might logically be called an ideal discrete-to-continuous (D-to-C) converter. The input to the D-to-C converter is the discrete-time signal containing the samples, and the D-to-C converter operates at a specified sampling rate f_s. Since ambiguity is inherent in the sampling process, it impossible to reconstruct the original signal except under the assumption that the maximum frequency in the output signal be less than half of the sampling frequency. If this is true, then the D-to-C converter will *interpolate* to create the correct values of $x(t)$ between the samples. If the signal is under-sampled, the ideal D-to-C converter will still reconstruct an output signal, but that output will correspond to the aliased spectral components that fall into the frequency band of $-\frac{1}{2} f_s < f < \frac{1}{2} f_s$. In Section 4.4 we will demonstrate that practical implementations of the D-to-C converter can approximate the ideal performance quite closely in some cases.

4.3   STROBE DEMONSTRATION

One effective means for demonstrating aliasing is to use a strobe light to illuminate a spinning object. In fact, this process is used routinely to set the timing of automobile engines and gives a practical example where aliasing is a desirable effect. In our case, we use a disk attached to the shaft of an electric motor that rotates at a constant angular speed (see Fig. 4.9). On this white disk is painted a black spot that is easily seen whenever the strobe light flashes. In our particular case, the rotation speed of the motor is approximately 750 rpm, and the flash rate of the strobe light is variable over a wide range.

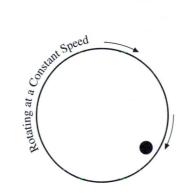

Figure 4.9 The disk attached to the shaft of a motor rotates clockwise at a constant speed.

Suppose that the flashing rate is very high, let's say nine times the rotation rate, i.e., $9 \times 750 = 6750$. In this case, the disk will not rotate very far between flashes. In fact, for the $9\times$ case, it will move only $360°/9 = 40°$ per flash, as in Fig. 4.10. Since the movement will be clockwise, the angular change from one flash to the next is $-40°$.

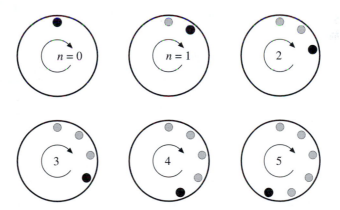

Figure 4.10 Six successive positions of the black spot for a very high flashing rate. Gray spots indicate previous locations of the black spot. The disk is rotating clockwise, and the spot appears to move in the same direction. Angular change is $-40°$ per flash.

If the flash rate of the strobe is set equal to the rotation rate of the disk, i.e., 750 flashes/min, then the black spot will appear to stand still. This is because the black spot makes exactly one revolution between flashes and therefore is always at the same position when illuminated by the strobe. This is exactly the situation illustrated in Fig. 4.7 where the frequency of the sinusoid aliased to zero. This is not the only flash rate for which the spot will stay still. In fact, a slower flash rate that permits two or three or any integer number of complete revolutions between flashes will create the same effect. In our case of a 750-rpm motor, flash rates of 375, 250, $187\frac{1}{2}$, 150, and 125 flashes/min will also work.

By using flashing rates that are close to these numbers, we can make the spot move slowly, and we can also control its direction of motion (clockwise or counter-clockwise). For example, if we set the strobe for a flashing rate that is just slightly higher than the rotation rate, we will observe another aliasing effect. Suppose that the flashing rate is 806 flashes/min; then the disk rotates slightly less than one full revolution between flashes. We can calculate the movement if we recognize that one complete revolution takes 1/750 min:

$$\Delta\theta = -360° \times \frac{1/806}{1/750} = -360° \times \frac{750}{806} = -335° = +25°$$

Once again, the minus sign indicates rotation in a clockwise direction, but since the angular change is almost $-360°$, we would observe a small positive angular change instead, and the black spot would appear to move in the counterclockwise direction, as shown in Fig. 4.11. The fact that one can distinguish between clockwise rotation and counterclockwise rotation is equivalent to saying that positive and negative frequencies have separate physical meanings.

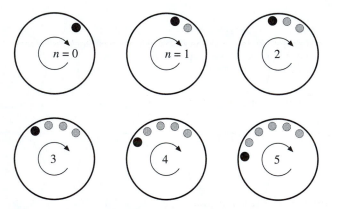

Figure 4.11   Six successive positions of the black spot for a flashing rate that aliases the spot motion. Gray spots indicate previous locations of the black spot. The disk is rotating clockwise, but the spot appears to move counter-clockwise. Angular change is $+25°$ per flash.

In order to analyze the strobe experiment mathematically, we need a notation for the motion of the black spot as a function of time. The spot moves in an x–y

coordinate system, so a succinct notation is given by a complex number whose real part is x and whose imaginary part is y. The position of the spot is

$$p(t) = x(t) + jy(t)$$

Furthermore, since the motion of the spot is on a circle of radius r, the correct formula for $p(t)$ is a complex exponential with constant frequency.

$$p(t) = re^{-j(2\pi f_m t - \phi)} \tag{4.3.1}$$

The minus sign in the exponent indicates clockwise rotation, and the initial phase ϕ specifies the location of the spot at $t = 0$ (see Fig. 4.12). The frequency of the motor rotation f_m is constant, as is the radius r. It will be convenient to set $r = 1$ in what follows so that the formulas will be less cluttered.

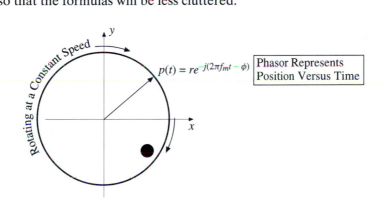

Figure 4.12   The position of the spot on the disk can be expressed as a rotating phasor $p(t) = x(t) + jy(t)$ versus time. The frequency f_m is the rotation rate of the motor in rpm.

The effect of the strobe light is to sample $p(t)$ at a fixed rate given by the flashing rate f_s. Thus the position of the spot at the nth flash can be expressed as a discrete-time signal $p[n]$:

$$p[n] = p(t)|_{t=nT_s} = p(nT_s) = p(n/f_s)$$

Substituting into the complex exponential (4.3.1) we get

$$p[n] = re^{-j(2\pi(f_m/f_s)n - \phi)} \tag{4.3.2}$$

If the constraint of the sampling theorem is met (i.e., $f_s > 2|f_m|$), then there will be no aliasing in the experiment. In fact, the angular change from one sample time to the next will be between $-180°$ and $0°$, so the spot will appear to rotate clockwise.

The more interesting cases occur when the flashing rate drops below $2|f_m|$. Then the disk may make one or more revolutions between flashes, which introduces the aliasing phenomenon. Using the sampled position formula (4.3.2), we can solve the following type of problem: Find *all* possible flashing rates so that the spot will move counter-clockwise at a rate of 25° per flash, which is equivalent to one revolution every 14.4 flashes. Assume a constant motor speed of f_m rpm. One twist to this problem is that the two rotation rates are specified in different units.

A systematic approach is possible if we use the property $e^{j2\pi\ell} = 1$, whenever ℓ is an integer. The desired rotation of the spot can be expressed as:

$$d[n] = re^{+j(2\pi(25/360)n+\psi)}$$

where the factor $2\pi(25/360)$ equals 25° converted to radians. The initial phase ψ is set equal to ϕ in $p[n]$, so that $d[0] = p[0]$. Then we can equate $p[n]$ and $d[n]$, but we throw in the factor $e^{j2\pi\ell n}$, which is just multiplying by one; i.e.,

$$p[n] = d[n]e^{j2\pi\ell n}$$

Now we can generate an equation that can be solved for the flashing rates:

$$re^{-j(2\pi(f_m/f_s)n-\phi)} = re^{+j(2\pi(25/360)n+\phi)}e^{j2\pi\ell n}$$

$$-\left(2\pi\frac{f_m}{f_s}n - \phi\right) = +\left(2\pi\frac{25}{360}n + \phi\right) + 2\pi\ell n$$

$$-\frac{f_m}{f_s} = \frac{25}{360} + \ell$$

$$\implies \quad -f_m = f_s\left(\frac{25}{360} + \ell\right)$$

So, finally we can solve for the flashing rate:

$$f_s = \frac{-f_m}{(5/72) + \ell} \tag{4.3.3}$$

where ℓ is any integer. Since we want positive values for the flashing rate, and since there is a minus sign associated with the clockwise rotation rate of the motor $(-f_m)$, we choose $\ell = -1, -2, -3, \ldots$ to generate the different solutions. For example, when the motor rpm is 750, the following flashing rates (in flashes/min) will give the desired spot movement:

ℓ	-1	-2	-3	-4
f_s	805.97	388.49	255.92	190.81

Exercise 4.1. To test your knowledge of this concept, try solving for *all* possible flashing rates so that the spot will move clockwise at a rate of one revolution every 9 flashes. Assume the same motor rotation speed of 750 rpm clockwise. What is the maximum flashing rate for this case?

4.3.1 Spectrum Interpretation

The strobe demo involves the sampling of complex exponentials, so we can present the results in the form of a spectrum diagram rather than using equations, as in the foregoing section. The rotating disk has an analog spectrum given by a single frequency component at $f = -f_m$ cycles/min (see Fig. 4.13).

Figure 4.13   Analog spectrum representing the disk spinning clockwise at f_m rpm. The units for f are cycles/min.

When the strobe light is applied at a rate of f_s flashes per minute, the resulting spectrum of the discrete-time signal $p[n]$ will contain an infinite number of frequency lines at the frequencies:

$$\hat{f}_\ell = \frac{-f_m + \ell f_s}{f_s} = -\hat{f}_m + \ell \qquad \ell = 0, \pm 1, \pm 2, \pm 3, \dots \qquad (4.3.4)$$

Equation (4.3.4) tells us that there are two steps needed to create the spectrum of a discrete-time signal:

1. Repeat each spectral line in the analog spectrum by shifting it by all integer multiples of the sampling frequency.

2. Then rescale the frequency axis by dividing by f_s. This will normalize the digital spectrum with respect to the sampling frequency, e.g., a digital frequency of $\hat{f} = \frac{1}{4}$ corresponds to $f_s/4$.

The discrete-time spectrum derived from Fig. 4.13 is shown in Fig. 4.14.

In the discrete-time spectrum, only the frequency component closest to $\hat{f} = 0$ counts in D-to-C reconstruction, so the strobed signal $p[n]$ appears to be rotating at that lowest normalized frequency. However, one last conversion must be made to give the perceived analog rotation rate in rpm. The discrete-time frequency $(\hat{f})$

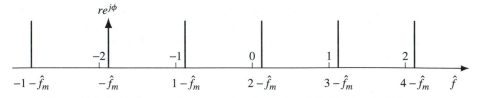

Figure 4.14   Digital spectrum representing the strobed disk spinning clockwise at f_m rpm, but sampled at f_s flashes per minute. The horizontal axis is normalized frequency: $\hat{f} = f/f_s$. The normalized motor frequency $\hat{f}_m = (f_m/f_s)$ appears at the aliases $\hat{f}_\ell = \ell - \hat{f}_m$, where ℓ is an integer.

must be converted back to analog frequency (f). In Fig. 4.14, the alias for $\ell = 2$ is the one closest to zero frequency, so the corresponding analog frequency is

$$f_{\text{spot}} = (\hat{f}_2)f_s = (2 - \hat{f}_m)f_s = 2f_s - f_m$$

This may seem like a roundabout way to say that the rotation rate of the spot differs from $-f_m$ by an integer multiple of the sampling rate, but it does provide a graphical picture of the relative frequency locations. Finally, the spectrum picture makes it clear that the lowest discrete-time frequency is the one that is "reconstructed." We will explain why this is true in Section 4.4.

The case where the sampling rate is variable and f_m is fixed is a bit harder to solve, but this is the actual situation for our strobed disk experiment. Nonetheless, a graphical approach is still possible because the desired spot frequency defines a line in the discrete-time spectrum, say, $\hat{f}_d$. This line is the one closest to the origin, so we must add an integer ℓ to $\hat{f}_d$ to match the normalized motor rotation frequency.

$$\hat{f}_d + \ell = \hat{f}_m = \frac{-f_m}{f_s}$$

This equation can be solved for the flashing rate f_s, but the final answer depends on the integer ℓ, which predicts, by the way, that there are many answers for f_s, as we already saw in (4.3.3).

4.4   DISCRETE-TO-CONTINUOUS CONVERSION

The ideal discrete-to-continuous (D-to-C) converter is depicted in Fig. 4.15. Its purpose is to interpolate a smooth continuous-time function through the input samples $y[n]$. Thus, in the special case when $y[n] = A\cos(2\pi f_0 n T_s + \phi)$, and if $f_0 < \frac{1}{2}f_s$, then according to the sampling theorem, the converter should produce

$$y(t) = A\cos(2\pi f_0 t + \phi) \qquad (4.4.1)$$

For sampled cosine signals *only*, the ideal D-to-C converter in effect replaces n by $f_s t$. On the other hand, if $f_0 > \frac{1}{2} f_s$, then we know that aliasing or folding distortion has occurred, and the ideal D-to-C converter will reconstruct a cosine wave with frequency equal to the alias frequency that is less than $\frac{1}{2} f_s$.

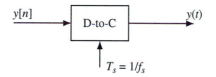

Figure 4.15   Block diagram of the ideal discrete-to-continuous (D-to-C) converter (interpolator) for sample spacing T_s.

4.4.1 Alias Frequencies Due to Sampling

The ideal continuous-to-discrete (C-to-D) converter is depicted again in Fig. 4.16. For cosine waves of the form

$$x(t) = A\cos(2\pi f_0 t + \phi) \tag{4.4.2}$$

the resulting sequence of samples is

$$x[n] = A\cos(2\pi f_0 n T_s + \phi) \tag{4.4.3}$$

Previously, we have shown that there are an infinite number of *alias frequencies*, $\pm f_0 + \ell f_s$, such that cosine waves with these frequencies have the same sample values when the sampling rate is f_s. For example, both

$$x_1(t) = A\cos(2\pi(f_s + f_0)t + \phi) \tag{4.4.4}$$

and

$$x_2(t) = A\cos(2\pi(f_s - f_0)t - \phi) \tag{4.4.5}$$

give the same samples as $x(t)$ in (4.4.2), i.e.,

$$x[n] = x(nT_s) = x_1(nT_s) = x_2(nT_s)$$

This ambiguity is caused by the C-to-D sampling, so the D-to-C converter is limited by this ambiguity, because it can reconstruct only one output, which will only be the right one if $f_0 < f_s/2$.

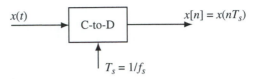

$$x(t) \quad \boxed{\text{C-to-D}} \quad x[n] = x(nT_s)$$

$$T_s = 1/f_s$$

Figure 4.16   Block diagram of the ideal continuous-to-discrete (C-to-D) converter.

4.4.2 Interpolation with Pulses

How does the D-to-C converter work? In this section, we explain how the D-to-C converter does interpolation, and then describe a practical system that is nearly the same as the ideal D-to-C converter. These actual hardware systems, called digital-to-analog (D-to-A) converters, approximate the behavior of the ideal D-to-C system.

A general formula that describes a broad class of D-to-C converters is given by the equation

$$y(t) = \sum_{n=-\infty}^{\infty} y[n]p(t - nT_s) \qquad (4.4.6)$$

where $p(t)$ is the characteristic pulse shape of the converter. Equation (4.4.6) states that the output signal is produced by adding together many pulses, each shifted in time. In other words, at each sample time $t_n = nT_s$, a pulse $p(t - nT_s)$ is emitted with an amplitude proportional to the sample value $y[n]$ corresponding to that time instant.[3] Note that all the pulses have a common waveshape specified by $p(t)$. If the pulse has duration greater than or equal to T_s, then the gaps between samples will be filled by adding overlapped pulses.

RECON-
STRUCT via
D-to-A

Obviously, the important issue is the choice of the pulse waveform $p(t)$. Unfortunately, we do not yet have the mathematical tools required to derive the optimal pulse shape required for exact reconstruction of a waveform $y(t)$ from its samples $y[n] = y(nT_s)$. This optimal pulse shape can be constructed during a mathematical

---

[3] A *pulse* is a continuous-time waveform that is concentrated in time. Typically, a pulse will be nonzero only over a finite interval of time.

proof of the sampling theorem. Nonetheless, we will demonstrate the plausibility of (4.4.6) by considering some simple (suboptimal) examples. Figure 4.17 shows four possible pulse waveforms for D-to-C conversion when $f_s = 200$ Hz.

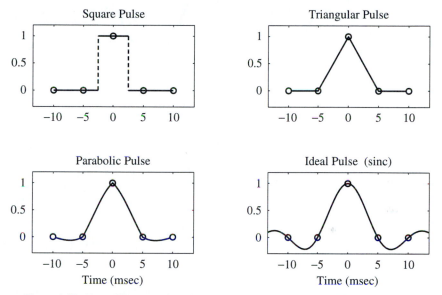

Figure 4.17   Four different pulses for D-to-C conversion. The sampling period is $T_s = 0.005$, i.e., $f_s = 200$ Hz. Note that the duration of each pulse is approximately one or two times T_s.

4.4.3 Zero-Order Hold Interpolation

The simplest pulse shape that we might propose is a symmetric square pulse of the form

$$p(t) = \begin{cases} 1 & -\tfrac{1}{2}T_s < t \leq \tfrac{1}{2}T_s \\ 0 & \text{otherwise} \end{cases} \tag{4.4.7}$$

This pulse is plotted in the upper left panel of Fig. 4.17.

From Fig. 4.17, we see that the total width of the square pulse is $T_s = 5$ ms and that its amplitude is 1. Therefore, each term $y[n]p(t - nT_s)$ in the sum (4.4.6) will create a flat region of amplitude $y[n]$ centered on $t = nT_s$. This is shown in the top part of Fig. 4.18, which shows the original 83 Hz cosine wave (solid curve), its samples taken at a sampling rate of 200 samples/sec, and the individual shifted and scaled pulses, $y[n]p(t - nT_s)$ (dashed curves). The sum of all the shifted and scaled pulses will be a "stairstep" waveform, as shown in the bottom panel of Fig. 4.18, where the dotted curve is the original cosine wave $x(t)$, and the solid curve shows the reconstructed waveform using the square pulse.

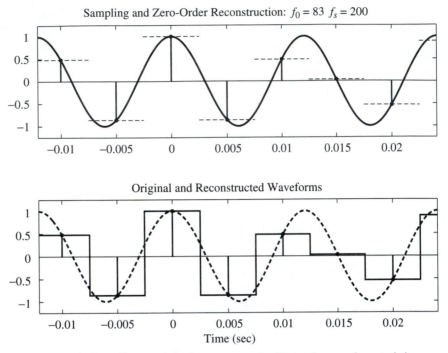

Figure 4.18 D-to-C conversion using a square pulse. Flat regions are characteristic of the zero-order hold.

The space between samples has indeed been filled with a continuous-time waveform; however, it is clear from the lower part of Fig. 4.18 that the reconstructed waveform for the square pulse is a poor approximation of the original cosine wave. Thus, using a square pulse in (4.4.6) is D-to-C conversion, but not *ideal D-to-C conversion.*[4] Even so, this is a useful model, since many physically realizable digital-to-analog (D-to-A) converters produce outputs that look exactly like this!

4.4.4 Linear Interpolation

The triangular pulse plotted in the upper right panel of Fig. 4.17 is defined as a pulse consisting of the first-order polynomial (straight line) segments

$$p(t) = \begin{cases} 1 - |t|/T_s & -T_s \le t \le T_s \\ 0 & \text{otherwise} \end{cases} \qquad (4.4.8)$$

Figure 4.19 (top) again shows the original 83-Hz cosine wave (solid curve) and its samples, together with the scaled pulses $y[n]p(t - nT_s)$ (dashed curves) for the

---

[4] Since a constant is a polynomial of zero order, and since the effect of the flat pulse is to "hold" or replicate each sample for T_s sec, the use of a flat pulse is called a *zero-order hold reconstruction.*

triangular pulse shape. In this case, the output $y(t)$ of the D-to-C converter at any time t is the sum of the scaled pulses that overlap at that time instant. Since the duration of the pulse is $2T_s$, and they are shifted by multiples of T_s, no more than two pulses can overlap at any given time. The resulting output is shown as the solid curve in the bottom panel of Fig. 4.19. Note that the result is that the samples are connected by straight lines. Also note that the values at $t = nT_s$ are exactly correct. This is because the triangular pulses are zero at $\pm T_s$, and only one scaled pulse (with value $y[n]$ at $t = nT_s$) contributes to the value at $t = nT_s$. In this case, we say that the continuous-time waveform has been obtained by *linear interpolation* between the samples. This is a smoother and better approximation of the original waveform (dotted curve), but there is still significant reconstruction error for this signal.

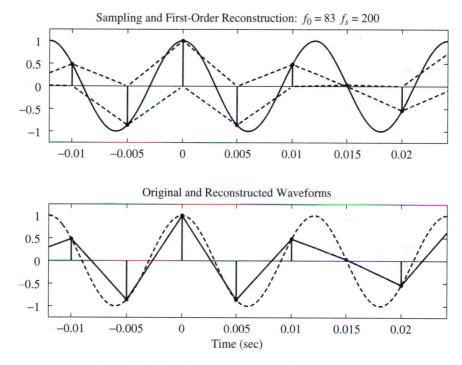

Figure 4.19 D-to-C conversion using a triangular pulse.

4.4.5 Parabolic Interpolation

A third pulse shape is shown in the lower left panel of Fig. 4.17. This pulse consists of four parabolic (second-order polynomial) segments. Note that it has a duration that is twice that of the triangular pulse and four times that of the square pulse. Also, note that this pulse has zeros at certain key locations:

$$p(t) = 0 \qquad \text{for } t = 0, \ \pm T_s, \ \pm 2T_s$$

The reconstruction with the parabolic pulses is shown in Fig. 4.20. The top panel of Fig. 4.20 shows the original waveform ($f_0 = 83$ Hz), its samples ($f_s = 200$ samples/sec), and the shifted and scaled pulses $y[n]p(t - nT_s)$. Now note that for values of t between two sample times, four pulses overlap and must be added together in the sum (4.4.6). This means that the reconstructed signal at a particular time instant, which is the sum of these overlapping pulses, depends on the two samples preceding and the two samples following that time instant. The lower panel of Fig. 4.20 shows the original waveform (dashed), the samples, and the output of the D-to-C converter with "parabolic pulse" interpolation (solid curve). Now we see that the approximation is getting smoother and better, but it is still far from perfect. We will see that this is because the sampling is only 2.4 times the highest frequency.

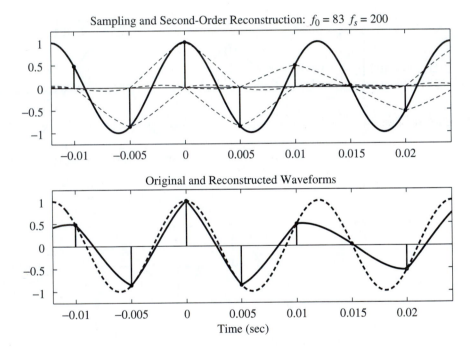

Figure 4.20   D-to-C conversion using a parabolic pulse.

4.4.6 Over-Sampling Aids Interpolation

From the previous three examples, it seems that one way to make a smooth reconstruction of a waveform such as a sinusoid is to use a pulse $p(t)$ that is smooth and has a long duration. Then the interpolation formula will involve several neighboring samples in the computation of each output value $y(t)$. However, the sampling rate is another important factor. If the original waveform does not vary much over the duration of $p(t)$, then we will also obtain a good reconstruction. One way to ensure this is by over-sampling, i.e., using a sampling rate that is much greater than the frequency of the cosine wave. This is illustrated in Figs. 4.21, 4.22, and 4.23, where the frequency of the cosine wave is still $f_0 = 83$ Hz, but the sampling frequency has been raised to $f_s = 800$ samples/sec. Now the reconstruction pulses are the same shape as in Fig. 4.17, but they are much shorter, since $T_s = 1.25$ msec instead of 5 msec. The signal changes much less over the duration of a single pulse, so the waveform appears "smoother" and is much easier to reconstruct accurately using only a few samples. Note that even for the case of the square pulse (Fig. 4.21) the reconstruction is better, but still discontinuous; the triangular pulse (Fig. 4.22) gives an excellent approximation; and the parabolic pulse gives a reconstruction that is indistinguishable from the original signal on the plotting scale of Fig. 4.23.

RECON-
STRUCT
MOVIES

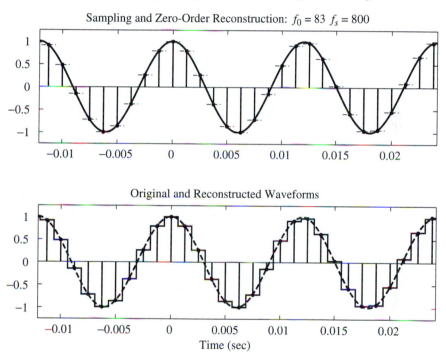

Figure 4.21   D-to-C conversion using a square pulse. The original 83-Hz sinusoid was over-sampled at $f_s = 800$ samples/sec.

Figures 4.21, 4.22, and 4.23 show that over-sampling can make it easier to re-construct a waveform from its samples. Indeed, this is why audio CD players have over-sampled D-to-A converters! In the CD case, $4\times$ or $2\times$ over-sampling is used to increase the sampling rate before sending the samples to the D-to-A converter. This makes it possible to use a simpler (and therefore less inexpensive) D-to-A converter to reconstruct an accurate output from a CD player.

4.4.7 Ideal Bandlimited Interpolation

So far in this section, we have demonstrated the basic principles of discrete-to-continuous conversion. We have shown that this process can be approximated very

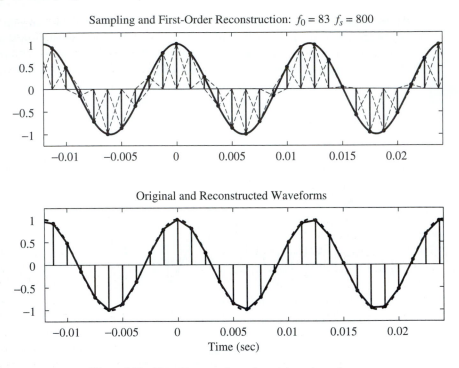

Figure 4.22   D-to-C conversion using a triangular pulse.

well by a sum of pulses (4.4.6). One question remains: What is the pulse shape that gives "ideal D-to-C conversion"? Well, it is given by the following equation:

$$p(t) = \frac{\sin \dfrac{\pi}{T_s} t}{\dfrac{\pi}{T_s} t} \qquad \text{for } -\infty < t < \infty \qquad (4.4.9)$$

This is a very long pulse, and its infinite length implies that to reconstruct a signal at time t exactly from its samples requires *all* the samples, not just those around that time. The lower right part of Fig. 4.17 shows this pulse over the interval $-2.6T_s < t < 2.6T_s$. It decays outside this interval, but never does reach and stay at zero. Since the pulse has zeros at multiples of T_s, this type of reconstruction is still an interpolation process, called *bandlimited interpolation*. Using this pulse to reconstruct from samples of a cosine wave will always produce a cosine wave exactly. If the sampling rate satisfies the conditions of the sampling theorem, the reconstructed cosine wave will be identical to the original signal that was sampled. If aliasing occurred in sampling, the ideal D-to-C converter reconstructs a cosine wave with the alias frequency that is less than $f_s/2$.

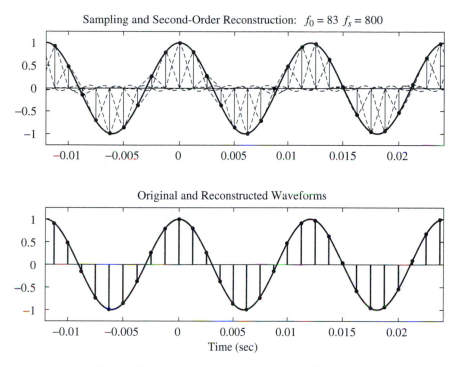

Figure 4.23   D-to-C conversion using a parabolic pulse.

4.5   THE SAMPLING THEOREM

This chapter has discussed the issues that arise in sampling continuous-time signals. Using the example of the continuous-time cosine signal, we have illustrated the phenomenon of aliasing, and we have shown how the original continuous-time cosine signal can be reconstructed by interpolation. All of the discussion of this chapter has been aimed at the goal of establishing confidence in the Shannon sampling theorem, which, because of its central importance to our study of digital signal processing, is repeated here.

Shannon Sampling Theorem

A continuous-time signal $x(t)$ with frequencies no higher than f_{max} can be reconstructed exactly from its samples $x[n] = x(nT_s)$, if the samples are taken at a rate $f_s = 1/T_s$ that is greater than $2f_{max}$.

A block diagram representation of the sampling theorem is shown in Fig. 4.24 in terms of the ideal C-to-D and D-to-C converters that we have defined in this chapter. The

sampling theorem states that for the sampling and reconstruction system shown in
Fig. 4.24, if the input is composed of sinusoidal signals limited to the set of frequencies
in the range $0 \leq f \leq f_{max}$, then the reconstructed signal is equal to the original
signal that was sampled; i.e., $y(t) = x(t)$.

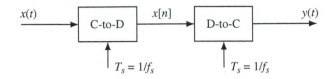

Figure 4.24   Sampling and reconstruction system.

Signals composed of sinusoids such that all frequencies are limited to a "band
of frequencies" of the form $0 \leq f \leq f_{max}$ are called *bandlimited signals*.[5] Such
signals could be represented as

$$x(t) = \sum_{k=0}^{N} x_k(t) \tag{4.5.1}$$

where each of the individual signals is of the form

$$x_k(t) = A_k \cos(2\pi f_k t + \phi_k) \tag{4.5.2}$$

and it is assumed that the frequencies are ordered so that $f_0 \geq 0$ and $f_N \leq f_{max}$.
As we have seen in Chapter 3, such an additive combination of cosine signals can
produce an infinite variety of both periodic and nonperiodic signal waveforms. If we
sample the signal represented by (4.5.1) and (4.5.2), we obtain

$$x[n] = x(nT_s) = \sum_{k=0}^{N} x_k(nT_s) = \sum_{k=0}^{N} x_k[n] \tag{4.5.3}$$

where $x_k[n] = A_k \cos(2\pi \hat{f}_k n + \phi_k)$ with $\hat{f}_k = f_k/f_s$. That is, if we sample a sum
of continuous-time cosines, we obtain a sum of sampled cosines each of which would
be subject to aliasing if the sampling rate is not high enough.

The final step in the process of sampling followed by reconstruction in Fig. 4.24
is discrete-to-continuous conversion by interpolation with

$$y(t) = \sum_{n=-\infty}^{\infty} x[n]p(t - nT_s) \tag{4.5.4}$$

---

[5] The corresponding complex exponential signals would be limited to the band $-f_{max} \leq f \leq f_{max}$.

where for perfect reconstruction, $p(t)$ would be given by (4.4.9). This expression for the reconstructed output is a linear operation of the samples $x[n]$. We can see this by substituting (4.5.3) into (4.5.4) as in

$$y(t) = \sum_{n=-\infty}^{\infty} \left(\sum_{k=0}^{N} x_k[n] \right) p(t - nT_s) = \sum_{k=0}^{N} \left(\sum_{n=-\infty}^{\infty} x_k[n] p(t - nT_s) \right) \quad (4.5.5)$$

Now since each individual sinusoid is assumed to satisfy the conditions of the sampling theorem, it follows that the D-to-C converter will reconstruct each component perfectly, and therefore we conclude that

$$y(t) = \sum_{k=0}^{N} x_k(t) = x(t)$$

Thus, we have shown that the Shannon sampling theorem applies to *any* signal that can be represented as a bandlimited sum of sinusoids, and since it can be shown that most real-world signals can be represented as bandlimited signals, it follows that the sampling theorem is a very general guide to sampling all kinds of signals.

4.6 SUMMARY AND LINKS

This chapter introduced the concept of sampling and the companion operation of reconstruction. With sampling, the possibility of aliasing always exists, and we have created the strobe demo to illustrate that concept in a direct and intuitive way.

None of the labs can be associated directly with this chapter. Some aspects of sampling have already been used in the music synthesis labs that are associated with Chapter 3, because the sounds must be produced by making their samples in a computer before playing them out through a D-to-A converter.

The CD-ROM also contains many demonstrations of sampling and aliasing:

1. Strobe movie filmed using the natural strobing of a video camera at 30 frames per second.

2. Synthetic strobe demos produced as MATLAB movies.

3. Reconstruction movies that show the interpolation process for different pulse shapes and different sampling rates.

Again, the CD-ROM contains a large number of solved homework problems available for review and practice.

PROBLEMS

4.1 Consider the cosine wave

$$x(t) = 10\cos(880\pi t + \phi)$$

Suppose that we obtain a sequence of numbers by sampling the waveform at equally spaced time instants nT_s. In this case, the resulting sequence would have values

$$x[n] = x(nT_s) = 10\cos(880\pi nT_s + \phi) \qquad -\infty < n < \infty$$

Suppose that $T_s = 0.0001$.

(a) How many samples will be taken in one period of the cosine wave?

(b) Now consider another waveform $y(t)$ such that

$$y(t) = 10\cos(\omega_0 t + \phi)$$

Find a frequency $\omega_0 > 880\pi$ such that $y(nT_s) = x(nT_s)$ for all integers n. *Hint:* Use the fact that $\cos(\theta + 2\pi n) = \cos(\theta)$ if n is an integer.

(c) For the frequency found in (b), what is the total number of samples taken in one period of $x(t)$?

4.2 A discrete-time signal $x[n]$ is known to be a sinusoid:

$$x[n] = A\cos(2\pi \hat{f}_0 n + \phi)$$

The values of $x[n]$ are tabulated for $n = 0, 1, 2, 3, 4,$ and 5.

n	1	2	3	4	5	6	7
$x[n]$	2.5000	0.5226	−1.5451	−3.3457	−4.5677	−5.0000	−4.5677

(a) Plot $x[n]$ versus n.

(b) Prove (with phasors, not trigonometry) the following identity for the cosine signal:

$$\beta = \frac{\cos(n+1)2\pi\hat{f}_0 + \cos(n-1)2\pi\hat{f}_0}{\cos n2\pi\hat{f}_0} \qquad \text{for all } n$$

Determine the value of the constant β.
Note: β does not depend on n, but it might be a function of $\hat{f}_0$.

(c) Now determine the numerical values of A, ϕ, and $\hat{f}_0$. (This is an easy calculation if you find $\hat{f}_0$ first.)

4.3 A discrete-time signal $x[n]$ is known to be a sinusoid:

$$x[n] = A \cos(2\pi \hat{f}_0 n + \phi)$$

The values of $x[n]$ are tabulated for $n = 0, 1, 2, 3, 4,$ and 5.

n	0	1	2	3	4	5
$x[n]$	2.4271	2.9002	2.9816	2.6603	1.9798	1.0318

Plot $x[n]$, then determine the numerical values of A, ϕ, and $\hat{f}_0$.

4.4 Let $x(t) = 7 \sin(11\pi t)$. In each of the following parts, the discrete-time signal $x[n]$ is obtained by sampling $x(t)$ at a rate f_s, and the resultant $x[n]$ can be written:

$$x[n] = A \cos(2\pi \hat{f}_0 n + \phi)$$

So, for each part below, determine the values of A, ϕ, and $\hat{f}_0$. In addition, state whether or not the signal has been over-sampled or under-sampled.

(a) Let the sampling frequency be $f_s = 10$ samples/sec.

(b) Let the sampling frequency be $f_s = 5$ samples/sec.

(c) Let the sampling frequency be $f_s = 15$ samples/sec.

4.5 Suppose that a discrete-time signal $x[n]$ is given by the formula:

$$x[n] = 2.2 \cos(0.3\pi n - \pi/3)$$

and that it was obtained by sampling a continuous-time signal $x(t) = A \cos(2\pi f_0 t + \phi)$ at a sampling rate of $f_s = 6000$ samples/sec. Determine three different continuous-time signals that could have produced $x[n]$. These continuous-time signals all should have a frequency less than 8 kHz. Write the mathematical formula for all three.

4.6 An amplitude-modulated (AM) cosine wave is represented by the formula

$$x(t) = [10 + \cos(2\pi (2000)t)] \cos(2\pi (10^4)t)$$

(a) Sketch the two-sided spectrum of this signal. Be sure to label important features of the plot.

(b) Is this waveform periodic? If so, what is the period?

(c) What relation must the sampling rate f_s satisfy so that $y(t) = x(t)$ for the following system?

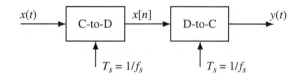

4.7 Suppose that a discrete-time signal $x[n]$ is given by the formula

$$x[n] = 10\cos(0.2\pi n - \pi/7)$$

and that it was obtained by sampling a continuous-time signal at a sampling rate of $f_s = 1000$ samples/sec.

(a) Determine two *different* continuous-time signals $x_1(t)$ and $x_2(t)$ whose samples are equal to $x[n]$; i.e., find $x_1(t)$ and $x_2(t)$ such that $x[n] = x_1(nT_s) = x_2(nT_s)$ if $T_s = 0.001$. Both of these signals should have a frequency less than 1000 Hz. Give a formula for each signal.

(b) If $x[n]$ is given by the equation above, what signal will be reconstructed by an ideal D-to-C converter operating at sampling rate of 2000 samples/sec? That is, what is the output $y(t)$ in the following figure if $x[n]$ is as given above?

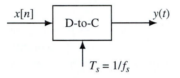

4.8 A non-ideal D-to-C converter takes a sequence $y[n]$ as input and produces a continuous-time output $y(t)$ according to the relation

$$y(t) = \sum_{n=-\infty}^{\infty} y[n]p(t - nT_s)$$

where $T_s = 0.1$ second. The input sequence is given by the formula

$$y[n] = \begin{cases} (0.8)^n & 0 \le n \le 5 \\ 0 & \text{otherwise} \end{cases}$$

(a) For the pulse shape

$$p(t) = \begin{cases} 1 & -0.05 \le t \le 0.05 \\ 0 & \text{otherwise} \end{cases}$$

carefully sketch the output waveform $y(t)$.

(b) For the pulse shape

$$p(t) = \begin{cases} 1 - 10|t| & -0.1 \le t \le 0.1 \\ 0 & \text{otherwise} \end{cases}$$

carefully sketch the output waveform $y(t)$.

4.9 Let $x[n]$ be the complex exponential

$$x[n] = 7e^{j(0.22\pi n - 0.25\pi)}$$

If we define a new signal $y[n]$ to be the second difference

$$y[n] = x[n+1] - 2x[n] + x[n-1]$$

it is possible to express $y[n]$ in the form

$$y[n] = Ae^{j(2\pi \hat{f}_0 n + \phi)}$$

Determine the numerical values of A, ϕ, and $\hat{f}_0$.

4.10 Suppose that MATLAB is used to plot a sinusoidal signal. The following MATLAB code generates a signal $x[n]$ and plots it. Unfortunately, the time axis of the plot is not labeled properly.

```
Ts = 0.01;
Duration = 0.3;
tt = 0 : Ts : Duration;
Fo = 394;
xx = 9*cos( 2*pi*Fo*tt  + pi/2 );
%
stem( xx )      %<--- OOPS! there is no time axis
```

   (a) Make the stem plot of the signal. Either sketch it or plot it using MATLAB.
   (b) For the plot above, determine the correct formula for the discrete-time signal in the form:

$$x[n] = A\cos(2\pi \hat{f}_0 n + \phi)$$

   (c) Explain how aliasing affects the plot that you see.

4.11 The spectrum diagram gives the frequency content of a signal.
   (a) Draw a sketch of the spectrum of

$$x(t) = \cos(50\pi t)\,\sin(700\pi t)$$

   Label the frequencies and complex amplitudes of each component.
   (b) Determine the minimum sampling rate that can be used to sample $x(t)$ without aliasing for any of the components.

4.12 The spectrum diagram gives the frequency content of a signal.
   (a) Draw a sketch of the spectrum of

$$x(t) = \sin^3(400\pi t)$$

   Label the frequencies and complex amplitudes of each component.
   (b) Determine the minimum sampling rate that can be used to sample $x(t)$ without aliasing for any of its components.

4.13 The intention of the following MATLAB program is to plot 13 cycles of a 13-Hz sinusoid, but it has a bug.

```
Fo = 13;
To = 1/Fo;     %-- Period
Ts = 0.07;
tt = 0 : Ts : (13*To);
xx = real( exp( j*(2*pi*Fo*tt  - pi/2) ) );
%
stem( tt, xx ),  xlabel('TIME    (sec)'),  grid
```

    (a) Draw a sketch of the plot that will be produced by MATLAB. Explain how aliasing or folding affects the plot that you see. In particular, how many periods do you observe?

    (b) Determine an acceptable value of Ts to get a very smooth plot of the desired 13 Hz signal.

4.14 An amplitude-modulated (AM) cosine wave is represented by the formula

$$x(t) = [3 + \sin(\pi t)]\cos(13\pi t + \pi/2)$$

    (a) Use *phasors* to show that $x(t)$ can be expressed in the form

$$x(t) = A_1 \cos(\omega_1 t + \phi_1) + A_2 \cos(\omega_2 t + \phi_2) + A_3 \cos(\omega_3 t + \phi_3)$$

    where $\omega_1 < \omega_2 < \omega_3$; i.e., find A_1, A_2, A_3, ϕ_1, ϕ_2, ϕ_3, ω_1, ω_2, ω_3.

    (b) Sketch the two-sided spectrum of this signal on a frequency axis. Be sure to label important features of the plot. Label your plot in terms of the numerical values of A_i, ϕ_i, and ω_i.

    (c) Determine the minimum sampling rate that can be used to sample $x(t)$ without aliasing of any of the components.

4.15 Consider the following system with ideal C-to-D and D-to-C converters.

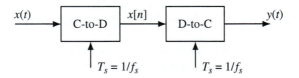

    (a) Suppose that the discrete-time signal $x[n]$ is given by the formula

$$x[n] = 10\cos(0.13\pi n + \pi/13)$$

    If the sampling rate is $f_s = 1000$ samples/sec, determine two *different* continuous-time signals $x(t) = x_1(t)$ and $x(t) = x_2(t)$ that could have been inputs to the above system; i.e., find $x_1(t)$ and $x_2(t)$ such that $x[n] = x_1(nT_s) = x_2(nT_s)$ if $T_s = 0.001$.

(b) If the input $x(t)$ is given by the two-sided spectrum representation shown below, determine a simple formula for $y(t)$ when $f_s = 700$ samples/sec (for both the C-to-D and D-to-C converters).

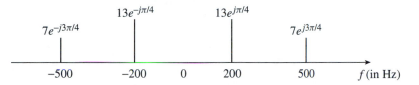

4.16 In the rotating disk and strobe demo, we observed that different flashing rates of the strobe light would make the spot on the disk stand still or move in different directions.

(a) Assume that the disk is rotating *clockwise* at a constant speed of 13 rev/sec. If the flashing rate is 15 times per second, express the movement of the spot on the disk as a complex phasor, $p[n]$, that gives the position of the spot at the nth flash. Assume that the spot is at the top when $n = 0$ (the first flash).

(b) For the conditions in (a), determine the apparent speed (in revolutions per second) and direction of movement of the "strobed" spot.

(c) *Now assume that the rotation speed of the disk is unknown.* If the flashing rate is 13 times per second, and the spot on the disk moves *counterclockwise* by 15 degrees with each flash, determine the rotation speed of the disk (in rev/sec). If the answer is not unique, give *all* possible rotation speeds.

4.17 In the rotating disk and strobe demo, we observed that different flashing rates of the strobe light would make the spot on the disk stand still.

(a) Assume that the disk is rotating in the counterclockwise direction at a constant speed of 720 rpm. Express the movement of the spot on the disk as a rotating complex phasor.

(b) If the strobe light can be flashed at a rate of n flashes per second where n is an integer greater than zero, determine all possible flashing rates such that the disk can be made to stand still.

Note: The only possible flashing rates are 1 per second, 2 per second, 3 per second, etc.

(c) If the flashing rate is 13 times per second, explain how the spot will move, and write a complex phasor that gives the position of the spot at each flash.

4.18 The following complex-valued signal is a phasor:

$$z[n] = e^{j\theta[n]}$$

where $\theta[n]$ is the phase.

(a) When the phase changes by a constant amount versus n, the phasor rotates at a constant speed. For the following phasor

$$z[n] = e^{j(0.08\pi n - 0.25\pi)}$$

make a plot of the phasor locations for $n = 0, 1, 2, 7, 10, 17, 20, 33, 50,$ and 99.

(b) What is the period of $z[n]$?

(c) Repeat for the complex phasor that corresponds to the chirp signal:

$$c[n] = e^{j0.1\pi n^2}$$

In this case, plot the phasor locations for $n = 0, 1, 2, 3, 4$, and 7.

4.19 A digital chirp signal is synthesized according to the following formula:

$$x[n] = \Re e\{e^{j\theta[n]}\} = \cos(\pi(0.7 \times 10^{-3})n^2) \qquad \text{for } n = 0, 1, 2, \ldots, 200$$

(a) Make a plot of the rotating phasor $e^{j\theta[n]}$ for $n = 10, 50$, and 100.

(b) If this signal is played out through a D-to-A converter whose sampling rate is 8 kHz, make a plot of the instantaneous analog frequency (in Hz) versus time for the analog signal.

(c) If the *constant frequency* digital signal $v[n] = \cos(0.7\pi n)$ is played out through a D-to-A converter whose sampling rate is 8 kHz, what (analog) frequency will be heard?

5

FIR Filters

Up to this point, we have focused our attention on signals and their mathematical representations. In this chapter, we begin to emphasize *systems* or *filters*. Strictly speaking, a filter is a system that is designed to remove some component or modify some characteristic of a signal, but often the two terms are used interchangeably. In this chapter, we introduce the class of FIR (finite impulse response) systems, or, as we will often refer to them, *FIR filters*. These filters are systems for which each output is the sum of a finite number of weighted samples of the input sequence. We will define the basic input–output structure of the FIR filter as a time-domain computation based on a *feed-forward* difference equation. The unit impulse response of the filter will be defined and shown to characterize the filter. The general concepts of linearity and time invariance will also be presented. These properties characterize a wide class of filters that are exceedingly important in both the continuous-time and the discrete-time cases.

Our purpose in this chapter is to introduce the basic ideas of discrete-time systems and to provide a starting point for further study. The analysis of both discrete-time and continuous-time systems is a rich subject of study that is necessarily based on mathematical representations and manipulations.[1] The systems that we introduce in this chapter are the simplest to analyze. The remainder of the text is concerned with extending the ideas of this chapter to other classes of systems and with developing tools for the analysis of other discrete-time systems.

5.1 DISCRETE-TIME SYSTEMS

A discrete-time system is a computational process for transforming one sequence, called the *input signal*, into another sequence called the *output signal*. As we have

---

[1] Systems can be analyzed effectively by the mathematical methods of Fourier analysis, which are introduced in Chapter 9 and are covered extensively in more advanced signals and systems texts.

already mentioned, systems are often depicted by block diagrams such as the one in Fig. 5.1. In Chapter 4, we used similar block diagrams to represent the operations of sampling and reconstruction. In the case of sampling, the input signal is a continuous-time signal and the output is a discrete-time signal, while for reconstruction the opposite is true. Now we want to begin to study discrete-time systems where the input and output are both discrete-time signals. Such systems are very interesting because they can be implemented with digital computation and because they can be designed so as to modify signals in many useful ways.

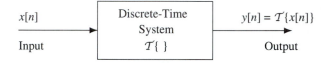

Figure 5.1   Block-diagram representation of a discrete-time system.

In general, we represent the operation of a system by the notation

$$y[n] = \mathcal{T}\{x[n]\}$$

which suggests that the output sequence is related to the input sequence by a process that can be described mathematically by an operator $\mathcal{T}$. Since a discrete-time signal is just a sequence of numbers, such operators can be described by giving a formula for computing the values of the output sequence from the values of the input sequence. For example, the relation

$$y[n] = (x[n])^2$$

defines a system for which the output sequence values are the square of the corresponding input sequence values. A more complicated example would be the following system definition:

$$y[n] = \max\{x[n], x[n-1], x[n-2]\}$$

In this case, the output depends on three consecutive input values. Obviously, infinite possibilities exist for defining discrete-time systems. In order to allow us to study discrete-time systems in a useful way, it is necessary to limit this range of possibilities by placing some restrictions on the properties of the systems that we study. Therefore we will begin our study of discrete-time systems in this chapter by introducing a very important class of discrete-time systems called "FIR filters." Specifically, we will discuss the representation, implementation, and analysis of discrete-time FIR systems, and illustrate how such systems can be used to modify signals.

5.2   THE RUNNING AVERAGE FILTER

A simple but useful transformation of a discrete-time signal is to compute a "moving average" or "running average" of two or more consecutive numbers of the sequence, thereby forming a new sequence of the average values. The FIR filter is a generalization of the idea of a running average. Averaging is commonly used whenever data fluctuate and must be smoothed prior to interpretation. For example, stock-market prices fluctuate noticeably from day to day, or hour to hour. Therefore, one might take an average of the stock price over several days before looking for any trend. Another everyday example concerns credit-card balances where interest is charged on the "average" daily balance.

In order to motivate the general definition of the class of FIR systems, let us consider the simple running average as an example of a system that processes an input sequence to produce an output sequence. To be specific, consider a 3-point averaging method; i.e., each value of the output sequence is the sum of three consecutive input sequence values divided by three. If we apply this algorithm to the triangularly shaped sequence shown in Fig. 5.2, we can compute a new sequence called $y[n]$, which is the output of the averaging operator. The sequence in Fig. 5.2 is an example of a *finite-length* signal. The *support* of such a sequence is the set of values over which the sequence is nonzero; in this case, the support of the sequence is the finite interval $0 \leq n \leq 4$. A 3-point average of the values $\{x[0], \ x[1], \ x[2]\} = \{2, \ 4, \ 6\}$ gives the answer $\frac{1}{3}(2 + 4 + 6) = 4$. This result defines one of the output values. The next output value is obtained by averaging $\{x[1], \ x[2], \ x[3]\} = \{4, \ 6, \ 4\}$ which, yields a value of 14/3. Before going any further, we should decide on the output indexing. For example, the values 4 and 14/3 could be assigned to $y[0]$ and $y[1]$, *but this is only one of many possibilities.* With this indexing, the equations for computing the output from the input are

$$y[0] = \tfrac{1}{3}(x[0] + x[1] + x[2])$$

$$y[1] = \tfrac{1}{3}(x[1] + x[2] + x[3])$$

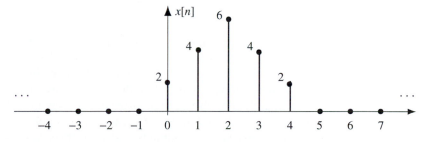

Figure 5.2   Finite-length input signal, $x[n]$.

which generalizes to the following input–output equation

$$y[n] = \tfrac{1}{3}(x[n] + x[n + 1] + x[n + 2]) \tag{5.2.1}$$

The equation given in (5.2.1) is called a *difference equation*. It is a complete description of the FIR system because we can use (5.2.1) to compute the entire output signal for all index values $-\infty < n < \infty$. For the triangular input of Fig. 5.2, the result is the signal $y[n]$ tabulated as follows:

n	$n < -2$	-2	-1	0	1	2	3	4	5	$n > 5$
$x[n]$	0	0	0	2	4	**6**	**4**	**2**	0	0
$y[n]$	0	$\frac{2}{3}$	2	4	$\frac{14}{3}$	**4**	2	$\frac{2}{3}$	0	0

Note that the values in bold type in the $x[n]$ row are the numbers involved in the computation of $y[2]$. Also note that $y[n] = 0$ outside of the finite interval $-2 \leq n \leq 4$; i.e., the output also has finite support. The output sequence is also plotted in Fig. 5.3. Observe that the output sequence is longer than the input sequence, and that the output appears to be a somewhat rounded-off version of the input; i.e. it is "smoother" than the input sequence. This behavior is characteristic of the running average FIR filter.

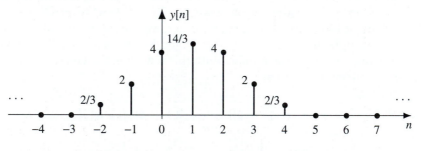

Figure 5.3   Output of running average, $y[n]$.

The choice of the output indexing is arbitrary, but it does matter when speaking about properties of the filter. For example, the filter defined in (5.2.1) has the property that its output starts (becomes nonzero) before the input starts. This would certainly be undesirable if the input signal values came directly from an A-to-D converter, as is common in audio signal-processing applications. In this case, n would stand for time, and we can interpret $y[n]$ in (5.2.1) as the computation of the "present" value of the output based on three input values. Since these inputs are indexed as $n, n+1$, and $n+2$, two of them are "in the future." In general, values from either the "past" or the "future" or both may be used in the computation, as shown in Fig. 5.4. In all cases of a 3-point running average, a "sliding window" of three samples determines which three samples are used in the computation of $y[n]$.

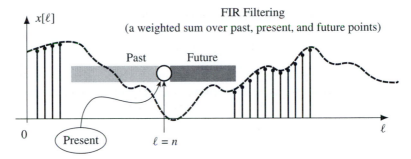

Figure 5.4   The running-average filter calculation at time index n uses values within a sliding window (shaded). Dark shading indicates the future ($\ell > n$); light shading, the past ($\ell < n$).

A filter that uses only the present and past values of the input is called a *causal* filter, implying that the cause does not precede the corresponding effect. Therefore, a filter that uses future values of the input is called *noncausal*. Noncausal systems cannot be implemented in a real-time application because the input is not yet available when the output has to be computed. In other cases, where stored data blocks are manipulated inside a computer, the issue of causality is not crucial.

An alternative output indexing scheme can produce a 3-point averaging filter that is causal. In this case, the output value $y[n]$ is the average of inputs at n (the present), $n - 1$ (one sample previous), and $n - 2$ (two samples previous). The difference equation for this filter is

$$y[n] = \tfrac{1}{3}(x[n] + x[n-1] + x[n-2]) \qquad (5.2.2)$$

The form given in (5.2.2) is a causal running averager, or it may well be called a *backward average*. Using the difference equation (5.2.2), we can make a table of all output values over the range $-\infty < n < \infty$. (Notice that now the boldface values of $x[n]$ are used in this case to compute $y[4]$ instead of $y[2]$.) The resulting signal $y[n]$ has the same values as before, but its support is now the index interval $0 \leq n \leq 6$. Observe that the output of the causal filter is simply a shifted version of the output of the previous noncausal filter. This filter is causal because the output depends on only the present and two previous (or past) values of the input. Therefore, the output does not change from zero before the input changes from zero.

n	$n < -2$	-2	-1	0	1	2	3	4	5	6	7	$n > 7$
$x[n]$	0	0	0	2	4	6	**4**	2	0	0	0	0
$y[n]$	0	0	0	$\frac{2}{3}$	2	4	$\frac{14}{3}$	**4**	2	$\frac{2}{3}$	0	0

Exercise 5.1. Determine the output of a *centralized averager*

$$y[n] = \tfrac{1}{3}(x[n+1] + x[n] + x[n-1])$$

for the input in Fig. 5.2. Is this filter causal or noncausal? What is the support of the output for this input? How would the plot of the output compare to Fig. 5.3?

5.3 THE GENERAL FIR FILTER

Note that (5.2.2) is a special case of the general difference equation

$$y[n] = \sum_{k=0}^{M} b_k\, x[n-k] \qquad (5.3.1)$$

That is, when $M = 2$ and $b_k = 1/3$ for $k = 0, 1, 2$, (5.3.1) reduces to the causal running average of (5.2.2). If the coefficients b_k are not all the same, then we might say that (5.3.1) defines a "weighted running average of $M+1$ samples." It is clear from (5.3.1) that the computation of $y[n]$ involves the samples $x[\ell]$ for $\ell = n, n - 1, n - 2, \ldots, n - M$; i.e., $x[n], x[n-1], x[n-2]$, etc. Since the filter in (5.3.1) does not involve future values of the input, the system is causal, and, therefore, the output cannot start before the input becomes non-zero.[2] Figure 5.5 shows that the causal FIR filter uses $x[n]$ and the past M points to compute the output. Figure 5.5 also shows that if the input has finite support ($0 \le \ell \le N - 1$), there will be an interval of M samples at the beginning, where the computation will involve fewer than $M+1$ samples as the sliding window of the filter engages with the input, and an interval of M samples at the end where the sliding window of the filter disengages from the input sequence. It also can be seen from Fig. 5.5 that the output sequence can be as much as M samples longer than the input sequence.

M-th Order FIR Filter Operation (Causal)

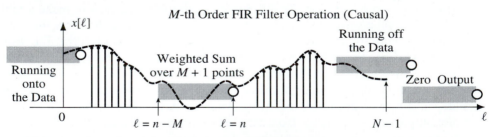

Figure 5.5 Operation of the Mth-order causal FIR filter showing various positions of the sliding window of $M + 1$ points under which the weighted average is calculated. When the input signal $x[\ell]$ is also finite length (N points), the sliding window will run onto and off the input data, so the resulting output signal will also have finite length.

---

[2] Note that a noncausal system can be represented by (5.3.1) by allowing negative values of the summation index k.

Example 5.1   The FIR filter is completely defined once the set of filter coefficients $\{b_k\}$ is known. For example, if the $\{b_k\}$ are

$$\{b_k\} = \{3, -1, 2, 1\}$$

then we have a length 4 filter with $M = 3$, and (5.3.1) expands into a 4-point difference equation:

$$y[n] = \sum_{k=0}^{3} b_k x[n - k] = 3x[n] - x[n - 1] + 2x[n - 2] + x[n - 3]$$

◇

The parameter M is the *order* of the FIR filter. The number of filter coefficients is also called the filter *length* (L). The length is one greater than the order; i.e., $L = M + 1$. This terminology will make more sense after we have introduced the z-transform in Chapter 7.

Exercise 5.2. Compute the output $y[n]$ for the length 4 filter whose coefficients are $\{b_k\} = \{3, -1, 2, 1\}$. Use the input signal given in Fig. 5.2. Verify that the answers tabulated here are correct, then fill in the missing values.

n	$n < 0$	0	1	2	3	4	5	6	7	8	$n > 8$
$x[n]$	0	2	4	6	4	2	0	0	0	0	0
$y[n]$	0	6	10	18	?	?	?	8	2	0	0

5.3.1 An Illustration of FIR Filtering

To illustrate some of the things that we have learned so far, and to show how FIR filters can modify sequences, consider a signal

$$x[n] = \begin{cases} (1.02)^n + 0.5\cos(2\pi n/8 + \pi/4) & 0 \le n \le 40 \\ 0 & \text{otherwise} \end{cases}$$

This signal is shown as the upper plot in Fig. 5.6. We often have real signals of this form; i.e., a component that is the signal of interest (in this case, it may be the slowly varying exponential component $(1.02)^n$) plus another component that is not of interest. Indeed, the second component is often considered to be *noise* that interferes with observation of the desired signal. In this case, we will consider the sinusoidal component $0.5\cos(2\pi n/8 + \pi/4)$ to be noise that we wish to remove. The solid, exponentially growing curve shown in each of the plots in Fig. 5.6 simply connects the sample values of the desired signal $(1.02)^n$ by straight lines for reference in the other two plots.

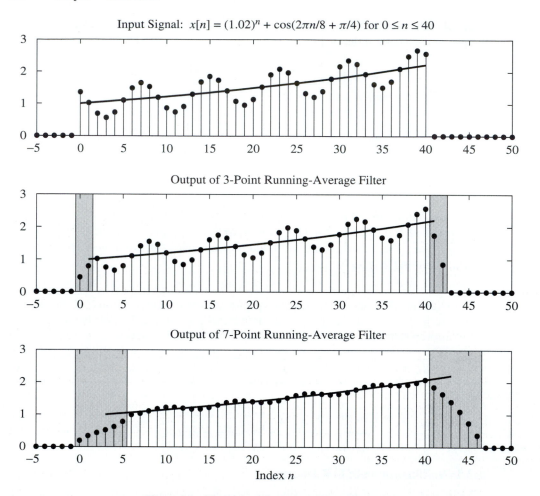

Figure 5.6   Illustration of running average filtering.

Now suppose that $x[n]$ is the input to a causal 3-point running averager, i.e.,

$$y_3[n] = \tfrac{1}{3}\left(\sum_{k=0}^{2} x[n-k]\right) \qquad (5.3.2)$$

In this case, $M = 2$ and all the coefficients are equal to 1/3. The output of this filter is shown in the middle plot in Fig. 5.6. We can notice several things about these plots.

1. Observe that the input sequence $x[n]$ is zero prior to $n = 0$, and from (5.3.2) it is clear that the output must be zero for $n < 0$.

2. The output becomes nonzero at $n = 0$, and the shaded interval of length $M = 2$ samples at the beginning of the nonzero part of the output sequence

is the interval where the 3-point averager "runs onto" the input sequence. For $2 \leq n \leq 40$, the input samples within the 3-point averaging window are all nonzero.

3. There is another shaded interval of length $M = 2$ samples at the end (after sample 40), where the filter window "runs off" the input sequence.

4. Observe that the size of the sinusoidal component has been reduced, but that the component is not eliminated by the filter. The solid line showing the values of the exponential component has been shifted to the right by $M/2 = 1$ sample to account for the shift introduced by the causal filter.

Clearly, the 3-point running averager has removed some of the fluctuations in the input signal, but we have not recovered the desired component. Intuitively, we might think that averaging over a longer interval might produce better results. The lower plot in Fig. 5.6 shows the output of a 7-point running-averager as defined by

$$y_7[n] = \frac{1}{7}\left(\sum_{k=0}^{6} x[n-k]\right) \tag{5.3.3}$$

In this case, since $M = 6$ and all the coefficients are equal to 1/7, we observe the following:

1. The shaded regions at the beginning and end of the output are now $M = 6$ samples long.

2. Now the size of the sinusoidal component is greatly reduced relative to the input sinusoid, and the exponential component is very close to the exponential component of the input (after a shift of $M/2 = 3$ samples).

What can we conclude from this example? First, it appears that FIR filtering can modify signals in ways that may be useful. Second, the length of the averaging interval seems to have a big effect on the resulting output. Third, the running-average filters appear to introduce a shift equal to $M/2$ samples. All of these observations can be shown to apply to more general FIR filters defined by (5.3.1). However, before we can fully appreciate the details of this example, we must explore the properties of FIR filters in greater detail. We will gain full appreciation of this example only upon the completion of Chapter 6.

5.3.2 The Unit Impulse Response

In this section, we will introduce three new ideas: the unit impulse sequence, the unit impulse response, and the convolution sum. We will show that the impulse response provides a complete characterization of the filter, because the convolution sum gives a formula for computing the output from the input when the unit impulse response is known.

5.3.2.1   Unit Impulse Sequence   The *unit impulse* is perhaps the simplest sequence because it has only one nonzero value, which occurs at $n = 0$. The mathematical notation is that of the Kronecker delta function, $\delta[n]$, where

$$\delta[n] = \begin{cases} 1 & n = 0 \\ 0 & n \neq 0 \end{cases} \tag{5.3.4}$$

It is tabulated in the second row of this table:

n	...	-2	-1	0	1	2	3	4	5	6	...
$\delta[n]$	0	0	0	1	0	0	0	0	0	0	0
$\delta[n-3]$	0	0	0	0	0	0	1	0	0	0	0

A shifted impulse such as $\delta[n-3]$ is nonzero when its argument is zero, i.e., $n-3 = 0$, or equivalently $n = 3$. The third row of the table gives the values of the shifted impulse $\delta[n-3]$, and Fig. 5.7 shows a plot of that sequence.

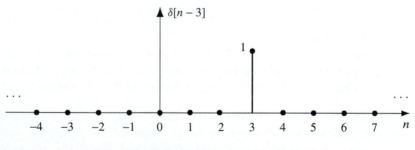

Figure 5.7   Shifted impulse sequence, $\delta[n-3]$.

The shifted impulse is a concept that is very useful in representing signals and systems. Consider, for example, the signal

$$x[n] = 2\delta[n] + 4\delta[n-1] + 6\delta[n-2] + 4\delta[n-3] + 2\delta[n-4] \tag{5.3.5}$$

To interpret (5.3.5), we must observe that the appropriate definition of multiplying a sequence by a number is to multiply each value of the sequence by that number; likewise, adding two or more sequences is defined as adding the sequence values at corresponding positions (times). The following table shows the individual sequences in (5.3.5) and their sum:

n	$\ldots$	-2	-1	0	1	2	3	4	5	6	$\ldots$
$2\delta[n]$	0	0	0	2	0	0	0	0	0	0	0
$4\delta[n-1]$	0	0	0	0	4	0	0	0	0	0	0
$6\delta[n-2]$	0	0	0	0	0	6	0	0	0	0	0
$4\delta[n-3]$	0	0	0	0	0	0	4	0	0	0	0
$2\delta[n-4]$	0	0	0	0	0	0	0	2	0	0	0
$x[n]$	0	0	0	2	4	6	4	2	0	0	0

Clearly, (5.3.5) is a compact representation of the signal in Fig. 5.2. Indeed, any sequence can be represented in this way. The equation

$$x[n] = \sum_k x[k]\delta[n-k]$$

$$= \ldots + x[-1]\delta[n+1] + x[0]\delta[n] + x[1]\delta[n-1] + \ldots \quad (5.3.6)$$

is obviously true if k ranges over all the nonzero values of the sequence $x[n]$. Equation (5.3.6) states the obvious: The sequence is formed by using scaled shifted impulses to place samples of the right size at the right positions.

5.3.2.2   Unit Impulse Response Sequence   When the input to the FIR filter (5.3.1) is a unit impulse sequence, $x[n] = \delta[n]$, the output is, by definition, the *unit impulse response*, which we will denote by $h[n]$.[3] This is depicted in the block diagram of Fig. 5.8. Substituting $x[n] = \delta[n]$ in (5.3.1) gives the output

$$y[n] = h[n] = \sum_{k=0}^{M} b_k \, \delta[n-k] = \begin{cases} b_n & n = 0, 1, 2, \ldots M \\ 0 & \text{otherwise} \end{cases}$$

Figure 5.8   Block diagram showing definition of impulse response.

As we have observed, the sum evaluates to a single term for each value of n because each $\delta[n-k]$ is nonzero only when $n-k = 0$ or $n = k$. In the tabulated form, the impulse response is

---

[3] We will usually shorten this to *impulse response*, with *unit* being understood.

n	$n < 0$	0	1	2	3	$\ldots$	M	$M+1$	$n > M+1$
$x[n] = \delta[n]$	0	1	0	0	0	0	0	0	0
$y[n] = h[n]$	0	b_0	b_1	b_2	b_3	$\ldots$	b_M	0	0

In other words, the impulse response $h[n]$ of the FIR filter is simply the sequence of difference equation coefficients. Since $h[n] = 0$ for $n < 0$ and for $n > M$, the length of the impulse response sequence $h[n]$ is finite. This is why the system (5.3.1) is called a *finite impulse response,* (FIR) system. Figure 5.9 illustrates a plot of the impulse response for the case of the causal 3-point running average filter.

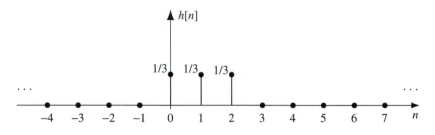

Figure 5.9   Impulse response of 3-point running average filter, $h[n]$.

Exercise 5.3. Determine and plot the impulse response of the FIR filter

$$y[n] = \sum_{k=0}^{10} kx[n - k]$$

5.3.2.3 The Unit-Delay System   One important system is the operator that performs a delay or shift by an amount n_0

$$y[n] = x[n - n_0] \tag{5.3.7}$$

When $n_0 = 1$, the system is called a unit delay. The output of the unit delay is particularly easy to visualize. In a plot, the values of $x[n]$ are moved to the right by one time interval. For example, $y[4]$ takes on the value of $x[3]$, $y[5]$ takes on the value of $x[4]$, $y[6]$ is $x[5]$, and so on.

The delay system is actually the simplest of FIR filters; it has only one nonzero coefficient. For example, a system that produces a "delay by 3" has filter coefficients $\{b_k\} = \{0, 0, 0, 1\}$. The order of this FIR filter is $M = 3$, and its difference equation is

$$y[n] = b_0 \cdot x[n] + b_1 \cdot x[n - 1] + b_2 \cdot x[n - 2] + b_3 \cdot x[n - 3]$$
$$= 0 \cdot x[n] + 0 \cdot x[n - 1] + 0 \cdot x[n - 2] + 1 \cdot x[n - 3]$$
$$= x[n - 3]$$

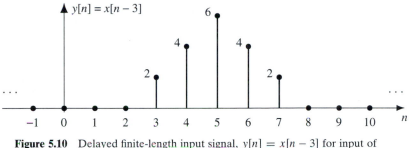

Figure 5.10   Delayed finite-length input signal, $y[n] = x[n-3]$ for input of Fig. 5.2.

Figure 5.10 shows the output of the delay system with a delay of 3 for the input of Fig. 5.2. The impulse response of the delay system is obtained by substituting $\delta[n]$ for $x[n]$ in (5.3.7). For the delay-by–3 case,

$$h[n] = \delta[n - n_0] = \delta[n - 3] = \begin{cases} 1 & n = 3 \\ 0 & n \neq 3 \end{cases}$$

This impulse response is the signal plotted previously in Fig. 5.7.

5.3.3 Convolution and FIR Filters

A general expression for the FIR filter's output (5.3.1) can be derived in terms of the impulse response. Since the filter coefficients in (5.3.1) are identical to the impulse response values, we can replace b_k in (5.3.1) by $h[k]$ to obtain

$$y[n] = \sum_{k=0}^{M} h[k]\, x[n - k] \tag{5.3.8}$$

When the relation between the input and the output of the FIR filter is expressed in terms of the input and the impulse response, as in (5.3.8), it is called a *finite convolution sum*, and we say that the output is obtained by *convolving* the sequences $x[n]$ and $h[n]$.

5.3.3.1 Computing the Output of a Convolution   The method of tabulating values for the output of an FIR filter works for short signals, but lacks the generality needed in more complicated problems. However, there is a simple interpretation of (5.3.8) that leads to a better algorithm for doing convolution. This algorithm can be implemented using the tableau in Fig. 5.11 that tracks the relative position of the signal values. The example in Fig. 5.11 shows how to convolve $x[n] = \{2,\ 4,\ 6,\ 4,\ 2\}$ with $h[n] = \{3,\ -1,\ 2,\ 1\}$. First of all, we write out the signals $x[n]$ and $h[n]$ on separate rows. Then we use a method similar to "synthetic polynomial multiplication" to form the output as the sum of shifted rows. Each shifted row is produced by multiplying the $x[n]$ row by one of the $h[k]$ values and shifting the result to the right so that it lines up with the $h[k]$ position. The final answer is obtained by summing down the columns.

n		0	1	2	3	4	5	6	7	8
x[n]		2	4	6	4	2				
h[n]		3	-1	2	1					
h[0]x[n-0]		6	12	18	12	6				
h[1]x[n-1]			-2	-4	-6	-4	-2			
h[2]x[n-2]				4	8	12	8	4		
h[3]x[n-3]					2	4	6	4	2	
y[n]		6	10	18	16	18	12	8	2	

Figure 5.11   Convolution of finite-length signals.

The justification of this algorithm for evaluating the convolution sum comes from writing out the sum in (5.3.8) as

$$y[n] = h[0]x[n] + h[1]x[n-1] + h[2]x[n-2] + h[3]x[n-3] + \ldots$$

A term such as $x[n-2]$ is the $x[n]$ signal with its values shifted two places to the right. The multiplier $h[2]$ scales the shifted signal $x[n-2]$ to produce the contribution $h[2]x[n-2]$.

Exercise 5.4. Use the "synthetic multiplication" convolution algorithm to compute the output $y[n]$ for the length 4 filter whose coefficients are $\{b_k\} = \{1, -2, 2, -1\}$. Use the input signal given in Fig. 5.2.

Later in this chapter, we will prove that convolution is the fundamental input–output algorithm for a large class of very useful filters that includes FIR filters as a special case. We will show that a general form of convolution that also applies to infinite-length signals is

$$y[n] = \sum_{k=-\infty}^{\infty} h[k]x[n-k] \tag{5.3.9}$$

This convolution sum (5.3.9) has infinite limits, but reduces to (5.3.8) when $h[n] = 0$ for $n < 0$ and $n > M$.

5.3.3.2   Convolution in MATLAB   In MATLAB, we implement FIR systems with the conv( ) function. For example, the following MATLAB statements

```
xx = sin(0.07*pi*(0:50));
hh = ones(11,1)/11;
yy = conv(hh, xx);
```

will evaluate the convolution of the 11-point sequence hh with the 51-point sinusoidal sequence xx. The particular choice for the MATLAB vector hh is actually the impulse response of an 11-point running average system:

$$h[n] = \begin{cases} 1/11 & n = 0, 1, 2, \ldots, 10 \\ 0 & \text{otherwise} \end{cases}$$

That is, all 11 filter coefficients are the same and equal to 1/11.

> **Exercise 5.5.** In MATLAB, we can compute only the convolution of finite-length signals. Determine the length of the output signal computed by the MATLAB convolution above.

We have already hinted that this operation called *convolution* is equivalent to polynomial multiplication. In Chapter 7, we will prove that this correspondence is true. At this point, we note that in MATLAB there is no function for multiplying polynomials. Instead, we must know that convolution is equivalent to polynomial multiplication. Then we can represent the polynomials by sequences of their coefficients and use the conv( ) function to convolve them, thereby doing polynomial multiplication.

> **Exercise 5.6.** Use MATLAB to compute the following product of polynomials:
>
> $$P(x) = (1 + 2x + 3x^2 + 5x^4)(1 - 3x - x^2 + x^3 + 3x^4)$$

5.4   IMPLEMENTATION OF FIR FILTERS

Recall that the general definition of an FIR filter is

$$y[n] = \sum_{k=0}^{M} b_k \, x[n - k] \tag{5.4.1}$$

We can see that to use (5.4.1) to compute the output of the FIR filter, we need the following: (1) a means for multiplying delayed-input signal values by the filter coefficients; (2) a means for adding the scaled sequence values; and (3) a means for obtaining delayed versions of the input sequence. We will find it useful to represent the operations of (5.4.1) as a block diagram. Such representations will lead to new insights about the properties of the system and about alternative ways to implement the system.

5.4.1 Building Blocks

The basic building-block systems we need are the multiplier, the adder, and the unit-delay operator as depicted in Fig. 5.12.

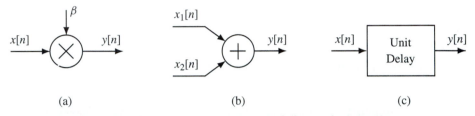

(a)   (b)   (c)

Figure 5.12   Building blocks for making any LTI discrete-time system: (a) multiplier, $y[n] = \beta x[n]$; (b) adder, $y[n] = x_1[n] + x_2[n]$; and (c) unit-delay, $y[n] = x[n-1]$.

5.4.1.1   Multiplier   The first elementary system performs multiplication of a signal by a constant (see Fig. 5.12(a)). The output signal $y[n]$ is given by the rule

$$y[n] = \beta\, x[n]$$

where the coefficient β is a constant. This system can be the hardware multiplier unit in a computer. For a DSP microprocessor, the speed of this multiplier is one of the fundamental limits on the throughput of the digital filtering process. In applications such as image convolution, billions of multiplications per second may have to be performed to implement a good filter, so quite a bit of engineering work has been directed at designing fast multipliers for DSP applications. Furthermore, since many filters require the same sequence of multiplications over and over, pipelining the multiplier also results in a dramatic speed-up of the filtering process.

Notice, by the way, that the simple multiplier is also an FIR filter, with $M = 0$, and $b_0 = \beta$ in (5.3.1). The impulse response of the multiplier system is simply $h[n] = \beta\delta[n]$.

5.4.1.2   Adder   The second elementary system in Fig. 5.12(b) performs the addition of two signals. This is a different sort of system because it has two inputs and one output. In hardware, the adder is simply the hardware adder unit in the computer. Since many DSP operations require a multiplication followed immediately by an addition, it is common in DSP microprocessors to build a special multiply-accumulate unit, often called a "MADD" or "MAC" unit.

Notice that the adder is a pointwise combination of the values of the two input sequences. It is not an FIR filter, because it has more than one input; however, it is a crucial building block of FIR filters. With many inputs, the adder could be drawn as a multi-input adder, but, in digital hardware, the additions are typically done two inputs at a time.

5.4.1.3   Unit Delay   The third elementary system performs a delay by one unit of time. It is represented by a block diagram, as in Fig. 5.12(c). In the case of discrete-time filters, the time dimension is indexed by integers, so this delay is by one "count" of the system clock. The hardware implementation of the unit delay is actually performed by acquiring a sample value, storing it in memory for one clock cycle, and then releasing it to the output. The delays by more than one time unit that are needed to implement (5.4.1) can be implemented (from the block-diagram point of view) by cascading several unit delays in a row. Therefore, an M-unit delay requires M memory cells configured as a shift register, which can be implemented as a circular buffer in computer memory.

5.4.2 Block Diagrams

In order to create a graphical representation that is useful for hardware structures, we use block-diagram notation, which defines the interconnection of the three basic building blocks to make more complex structures. In such a directed graph, the nodes (i.e., junction points) are either summing nodes, splitting nodes, or input–output nodes. The connections between nodes are either delay branches or multiplier branches. Figure 5.13 shows the general block diagram for a third-order FIR digital filter ($M = 3$). This structure shows why the FIR filter is also called a *feed-forward difference equation,* since all paths lead forward from the input to the output; there are no loops in the block diagram. In Chapter 8 we will discuss filters with *feedback,* where both input and past output values are involved in the computation of the output.

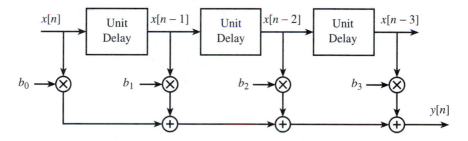

Figure 5.13   Block-diagram structure for the Mth order FIR filter.

Strictly speaking, the structure of Fig. 5.13 is a block-diagram representation of the equation

$$y[n] = (((b_0x[n] + b_1x[n-1]) + b_2x[n-2]) + b_3x[n-3])$$

that clearly expands into an equation that can be represented as in (5.4.1). The input signal is delayed by the cascaded unit delays, each delayed signal is multiplied by a filter coefficient, and the products are accumulated to form the sum. Thus, it is easy to see that there is a one-to-one correspondence between the block diagram and the difference equation (5.4.1) of the FIR filter, because both are defined by the filter coefficients $\{b_k\}$. A useful skill is to start with one representation and then produce the other. The structure in Fig. 5.13 displays a regularity that makes it simple to define longer filters; the number of cascaded delay elements is increased to M, and then the filter coefficients $\{b_k\}$ are substituted into the diagram. This standard structure is called the *direct form*. Going from the block diagram back to the difference equation is just as easy, as long as we stick to direct form. Here is a simple exercise to make the correspondence.

Exercise 5.7. Determine the difference equation for the following block diagram:

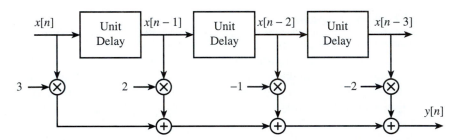

5.4.2.1   Other Block Diagrams   Many block diagrams will "implement" the same FIR filter, in the sense that the external behavior from input to output will be the same. Direct form is just one possibility. Other block diagrams would represent a different internal computation, or a different order of computation. In some cases, the internal multipliers might use different coefficients. After we have studied the z-transform in Chapters 7 and 8, we will have the tools to produce many different implementations.

When faced with an arbitrary block diagram, the following four-step procedure may be used to derive the difference equation from the block diagram.

1. Give a unique signal name to the input of each unit-delay block.
2. Notice that the output of a unit delay can be written in terms of its input.
3. At each summing node of the structure, write a signal equation. Use the signal names introduced in steps 1 and 2.
4. At this point, you will have several equations involving $x[n]$, $y[n]$, and the internal signal names. These can be reduced to one equation involving only $x[n]$ and $y[n]$ by eliminating variables, as is done with simultaneous equations.

Let us try this procedure on the simple but useful example shown in Fig. 5.14. First of all, we observe that the internal signal variables $\{v_1[n],\ v_2[n],\ v_3[n]\}$ have been defined in Fig. 5.14 as the inputs to the three delays. Then the delay outputs are, from left to right, $v_3[n-1]$, $v_2[n-1]$, and $v_1[n-1]$. We also notice that $v_3[n]$ is a scaled version of the input $x[n]$. In addition, we write the three equations at the output of the three summing nodes:

$$y[n] = b_0 x[n] + v_1[n-1]$$
$$v_1[n] = b_1 x[n] + v_2[n-1]$$
$$v_2[n] = b_2 x[n] + v_3[n-1]$$
$$v_3[n] = b_3 x[n]$$

Now we have four equations in five "unknowns." To show that this set of equations is equivalent to the direct form, we can eliminate the $v_i[n]$ by combining the equations in a pairwise fashion.

$$y[n] = b_0 x[n] + b_1 x[n-1] + v_2[n-2]$$
$$v_2[n] = b_2 x[n] + b_3 x[n-1]$$
$$\implies \quad y[n] = b_0 x[n] + b_1 x[n-1] + b_2 x[n-2] + b_3 x[n-3]$$

Thus, we have derived the same difference equation as before, so this new structure must be a different way of computing the same thing. In fact, it is widely used and is called the *transposed form* for the FIR filter.

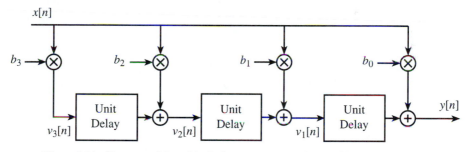

Figure 5.14   Transposed form block diagram structure for the Mth order FIR filter.

Exercise 5.8. Why do you think that Fig. 5.14 is called the *transposed form* of the direct form in Fig. 5.13?

5.4.2.2   Internal Hardware Details   A block diagram shows dependencies among the different signal variables. Therefore, different block diagrams that implement the same input–output operation may have dramatically different characteristics as far as their internal behavior goes. Several issues come to mind:

1. The order of computation is specified by the block diagram. In high-speed applications where parallelism or pipelining must be exploited, dependencies in the block diagram represent constraints on the computation.

2. Partitioning a filter for a VLSI chip would be done (at the highest level) in terms of the block diagram. Likewise, algorithms that must be mapped onto special DSP architectures can be managed by using the block diagram with special compilers that translate the block diagram into optimized code for the DSP chip.

3. Finite word-length effects are important if the filter is constructed using fixed-point arithmetic. In this case, round-off noise and overflow are important real-world problems that depend on the internal order of computation.

Now that we know something about convolution and the implementation of FIR filters, it is time to consider discrete-time systems in a more general way. In the next section, we will show that FIR filters are a special case of the general class of linear time-invariant systems. Much of what we have learned about FIR filters will apply to this more general class of systems.

5.5   LINEAR TIME-INVARIANT (LTI) SYSTEMS

In this section, we discuss two general properties of systems. These properties, *linearity* and *time invariance*, lead to simplifications of mathematical analysis and greater insight and understanding of system behavior. To facilitate the discussion of these properties, it is useful to recall that the block-diagram representation of a general discrete-time system is shown in Fig. 5.15. This block diagram depicts a transformation of the input signal $x[n]$ into an output signal $y[n]$. It will also be useful to introduce the notation

$$x[n] \quad \longmapsto \quad y[n] \tag{5.5.1}$$

to represent this transformation. In a specific case, the system is defined by giving a formula or algorithm for computing all values of the output sequence from the values of the input sequence. A specific example is the "square-law" system defined by the rule:

$$y[n] = (x[n])^2 \tag{5.5.2}$$

Another example is (5.3.7), which defines the general delay system; still another is (5.4.1), which defines the general FIR filter and includes the delay system as a special case. We will see that these FIR filters are both linear and time-invariant, while the square-law system is not linear.

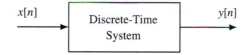

Figure 5.15   Discrete-time system.

5.5.1 Time Invariance

A discrete-time system is said to be *time-invariant* if, when an input is delayed (shifted) by n_0, the output is delayed by the same amount. Using the notation introduced above, we can express this condition as

$$x[n - n_0] \longmapsto y[n - n_0] \qquad (5.5.3)$$

where $x[n] \longmapsto y[n]$. The condition must be true for any choice of n_0, the integer that determines the amount of shift.

A block-diagram view of the time-invariance property is given in Fig. 5.16. In the upper branch, the input is shifted prior to the system; in the lower branch, the output is shifted. Thus, a system can be tested for time-invariance by checking whether or not $w[n] = y[n - n_0]$ in Fig. 5.16.

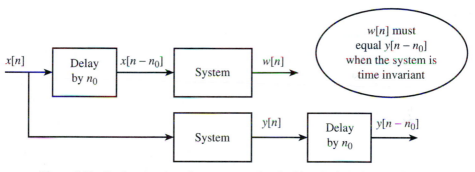

Figure 5.16   Testing time-invariance property by checking the interchange of operations.

Consider the example of the square-law system defined by (5.5.2). If we use the delayed input as the input to the square-law system, we obtain

$$w[n] = (x[n - n_0])^2$$

If $x[n]$ is the input to the square-law system, then $y[n] = (x[n])^2$ and

$$y[n - n_0] = (x[n - n_0])^2 = w[n]$$

so the square-law system is time-invariant.

A second simple example is the "time-flip" system, defined by the equation

$$y[n] = x[-n]$$

This system simply reverses the order of ("flips") the input sequence about the origin. If we delay the input and then reverse the order about the origin, we obtain

$$w[n] = x[(-n) - n_0] = x[-n - n_0]$$

However, if we first flip the input sequence and then delay it, we obtain a different sequence from $w[n]$; i.e., since $y[n] = x[-n]$

$$y[n - n_0] = x[-(n - n_0)] = x[-n + n_0]$$

Thus, the time-flip system is *not* time-invariant.

Exercise 5.9. Test the system defined by the equation

$$y[n] = nx[n]$$

to determine whether it is a time-invariant system.

5.5.2 Linearity

Linear systems have the property that if $x_1[n] \longmapsto y_1[n]$ and $x_2[n] \longmapsto y_2[n]$, then

$$x[n] = \alpha x_1[n] + \beta x_2[n] \quad \longmapsto \quad y[n] = \alpha y_1[n] + \beta y_2[n] \tag{5.5.4}$$

This mathematical condition must be true for any choice of the constants α and β. Equation (5.5.4) states that if the input consists of a sum of scaled sequences, then the corresponding output is a sum of scaled outputs corresponding to the individual input sequences. A block diagram view of the linearity property is given in Fig. 5.17, which shows that a system can be tested for the linearity property by checking whether or not $w[n] = y[n]$.

The linearity condition in (5.5.4) is equivalent to the *principle of superposition*: If the input is the sum (superposition) of two or more scaled sequences, we can find the output due to each sequence acting alone and then add (superimpose) the separate scaled outputs. Sometimes it is useful to separate (5.5.4) into two conditions. Setting $\alpha = \beta = 1$ we get the condition

$$x[n] = x_1[n] + x_2[n] \quad \longmapsto \quad y[n] = y_1[n] + y_2[n] \tag{5.5.5}$$

and using only one scaled input gives

$$x[n] = \alpha x_1[n] \quad \longmapsto \quad y[n] = \alpha y_1[n] \tag{5.5.6}$$

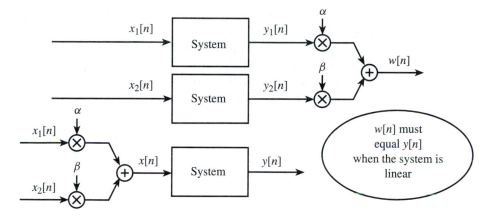

Figure 5.17   Testing linearity by checking the interchange of operations.

Both (5.5.5) and (5.5.6) must be true in order for (5.5.4) to be true.

Reconsider the example of the square-law system defined by (5.5.2). The output $w[n]$ in Fig. 5.17 for this system is

$$w[n] = \alpha(x_1[n])^2 + \beta(x_2[n])^2$$

while

$$y[n] = (\alpha x_1[n] + \beta x_2[n])^2 = \alpha^2(x_1[n])^2 + 2\alpha\beta x_1[n]x_2[n] + \beta^2(x_2[n])^2$$

Thus, $w[n] \neq y[n]$, and the square-law system has been shown not to be linear. Systems that are not linear are called *nonlinear* systems.

Exercise 5.10. Show that the time-flip system $y[n] = x[-n]$ is a linear system.

5.5.3 The FIR Case

Systems described by (5.3.1) satisfy both the linearity and time invariance conditions. A mathematical proof of time-invariance can be constructed using the procedure depicted in Fig. 5.16. If we define the signal $v[n]$ to be $x[n - n_0]$, then the difference equation relating $v[n]$ to $w[n]$ in the upper branch of Fig. 5.16 is

$$w[n] = \sum_{k=0}^{M} b_k v[n - k] = \sum_{k=0}^{M} b_k x[(n - k) - n_0] = \sum_{k=0}^{M} b_k x[(n - n_0) - k]$$

For comparison, we construct $y[n - n_0]$ in the lower branch

$$y[n] = \sum_{k=0}^{M} b_k x[n - k] \qquad \Rightarrow \qquad y[(n - n_0)] = \sum_{k=0}^{M} b_k x[(n - n_0) - k]$$

These two expressions are identical, so $w[n] = y[n - n_0]$, and we have proved that the FIR filter is time-invariant.

The linearity condition is simpler to prove. Just substitute into the difference equation (5.3.1) and collect terms:

$$y[n] = \sum_{k=0}^{M} b_k x[n - k]$$

$$= \sum_{k=0}^{M} b_k (\alpha x_1[n - k] + \beta x_2[n - k])$$

$$= \alpha \sum_{k=0}^{M} b_k x_1[n - k] + \beta \sum_{k=0}^{M} b_k x_2[n - k] = \alpha y_1[n] + \beta y_2[n]$$

Thus, the FIR filter obeys the principle of superposition; therefore, it is a linear system.

A system that satisfies both properties is called a *linear time-invariant* system, or simply LTI. It should be emphasized that the LTI condition is a general condition. The FIR filter is an example of an LTI system. Not all LTI systems are described by (5.3.1), but all systems described by (5.3.1) are LTI systems.

5.6   CONVOLUTION AND LTI SYSTEMS

Consider an LTI discrete-time system as depicted in Fig. 5.18. The impulse response $h[n]$ of the LTI system is simply the output when the input is the unit impulse sequence $\delta[n]$. In this section, we will show that the impulse response is a complete characterization for any LTI system, and that convolution is the general formula that allows us to compute the output from the input *for any LTI system*. In our initial discussion of convolution, we considered only finite-length input sequences and FIR filters. Now, we will give a completely general presentation.

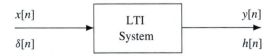

Figure 5.18   Linear time-invariant system with impulse input.

5.6.1   Derivation of the Convolution Sum

We begin by recalling from the discussion in Section 5.3.2 that *any* signal $x[n]$ can be represented as a sum of scaled and shifted impulse signals. Each nonzero sample of the signal $x[n]$ multiplies an impulse signal that is shifted to the index of that sample. Specifically, we can write $x[n]$ as follows:

$$x[n] = \sum_{\ell} x[\ell]\delta[n - \ell] \tag{5.6.1}$$

$$= \ldots + x[-2]\delta[n + 2] + x[-1]\delta[n + 1] + x[0]\delta[n]$$
$$+ \; x[1]\delta[n - 1] + x[2]\delta[n - 2] + \ldots$$

In the most general case, the range of summation in (5.6.1) could be from $-\infty$ to $+\infty$. In (5.6.1) we have left the range indefinite, realizing that the sum would include all nonzero samples of the input sequence.

A sum of scaled sequences such as (5.6.1) is commonly referred to as a "linear combination" or superposition of scaled sequences. Thus, (5.6.1) is a representation of the sequence $x[n]$ as a linear combination of scaled, shifted impulses. Since LTI systems respond in simple and predictable ways to sums of signals and to shifted signals, this representation is particularly useful for our purpose of deriving a general formula for the output of a LTI system.

Figure 5.18 reminds us that the response to the input $\delta[n]$ is, by definition, the impulse response $h[n]$. Time invariance gives us additional information; the response due to $\delta[n - 1]$ is $h[n - 1]$. In fact, we can write a whole family of input–output pairs:

$$\delta[n] \longmapsto h[n] \Rightarrow \delta[n - 1] \qquad \longmapsto h[n - 1]$$
$$\Rightarrow \delta[n - 2] \qquad \longmapsto h[n - 2]$$
$$\Rightarrow \delta[n - (-1)] \longmapsto h[n - (-1)] = h[n + 1]$$
$$\Rightarrow \delta[n - \ell] \qquad \longmapsto h[n - \ell] \quad \text{for any integer } \ell$$

Now we are in a position to use linearity, because (5.6.1) expresses a general input signal as a linear combination of shifted impulse signals. We can write out a few of the cases:

$$x[0]\delta[n] \qquad \longmapsto x[0]h[n]$$
$$x[1]\delta[n - 1] \qquad \longmapsto x[1]h[n - 1]$$
$$x[2]\delta[n - 2] \qquad \longmapsto x[2]h[n - 2]$$
$$x[\ell]\delta[n - \ell] \qquad \longmapsto x[\ell]h[n - \ell] \quad \text{for any integer } \ell$$

Then we use superposition to put it all together:

$$x[n] = \sum_{\ell} x[\ell]\delta[n - \ell] \longmapsto y[n] = \sum_{\ell} x[\ell]h[n - \ell] \tag{5.6.2}$$

The derivation of (5.6.2) did not assume that either $x[n]$ or $h[n]$ was of finite duration, so, in general, we may need infinite limits on the sum. With this modification we obtain

---

The Convolution Sum Formula

$$y[n] = \sum_{\ell=-\infty}^{\infty} x[\ell]h[n-\ell] \qquad (5.6.3)$$

This expression represents the convolution operation in the most general sense, so we have proved that *all LTI systems can be represented by a convolution sum*. The infinite limits take care of all possibilities, including the cases where either or both of the sequences are finite in length.

Example 5.2   For example, if $h[n]$ is nonzero only in the interval $0 \le n \le M$, then (5.6.3) reduces to

$$y[n] = \sum_{\ell=n-M}^{n} x[\ell]h[n-\ell] \qquad (5.6.4)$$

because the argument $n-\ell$ must lie in the range $0 \le n-\ell \le M$, so the range for ℓ in (5.6.3) is restricted to $(n-M) \le \ell \le n$. ◇

Exercise 5.11. By making the substitution $k = n - \ell$ in (5.6.4), show that $y[n]$ can also be expressed in the more familiar form

$$y[n] = \sum_{k=0}^{M} h[k]x[n-k]$$

5.6.2 Some Properties of LTI Systems

CASCADING LTI SYSTEM

The properties of convolution are the properties of LTI systems. Thus, it is of interest to explore these properties and relate them to properties of LTI systems.

5.6.2.1   Convolution as an Operator   An interesting aspect of convolution is its algebraic character as a operation between two signals. The convolution of $x[n]$ and $h[n]$ is an operation that will be denoted by $*$, i.e.,

$$y[n] = x[n] * h[n] = \sum_{\ell=-\infty}^{\infty} x[\ell]h[n-\ell] \qquad (5.6.5)$$

We say that the sequence $x[n]$ is *convolved* with the sequence $h[n]$ to produce the output $y[n]$.

The notation $x[n]*h[n]$ is useful because it allows us to think about convolution problems in an operational way. As a simple example, recall that the ideal delay system has impulse response $h[n] = \delta[n - n_0]$. We know that the output of the ideal delay system is $y[n] = x[n - n_0]$. Therefore, it follows that

---

Convolution with an Impulse

$$x[n] * \delta[n - n_0] = x[n - n_0] \qquad (5.6.6)$$

This is a very important and useful result. Equation (5.6.6) is stating that "to convolve any sequence $x[n]$ with an impulse located at $n = n_0$, all we need to do is translate the origin of $x[n]$ to n_0."

Furthermore, as an algebraic operator, convolution in (5.6.5) satisfies commutative and associative properties similar to the commutative and associative properties of multiplication, i.e.,

$$x[n] * h[n] = h[n] * x[n] \qquad \text{(Commutative)}$$

$$(x_1[n] * x_2[n]) * x_3[n] = x_1[n] * (x_2[n] * x_3[n]) \qquad \text{(Associative)}$$

We will reconsider both of these properties when we introduce the z-transform in Chapter 7.

5.6.2.2   Commutative Property of Convolution   It is relatively easy to prove that convolution is a commutative operation between two sequences. In fact, the truth of the commutative property is clear from the computational algorithm in Fig. 5.11, where it is obvious that either $x[n]$ or $h[n]$ can be written on the first row, so $x[n]$ and $h[n]$ are interchangeable. In the following algebraic manipulation, we make the change of variables $k = n - \ell$ and sum on the new dummy variable k.

$$y[n] = x[n] * h[n] = \sum_{\ell=-\infty}^{\infty} x[\ell]h[n - \ell]$$

$$= \sum_{k=+\infty}^{-\infty} x[n - k]h[k] \qquad (\text{Note: } k = n - \ell)$$

$$= \sum_{k=-\infty}^{\infty} h[k]x[n - k] = h[n] * x[n]$$

On the second line in this set of equations, the limits on the sum can be swapped without consequence, because a set of numbers can be added in any order. Thus, we have proved that convolution is a commutative operation, i.e.,

$$y[n] = x[n] * h[n] = h[n] * x[n]$$

5.6.2.3 Associative Property of Convolution   The associative property which is perhaps less obvious than the commutative property states that, when we are convolving three signals, we can convolve two of them and then convolve that result with the third signal. The algebraic manipulation needed to prove the associative property is tedious, but we give it here to illustrate how convolution expressions can be manipulated into different forms. We assume that $x_1[n]$, $x_2[n]$, and $x_3[n]$ are three arbitrary sequences to be convolved.

$$x_1[n] * (x_2[n] * x_3[n]) = \sum_{\ell=-\infty}^{\infty} x_1[\ell] \left(\sum_{k=-\infty}^{\infty} x_2[k]x_3[(n-\ell)-k] \right)$$

$$= \sum_{\ell=-\infty}^{\infty} x_1[\ell] \sum_{q=-\infty}^{\infty} x_2[q-\ell]x_3[n-q]$$

$$= \sum_{q=-\infty}^{\infty} \sum_{\ell=-\infty}^{\infty} x_1[\ell]x_2[q-\ell]x_3[n-q]$$

$$= \sum_{q=-\infty}^{\infty} \left(\sum_{\ell=-\infty}^{\infty} x_1[\ell]x_2[q-\ell] \right) x_3[n-q]$$

$$= (x_1[n] * x_2[n]) * x_3[n]$$

In the second line, we used the change of variables: $q = \ell + k$. This proves that convolution is an associative operation. The implication of this property for cascaded systems is explored in Section 5.7.

5.7 CASCADED LTI SYSTEMS

In a cascade connection of two systems, the output of the first system is the input to the second system, and the overall output of the cascade system is taken to be the output of the second system. Figure 5.19 shows two LTI systems (LTI 1 and LTI 2) in cascade. LTI systems have the remarkable property that two LTI systems in cascade can be implemented in either order. This property is a direct consequence of the commutative and associative properties of convolution, as demonstrated by the following three equivalent expressions that can be obtained by applying the commutative and associative properties of convolution:

$$y[n] = (x[n] * h_1[n]) * h_2[n] \tag{5.7.1}$$

$$= x[n] * (h_1[n] * h_2[n]) \tag{5.7.2}$$

$$= x[n] * (h_2[n] * h_1[n]) \tag{5.7.3}$$

$$= (x[n] * h_2[n]) * h_1[n] \tag{5.7.4}$$

Equation (5.7.1) is a mathematical statement of the fact that the second system processes the output of the first, which is $w[n] = x[n]*h_1[n]$. Equation (5.7.2) shows that the output $y[n]$ is the convolution of the input with a new impulse response $h_1[n] * h_2[n]$. This corresponds to the upper diagram in Fig. 5.20 with $h[n] = h_1[n] * h_2[n]$. Equation (5.7.3) uses the commutative property of convolution to show that $h[n] = h_1[n] * h_2[n] = h_2[n] * h_1[n]$. Applying the associative property leads to (5.7.4), which implies the cascade connection in the lower part of Fig. 5.20. Notice that reordering the LTI systems in cascade gives the same final output, i.e., it is correct to label the outputs of all three systems in Figs. 5.19 and 5.20 with the same symbol, $y[n]$, even though the intermediate signals, $w[n]$ and $v[n]$, are different.

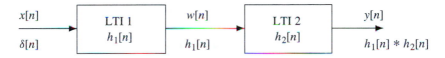

Figure 5.19   A Cascade of Two LTI Systems.

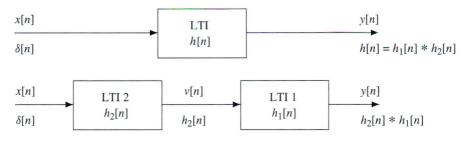

Figure 5.20   Switching the order of cascaded LTI systems.

Another way to show that the order of cascaded LTI systems does not affect the overall system response is to prove that the impulse response of the two cascade systems is the same. In Fig. 5.19, the impulse input and the corresponding outputs are shown below the arrows. When the input to the first system is an impulse, the output of LTI 1 is its impulse response, $h_1[n]$, which becomes the input to LTI 2. The output of LTI 2 is, therefore, just the convolution of its input $h_1[n]$ with its impulse response $h_2[n]$. Therefore the overall impulse response of Fig. 5.19 is $h_1[n] * h_2[n]$. In the same way, we can easily show that the overall impulse response of the other cascade system in Fig. 5.20 is $h_2[n] * h_1[n]$. Since convolution is commutative, the two cascade systems have the same impulse response

$$h[n] = h_1[n] * h_2[n] = h_2[n] * h_1[n] \tag{5.7.5}$$

which is also the impulse response $h[n]$ of the equivalent system. Again, since the overall impulse response is the same for each of the three systems in Figs. 5.19 and 5.20, the output is the same for all three systems for the same input.

Example 5.3   To illustrate the utility of the results that we have obtained for cascaded LTI systems, consider the cascade of two systems defined by

$$h_1[n] = \begin{cases} 1 & 0 \leq n \leq 3 \\ 0 & \text{otherwise} \end{cases} \quad \text{and} \quad h_2[n] = \begin{cases} 1 & 0 \leq n \leq 2 \\ 0 & \text{otherwise} \end{cases}$$

The results of this section show that the overall cascade system has impulse response

$$h[n] = h_1[n] * h_2[n]$$

Therefore, to find the overall impulse response we must convolve $h_1[n]$ with $h_2[n]$. This can be done by using the polynomial multiplication algorithm of Section 5.3.3.1. In this case, the computation is as follows:

```
    n    |   0    1    2    3    4    5
---------------+-------------------------------------
 h_1[n] |   1    1    1    1
 h_2[n] |   1    1    1
        |   -----------------------------------------
h_1[0]h_2[n-0] |   1    1    1    1
h_1[1]h_2[n-1] |        1    1    1    1
h_1[2]h_2[n-2] |             1    1    1    1
        |   -----------------------------------------
   h[n] |   1    2    3    3    2    1
```

Therefore, the equivalent impulse response is

$$h[n] = \sum_{k=0}^{5} b_k \delta[n - k]$$

where $\{b_k\}$ is the sequence $\{1, 2, 3, 3, 2, 1\}$.

This result means that a system with impulse response $h[n]$ can be implemented either by the single difference equation

$$y[n] = \sum_{k=0}^{5} b_k x[n - k] \tag{5.7.6}$$

where $\{b_k\}$ is the above sequence, or by the pair of difference equations

$$w[n] = \sum_{k=0}^{3} x[n-k] \quad \text{and} \quad y[n] = \sum_{k=0}^{2} w[n-k] \tag{5.7.7}$$

$\diamond$

Example 5.3 illustrates an important point. There is a significant difference between (5.7.6) and (5.7.7). It can be seen that the implementation in (5.7.7) requires a total of only five additions to compute each value of the output sequence, while (5.7.6) requires five additions and an additional four multiplications by coefficients that are not equal to unity. On a larger scale (longer filters), such differences in the amount and type of computation can be very important in practical applications of FIR filters. Thus, the existence of alternative equivalent implementations of the same filter is significant.

5.8 EXAMPLE OF FIR FILTERING

We conclude this chapter with an example of the use of FIR filtering on a real signal. An example of sampled data that can be viewed as a signal is the Dow-Jones Industrial Average. The DJIA is a sequence of numbers obtained by averaging the closing prices of a selected number of representative stocks. It has been computed since 1897, and is used in many ways by investors and economists. The entire sequence dating back to 1897 makes up a signal that has positive values and is exponentially growing. In order to obtain an example where fine detail is visible, we have selected the segment from 1950 to 1970 of the weekly closing price average to be our signal $x[n]$.[4] This signal is the one showing high variability in Fig. 5.21, where each weekly value is plotted as a distinct point. The smoother curve in Fig. 5.21 is the output of a 51-point causal running averager; i.e.,

$$y[n] = \frac{1}{51} \sum_{k=0}^{50} x[n-k] = x[n] * h[n]$$

where $h[n]$ is the impulse response of the causal 51-point running averager.

---

[4] The complete set of weekly sampled data for the period 1897–1997 would involve over 5000 samples and would range from about 40 to over 8000. A plot of the entire sequence would not show the effects that we wish to illustrate.

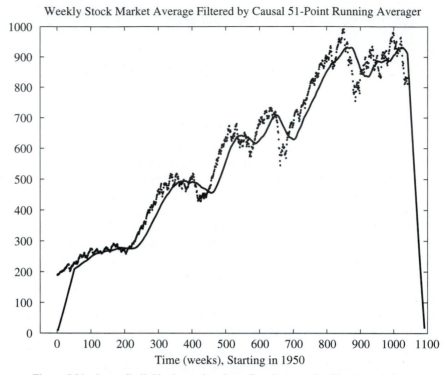

Weekly Stock Market Average Filtered by Causal 51-Point Running Averager

Figure 5.21 Input (individual samples plotted) and output for 51-point running averager (sample points connected by straight lines to distinguish them from the input).

Notice that, as we have observed in Section 5.3.1, there is a region at the beginning and end (50 samples, in this case) where the filter is engaging and disengaging the input signal. Also notice that much of the fine-scale variation has been removed by the filter. Finally, notice that the output is shifted relative to the input. In Chapter 6, we will develop techniques that will allow us to show that the shift introduced by this filter is exactly $M/2 = 25$ samples.

In this example, it is important to be able to compare the input and output without the shift. It would be better use the noncausal "centralized" running averager

$$\tilde{y}[n] = \frac{1}{51} \sum_{k=-25}^{25} x[n-k] = x[n] * \tilde{h}[n]$$

where $\tilde{h}[n]$ is the impulse response of the centralized running averager. The output $\tilde{y}[n]$ can be obtained by shifting (advancing) the output of the causal running averager by 25 samples to the left. In terms of our previous discussion of cascaded FIR systems, we could think of the centralized system as a cascade of the causal system with a system whose impulse response is $\delta[n + 25]$, i.e.,[5]

$$
\begin{aligned}
\tilde{y}[n] &= y[n] * \delta[n + 25] \\
&= (x[n] * h[n]) * \delta[n + 25] \\
&= x[n] * (h[n] * \delta[n + 25]) \\
&= x[n] * h[n + 25]
\end{aligned}
$$

Thus, we find that $\tilde{h}[n] = h[n + 25]$, i.e., the impulse response of the centralized running averager is a shifted version of the impulse response of the causal running averager. By shifting the impulse response, we remove the delay introduced by the causal system.

> **Exercise 5.12.** Determine the impulse response $h[n]$ of the 51-point causal running averager and determine the impulse response $\tilde{h}[n]$ for the 51-point centralized running averager.

The cascade representation of the centralized running averager is depicted in Fig. 5.22. Another way to describe the system in Fig. 5.22 is that the second system *compensates* for the delay of the first, or we might describe the centralized running averager as a *delay-compensated* running average filter.

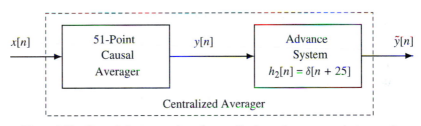

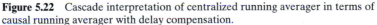

Figure 5.22   Cascade interpretation of centralized running averager in terms of causal running averager with delay compensation.

Figure 5.23 shows the input $x[n]$ and output $\tilde{y}[n]$ for the delay-compensated running average filter. It now can be seen that corresponding features of the input and output are well aligned.

As the example in this section illustrates, FIR filters can be used to remove rapid fluctuations in signals. Furthermore, the example shows that it is worthwhile to develop the fundamental mathematical properties of such systems because these properties can be useful in helping us to understand the way such systems work. In Chapter 6, we will further develop our understanding of FIR systems.

---

[5] Recall from (5.6.6) that convolution of $y[n]$ with an impulse simply shifts the origin of the sequence to the location of the impulse (in this case, to $n = -25$).

Weekly Stock Market Average Filtered by Noncausal 51-Point Running Averager

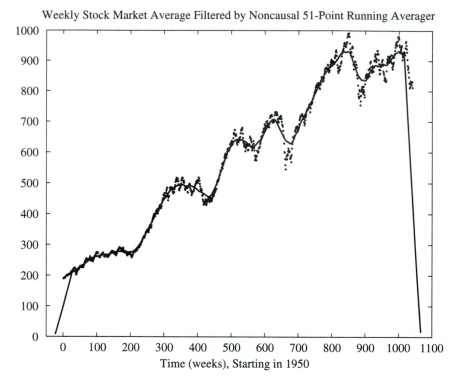

Figure 5.23   Input and output for delay-compensated 51-point running averager.

5.9   SUMMARY AND LINKS

This chapter introduced the concept of FIR filtering. Among the laboratory projects in Appendix C, there are three (Labs C.5, C.6, and C.7) that deal with filtering sinusoids and the frequency response of filters. Since the frequency response is necessary for understanding the behavior of FIR filters, these labs are best done after Chapter 6. The CD-ROM also contains two demonstrations where the properties of linearity and time invariance are illustrated by using combinations of filters, such as cascades. Finally, the reader is once again reminded of the large number of solved homework problems that are available for review and practice on the CD-ROM.

PROBLEMS

5.1   Evaluate the "running" average

$$y[n] = \frac{1}{L} \sum_{k=0}^{L-1} x[n-k]$$

for the *unit-step* input signal, i.e.,

$$x[n] = u[n] = \begin{cases} 0 & \text{for } n < 0 \\ 1 & \text{for } n \geq 0 \end{cases}$$

(a) Make a plot of $u[n]$ before working out the answer for $y[n]$.

(b) Now compute the numerical values of $y[n]$ over the range $-5 \leq n \leq 10$, assuming that $L = 5$.

(c) Make a sketch of the output over the range $-5 \leq n \leq 10$, assuming that $L = 5$. Use MATLAB if necessary, but learn to do it by hand also.

(d) Finally, derive a general formula for $y[n]$ that will apply for any length L and for the index range $n \geq 0$.

5.2 A linear time-invariant system is described by the difference equation

$$y[n] = 2x[n] - 3x[n-1] + 2x[n-2]$$

(a) When the input to this system is

$$x[n] = \begin{cases} 0 & n < 0 \\ n+1 & n = 0, 1, 2 \\ 5-n & n = 3, 4 \\ 1 & n \geq 5 \end{cases}$$

Compute the values of $y[n]$, over the range $0 \leq n \leq 10$.

(b) For the previous part, plot both $x[n]$ and $y[n]$.

(c) Determine the response of this system to a unit impulse input; i.e., find the output $y[n] = h[n]$ when the input is $x[n] = \delta[n]$. Plot $h[n]$ as a function of n.

5.3 A linear time-invariant system is described by the difference equation

$$y[n] = 2x[n] - 3x[n-1] + 2x[n-2]$$

(a) Draw the implementation of this system as a block diagram in direct form.

(b) Give the implementation as a block diagram in transposed direct form.

5.4 Consider a system defined by

$$y[n] = \sum_{k=0}^{M} b_k x[n-k]$$

(a) Suppose that the input $x[n]$ is nonzero only for $0 \leq n \leq N - 1$; i.e., it has a support of N samples. Show that $y[n]$ is nonzero at most over a finite interval of the form $0 \leq n \leq P - 1$. Determine P and the support of $y[n]$ in terms of M and N.

(b) Suppose that the input $x[n]$ is nonzero only for $N_1 \leq n \leq N_2$. What is the support of $x[n]$? Show that $y[n]$ is nonzero at most over a finite interval of the form $N_3 \leq n \leq N_4$. Determine N_3 and N_4 and the support of $y[n]$ in terms of N_1, N_2, and M.

Hint: Draw a sketch similar to Fig. 5.3.

5.5 The *unit-step* signal turns on at $n = 0$, and is usually denoted by $u[n]$. It is defined by the formula

$$u[n] = \begin{cases} 0 & n < 0 \\ 1 & n \geq 0 \end{cases}$$

(a) Make a plot of $u[n]$.

(b) We can use the unit-step sequence to represent other sequences that are zero for $n < 0$. Plot the sequence $x[n] = (0.5)^n u[n]$.

(c) The L-point running average is defined as

$$y[n] = \frac{1}{L} \sum_{k=0}^{L-1} x[n-k]$$

For the input sequence $x[n] = (0.5)^n u[n]$, compute the numerical value of $y[n]$ over the range $-5 \leq n \leq 10$, assuming that $L = 4$.

(d) For the input sequence $x[n] = a^n u[n]$, derive a general formula for $y[n]$ that will apply for any value a, for any length L, and for the index range $n \geq 0$. In doing so, you may have use for the formula:

$$\sum_{k=M}^{N} \alpha^k = \frac{\alpha^M - \alpha^{N+1}}{1 - \alpha}$$

5.6 Answer the following questions about the time-domain response of FIR digital filters:

$$y[n] = \sum_{k=0}^{M} b_k \, x[n-k]$$

(a) When tested with an input signal that is an impulse, $x[n] = \delta[n]$, the observed output from the filter is the signal $h[n]$ shown here

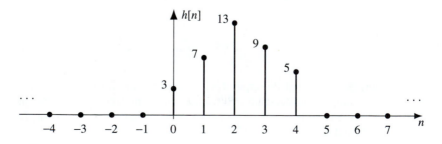

Determine the filter coefficients $\{b_k\}$ of the difference equation for the FIR filter.

(b) If the filter coefficients are $\{b_k\} = \{13, -13, 13\}$ and the input signal is

$$x[n] = \begin{cases} 0 & \text{for } n \text{ even} \\ 1 & \text{for } n \text{ odd} \end{cases}$$

determine the output signal $y[n]$ for all n. Give your answer as either a plot or a formula.

5.7 For each of the following systems, determine whether or not the system is (1) linear, (2) time-invariant, and (3) causal.

   (a) $y[n] = x[n]\cos(0.2\pi n)$

   (b) $y[n] = x[n] - x[n-1]$

   (c) $y[n] = |x[n]|$

   (d) $y[n] = Ax[n] + B$, where A and B are constants.

5.8 Suppose that S is a linear, time-invariant system whose exact form is unknown. It is tested by running some inputs into the system, and then observing the output signals. Suppose that the following input–output pairs are the result of the tests:

$$x[n] = \delta[n] - \delta[n-1] \longmapsto y[n] = \delta[n] - \delta[n-1] + 2\delta[n-3]$$

$$x[n] = \cos(\pi n/2) \longmapsto y[n] = 2\cos(\pi n/2 - \pi/4)$$

   (a) Make a plot of the signal: $y[n] = \delta[n] - \delta[n-1] + 2\delta[n-3]$.

   (b) Use linearity and time invariance to find the output of the system when the input is

$$x[n] = 7\delta[n] - 7\delta[n-2]$$

5.9 A linear time-invariant system has impulse response

$$h[n] = 3\delta[n] - 2\delta[n-1] + 4\delta[n-2] + \delta[n-4]$$

   (a) Draw the implementation of this system as a block diagram in direct form.

   (b) Give the implementation as a block diagram in transposed direct form.

5.10 For a particular LTI system, when the input is

$$x_1[n] = u[n] = \begin{cases} 1 & n \geq 0 \\ 0 & n < 0 \end{cases}$$

the corresponding output is

$$y_1[n] = \delta[n] + 2\delta[n-1] - \delta[n-2] = \begin{cases} 0 & n < 0 \\ 1 & n = 0 \\ 2 & n = 1 \\ -1 & n = 2 \\ 0 & n \geq 3 \end{cases}$$

Determine the output when the input to the LTI system is $x_2[n] = 3u[n] - 2u[n - 4]$. Give your answer as a formula expressing $y_2[n]$ in terms of known sequences, or give a list of values for $-\infty < n < \infty$.

5.11 Suppose that three systems are connected in cascade. In other words, the output of S_1 is the input to S_2, and the output of S_2 is the input to S_3. The three systems are specified as follows:

$$S_1: \quad y_1[n] = x_1[n] - x_1[n - 1]$$

$$S_2: \quad y_2[n] = x_2[n] + x_2[n - 2]$$

$$S_3: \quad y_3[n] = x_3[n - 1] + x_3[n - 2]$$

Thus $x_1[n] = x[n]$, $x_2[n] = y_1[n]$, $x_3[n] = y_2[n]$, and $y[n] = y_3[n]$. Determine the equivalent system that is a single operation from the input $x[n]$ (into S_1) to the output $y[n]$, which is the output of S_3.

(a) Determine the impulse response $h_i[n]$ for each individual subsystem S_i.

(b) Determine the impulse response $h[n]$ of the overall system, i.e., find $h[n]$ so that $y[n] = x[n] * h[n]$.

(c) Write one difference equation that defines the overall system in terms of $x[n]$ and $y[n]$ only.

5.12 Consider a system implemented by the following MATLAB program:

```
load xx    % xx.mat file contains vector of input samples
yy1 = conv(ones(1,4),xx);    % xx is name of input vector
yy2 = conv([1, -1, -1, 1],xx);
ww = yy1 + yy2;
yy = conv(ones(1,3),ww);
```

The overall system from input xx to output yy is an LTI system composed of three LTI systems.

(a) Draw a block diagram of the system that is implemented by the program given above. Be sure to indicate the impulse responses and difference equations of each of the component systems.

(b) The overall system is an LTI system. What is its impulse response and what is the difference equation that is satisfied by the input $x[n]$ and the output $y[n]$?

6

Frequency Response of FIR Filters

Chapter 5 introduced the class of FIR discrete-time systems. We showed that the weighted running average of a finite number of input sequence values defines a discrete-time system, and we showed that such systems are linear and time-invariant. We also showed that the impulse response of an FIR system completely defines the system. In this chapter, we introduce the concept of the *frequency response* of a linear time-invariant FIR filter and show that the frequency response and impulse response are uniquely related. It is remarkable that for linear time-invariant systems, when the input is a complex sinusoid, the corresponding output signal is another complex sinusoid of exactly the same frequency, but with different magnitude and phase. The frequency-response function over all frequencies summarizes the response of an LTI system by giving the magnitude and phase change experienced by all possible sinusoids. Furthermore, since linear time-invariant systems obey the principle of superposition, the frequency-response function is a complete characterization of the behavior of the system for any input that can be represented as a sum of sinusoids. Since almost any discrete-time signal can be represented by a superposition of sinusoids, the frequency response is sufficient therefore to represent the system for almost any signal.

6.1 SINUSOIDAL RESPONSE OF FIR SYSTEMS

Linear time-invariant systems behave in a particularly simple way when the input is a discrete-time complex exponential. To see this, consider the FIR system

$$y[n] = \sum_{k=0}^{M} b_k x[n-k] = \sum_{k=0}^{M} h[k]x[n-k] \qquad (6.1.1)$$

and assume that the input is a complex exponential signal with normalized radian frequency $\hat{\omega}$

$$x[n] = Ae^{j\phi}e^{j\hat{\omega}n} \qquad -\infty < n < \infty$$

Recall that this discrete-time signal could have been obtained by sampling the continuous-time signal

$$x(t) = Ae^{j\phi}e^{j\omega t}$$

If $x[n] = x(nT_s)$, then ω and $\hat{\omega}$ are related by $\hat{\omega} = \omega T_s$, where T_s is the sampling period. For such inputs, the corresponding output is

$$y[n] = \sum_{k=0}^{M} b_k \, Ae^{j\phi}e^{j\hat{\omega}(n-k)}$$

$$= \left(\sum_{k=0}^{M} b_k e^{-j\hat{\omega}k} \right) Ae^{j\phi}e^{j\hat{\omega}n}$$

$$= \mathcal{H}(\hat{\omega})Ae^{j\phi}e^{j\hat{\omega}n} \qquad -\infty < n < \infty \qquad (6.1.2)$$

where

$$\mathcal{H}(\hat{\omega}) = \sum_{k=0}^{M} b_k e^{-j\hat{\omega}k} \qquad (6.1.3)$$

Because we have represented the frequency of the complex exponential signal as the general symbol $\hat{\omega}$, we have obtained an expression (6.1.3) that is a function of $\hat{\omega}$. Thus, (6.1.3) describes the response of the LTI system to a complex exponential signal of *any* frequency $\hat{\omega}$. The quantity $\mathcal{H}(\hat{\omega})$ defined by (6.1.3) is therefore called the *frequency-response function* for the system. (Generally, we shorten this to *frequency response*.)

Note that since the impulse response sequence of an FIR filter is the same as the sequence of filter coefficients, we can express the frequency response in terms of either the filter coefficients b_k or the impulse response $h[k]$; i.e.,

---

The frequency response of an LTI system

$$\mathcal{H}(\hat{\omega}) = \sum_{k=0}^{M} b_k e^{-j\hat{\omega}k} = \sum_{k=0}^{M} h[k]e^{-j\hat{\omega}k} \qquad (6.1.4)$$

Several important points can be made about (6.1.2) and (6.1.4). First of all, the precise interpretation of (6.1.2) is as follows: When the input is a discrete-time complex exponential signal, the output of an LTI FIR filter is also a discrete-time complex exponential signal with a different complex amplitude, but the same frequency $\hat{\omega}$. The frequency response multiplies the signal, thereby changing the complex amplitude. While it is tempting to express this fact by the mathematical statement $y[n] = \mathcal{H}(\hat{\omega})x[n]$, it is strongly recommended that this never be done because it is too easy to forget that the mathematical statement is true *only* for complex exponential signals of frequency $\hat{\omega}$. The statement $y[n] = \mathcal{H}(\hat{\omega})x[n]$ is *meaningless* for any signal other than signals of precisely the form $x[n] = Ae^{j\phi}e^{j\hat{\omega}n}$. *It is very important to understand this point.*

A second important point is that the frequency response $\mathcal{H}(\hat{\omega})$ is complex-valued so it can be expressed either as $\mathcal{H}(\hat{\omega}) = |\mathcal{H}(\hat{\omega})|e^{j\angle \mathcal{H}(\hat{\omega})}$ or as $\mathcal{H}(\hat{\omega}) = \Re e\{\mathcal{H}(\hat{\omega})\} + j\Im m\{\mathcal{H}(\hat{\omega})\}$. The effect of the LTI system on the magnitude and phase of the input complex exponential signal is determined completely by the frequency response function $\mathcal{H}(\hat{\omega})$. Specifically, if the input is $x[n] = Ae^{j\phi}e^{j\hat{\omega}n}$, then using the polar form of $\mathcal{H}(\hat{\omega})$ gives the result

$$y[n] = |\mathcal{H}(\hat{\omega})|e^{j\angle \mathcal{H}(\hat{\omega})} \cdot Ae^{j\phi}e^{j\hat{\omega}n}$$

$$= (|\mathcal{H}(\hat{\omega})|A) \cdot e^{j(\angle \mathcal{H}(\hat{\omega})+\phi)}e^{j\hat{\omega}n} \tag{6.1.5}$$

The magnitude and angle form of the frequency response is the most convenient form, since multiplication is most conveniently accomplished in polar form. The angle of the frequency response simply adds to the phase of the input, thereby producing additional phase shift in the complex exponential signal. Since the magnitude of the frequency response multiplies the magnitude of the complex exponential signal, this part of the frequency response controls the size of the output. Thus, $|\mathcal{H}(\hat{\omega})|$ is also referred to as the *gain* of the system.

Example 6.1   Consider an LTI system for which the difference equation coefficients are $\{b_k\} = \{1, 2, 1\}$. Substituting into (6.1.3) gives

$$\mathcal{H}(\hat{\omega}) = 1 + 2e^{-j\hat{\omega}} + e^{-j\hat{\omega}2}$$

To obtain formulas for the magnitude and phase of the frequency response of this FIR filter, we can manipulate the equation as follows:

$$\mathcal{H}(\hat{\omega}) = 1 + 2e^{-j\hat{\omega}} + e^{-j\hat{\omega}2}$$

$$= e^{-j\hat{\omega}}\left(e^{j\hat{\omega}} + 2 + e^{-j\hat{\omega}}\right)$$

$$= e^{-j\hat{\omega}}\left(2 + 2\cos\hat{\omega}\right)$$

Since $(2 + 2\cos\hat{\omega}) \geq 0$ for frequencies $-\pi < \hat{\omega} \leq \pi$, the magnitude is $|\mathcal{H}(\hat{\omega})| = (2 + 2\cos\hat{\omega})$ and the phase is $\angle \mathcal{H}(\hat{\omega}) = -\hat{\omega}$. $\diamond$

Example 6.2   Consider the complex input $x[n] = 2e^{j\pi/4}e^{j\pi n/3}$. If this signal is the input to the system of Example 6.1, then $|\mathcal{H}(\pi/3)| = 2 + 2\cos(\pi/3) = 3$ and $\angle\mathcal{H}(\hat{\omega}) = -\pi/3$. Therefore, the output of the system for the given input is

$$
\begin{aligned}
y[n] &= 3e^{-j\pi/3} \cdot 2e^{j\pi/4}e^{j\pi n/3} \\
&= (3 \cdot 2) \cdot e^{(j\pi/4 - j\pi/3)}e^{j\pi n/3} \\
&= 6e^{-j\pi/12}e^{j\pi n/3} = 6e^{j\pi/4}e^{j\pi(n-1)/3}
\end{aligned}
$$

Thus, for this system and the given input $x[n]$, the output is equal to the input multiplied by 3, and the phase shift corresponds to a delay of one sample.   ◇

Exercise 6.1. When the sequence of coefficients is symmetrical ($b_0 = b_M$, $b_1 = b_{M-1}$, etc.), the frequency response can be manipulated as in Example 6.1. Following the style of that example, show that the frequency response of an FIR filter with coefficients $\{b_k\} = \{1, -2, 4 - 2, 1\}$ can be expressed as

$$
\mathcal{H}(\hat{\omega}) = [4 - 4\cos(\hat{\omega}) + 2\cos(2\hat{\omega})]e^{-j2\hat{\omega}}
$$

6.2   SUPERPOSITION AND THE FREQUENCY RESPONSE

The principle of superposition makes it very easy to find the output of a linear time-invariant system if the input is a sum of complex exponential signals. This is why the frequency response is so important in the analysis and design of LTI systems.

As an example, suppose that the input to an LTI system is a cosine wave with a specific normalized frequency $\hat{\omega}_1$ plus a DC level,

$$
x[n] = A_0 + A_1 \cos(\hat{\omega}_1 n + \phi_1)
$$

If we represent the signal in terms of complex exponentials, the signal is composed of three complex exponential signals,

$$
x[n] = A_0 e^{j0n} + \frac{A_1}{2}e^{j\phi_1}e^{j\hat{\omega}_1 n} + \frac{A_1}{2}e^{-j\phi_1}e^{-j\hat{\omega}_1 n}
$$

with frequencies $\hat{\omega} = 0$, $\hat{\omega}_1$, and $-\hat{\omega}_1$. By superposition, we can determine the output due to each term separately and then add them to obtain the output $y[n]$ corresponding to $x[n]$. Because the components of the input signal are all complex exponential signals, it is easy to find their respective outputs if we know the frequency response of the system; we just multiply each component by $\mathcal{H}(\hat{\omega})$ evaluated at the corresponding frequency, i.e.,

$$y[n] = \mathcal{H}(0)A_0 e^{j0n} + \mathcal{H}(\hat{\omega}_1)\frac{A_1}{2}e^{j\phi_1}e^{j\hat{\omega}_1 n} + \mathcal{H}(-\hat{\omega}_1)\frac{A_1}{2}e^{-j\phi_1}e^{-j\hat{\omega}_1 n}$$

Note that we have used the fact that a constant signal is a complex exponential with $\hat{\omega} = 0$. If we express $\mathcal{H}(\hat{\omega}_1)$ as $\mathcal{H}(\hat{\omega}_1) = |\mathcal{H}(\hat{\omega}_1)|e^{j\angle\mathcal{H}(\hat{\omega}_1)}$, then the following algebraic steps[1] show that $y[n]$ can finally be expressed as a cosine signal:

$$y[n] = \mathcal{H}(0)A_0 e^{j0n} + \mathcal{H}(\hat{\omega}_1)\frac{A_1}{2}e^{j\phi_1}e^{j\hat{\omega}_1 n} + \mathcal{H}^*(\hat{\omega}_1)\frac{A_1}{2}e^{-j\phi_1}e^{-j\hat{\omega}_1 n}$$

$$= \mathcal{H}(0)A_0 + |\mathcal{H}(\hat{\omega}_1)|e^{j\angle\mathcal{H}(\hat{\omega}_1)}\frac{A_1}{2}e^{j\phi_1}e^{j\hat{\omega}_1 n}$$

$$+ |\mathcal{H}(\hat{\omega}_1)|e^{-j\angle\mathcal{H}(\hat{\omega}_1)}\frac{A_1}{2}e^{-j\phi_1}e^{-j\hat{\omega}_1 n}$$

$$= \mathcal{H}(0)A_0 + |\mathcal{H}(\hat{\omega}_1)|\frac{A_1}{2}e^{j(\hat{\omega}_1 n + \phi_1 + \angle\mathcal{H}(\hat{\omega}_1))}$$

$$+ |\mathcal{H}(\hat{\omega}_1)|\frac{A_1}{2}e^{-j(\hat{\omega}_1 n + \phi_1 + \angle\mathcal{H}(\hat{\omega}_1))}$$

$$= \mathcal{H}(0)A_0 + |\mathcal{H}(\hat{\omega}_1)|A_1 \cos\left(\hat{\omega}_1 n + \phi_1 + \angle\mathcal{H}(\hat{\omega}_1)\right)$$

Notice that the magnitude and phase change of the cosine input signal are taken from the positive frequency part of $\mathcal{H}(\hat{\omega})$, but also notice that it was crucial to express $x[n]$ as a sum of complex exponentials and then use the frequency response to find the output due to each component separately.

Example 6.3   For the FIR filter with coefficients $\{b_k\} = \{1, 2, 1\}$, find the output when the input is

$$x[n] = 2\cos\left(\frac{\pi}{3}n - \frac{\pi}{2}\right)$$

The frequency response of the system was determined in Example 6.1 to be

$$\mathcal{H}(\hat{\omega}) = (2 + 2\cos\hat{\omega})e^{-j\hat{\omega}}$$

---

[1] It is assumed that $\mathcal{H}(\hat{\omega})$ has the "conjugate-symmetry" property $\mathcal{H}(-\hat{\omega}) = \mathcal{H}^*(\hat{\omega})$, which is always true when the filter coefficients are real. (See Section 6.4.3.)

Note that $\mathcal{H}(-\hat{\omega}) = \mathcal{H}^*(\hat{\omega})$; i.e., $\mathcal{H}(\hat{\omega})$ has conjugate symmetry. Solution of this problem requires just one evaluation of $\mathcal{H}(\hat{\omega})$ at the frequency $\hat{\omega} = \pi/3$:

$$\mathcal{H}(\pi/3) = e^{-j\pi/3} (2 + 2\cos(\pi/3))$$
$$= e^{-j\pi/3} \left(2 + 2(\tfrac{1}{2})\right) = 3e^{-j\pi/3}$$

Therefore, the magnitude is $|\mathcal{H}(\pi/3)| = 3$ and the phase is $\angle \mathcal{H}(\pi/3) = -\pi/3$, so the output is

$$y[n] = (3)(2)\cos\left(\frac{\pi}{3}n - \frac{\pi}{3} - \frac{\pi}{2}\right) = 6\cos\left(\frac{\pi}{3}n - \frac{5\pi}{6}\right)$$
$$= 6\cos\left(\frac{\pi}{3}(n-1) - \frac{\pi}{2}\right)$$

Notice that the magnitude of the frequency response multiplies the amplitude of the cosine signal, and the phase angle of the frequency response adds to the phase of the cosine signal.   ◇

If the input signal consists of many complex exponential signals, the frequency response can be applied to find the output due to each component separately, and the results added to determine the total output. This is the principle of superposition at work. If we can find a representation for a signal in terms of complex exponentials, the frequency response gives a simple and highly intuitive means for determining what an LTI system does to that input signal. For example, if the input to an LTI system is a real signal and can be represented as

$$x[n] = X_0 + \sum_{k=1}^{N} \left(\frac{X_k}{2}e^{j\hat{\omega}_k n} + \frac{X_k^*}{2}e^{-j\hat{\omega}_k n}\right)$$
$$= X_0 + \sum_{k=1}^{N} |X_k|\cos(\hat{\omega}_k n + \angle X_k)$$

then it follows that if $\mathcal{H}(-\hat{\omega}) = \mathcal{H}^*(\hat{\omega})$, the corresponding output is

$$y[n] = \mathcal{H}(0)X_0 + \sum_{k=1}^{N} \left(\mathcal{H}(\hat{\omega}_k)\frac{X_k}{2}e^{j\hat{\omega}_k n} + \mathcal{H}(-\hat{\omega}_k)\frac{X_k^*}{2}e^{-j\hat{\omega}_k n}\right)$$
$$= \mathcal{H}(0)X_0 + \sum_{k=1}^{N} |\mathcal{H}(\hat{\omega}_k)||X_k|\cos\left(\hat{\omega}_k n + \angle X_k + \angle \mathcal{H}(\hat{\omega}_k)\right)$$

That is, each individual complex exponential component is modified by the frequency response evaluated at the frequency of that component.

Example 6.4   For the FIR filter with coefficients $\{b_k\} = \{1, 2, 1\}$, find the output when the input is

$$x[n] = 4 + 3\cos\left(\frac{\pi}{3}n - \frac{\pi}{2}\right) + 3\cos\left(\frac{20\pi}{21}n\right) \tag{6.2.1}$$

The frequency response of the system was determined in Example 6.1, and is the same as the frequency response in Example 6.3. The input in this example differs from that of Example 6.3 by the addition of a constant (DC) term and an additional cosine signal of frequency $20\pi/21$. The solution by superposition therefore requires that we evaluate $\mathcal{H}(\hat{\omega})$ at frequencies 0, $\pi/3$, and $20\pi/21$, giving

$$\mathcal{H}(0) = 4$$

$$\mathcal{H}(\pi/3) = 3e^{-j\pi/3}$$

$$\mathcal{H}(20\pi/21) = 0.0223e^{-j20\pi/21}$$

Therefore, the output is

$$y[n] = 4 \cdot 4 + 3 \cdot 3\cos\left(\frac{\pi}{3}n - \frac{\pi}{3} - \frac{\pi}{2}\right) + 0.0223 \cdot 3\cos\left(\frac{20\pi}{21}n - \frac{20\pi}{21}\right)$$

$$= 16 + 9\cos\left(\frac{\pi}{3}(n-1) - \frac{\pi}{2}\right) + 0.067\cos\left(\frac{20\pi}{21}(n-1)\right)$$

Notice that, in this case, the DC component is multiplied by 4, the component at frequency $\hat{\omega} = \pi/3$ is multiplied by 3, but the component at frequency $\hat{\omega} = 20\pi/21$ is multiplied by 0.0223. Because the frequency-response magnitude (gain) is so small at frequency $\hat{\omega} = 20\pi/21$, the component at this frequency is essentially "filtered out" of the input signal.   ◇

The examples of this section illustrate an approach to solving problems that is often called the *frequency-domain* approach. As these examples show, we do not need to deal with the *time-domain* description (i.e., the difference equation or impulse response) of the system when the input is a complex exponential signal. We can work exclusively with the frequency-domain description (i.e., the frequency-response function), if we think about how the spectrum of the signal is modified by the system rather than considering what happens to the individual samples of the input signal. We will have ample opportunity to visit both the time-domain and the frequency-domain in the remainder of this chapter.

6.3 STEADY STATE AND TRANSIENT RESPONSE

In Section 6.1, we showed that if the input is

$$x[n] = Xe^{j\hat{\omega}n} \qquad -\infty < n < \infty \tag{6.3.1}$$

where $X = Ae^{j\phi}$, then the corresponding output of an LTI FIR system is

$$y[n] = \mathcal{H}(\hat{\omega})Xe^{j\hat{\omega}n} \qquad -\infty < n < \infty \tag{6.3.2}$$

where

$$\mathcal{H}(\hat{\omega}) = \sum_{k=0}^{M} b_k e^{-j\hat{\omega}k} \tag{6.3.3}$$

In (6.3.1), the condition that $x[n]$ be a complex exponential signal existing over $-\infty < n < \infty$ is important. Without this condition, we will not obtain the simple result of (6.3.2). However, this condition appears to be somewhat impractical. In any practical implementation, we surely would not have actual input signals that exist back to $-\infty$! Fortunately, we can relax the condition that the complex exponential be defined over the doubly infinite interval and still take advantage of the convenience of (6.3.2). To see this, consider the following "suddenly applied" complex exponential signal that starts at $n = 0$ and is nonzero only for $0 \le n$:

$$x[n] = Xe^{j\hat{\omega}n}u[n] = \begin{cases} Xe^{j\hat{\omega}n} & 0 \le n \\ 0 & n < 0 \end{cases} \tag{6.3.4}$$

Note that multiplication by the unit-step signal is a convenient way to impose the suddenly applied condition. The output of an LTI FIR system for this input is

$$y[n] = \sum_{k=0}^{M} b_k Xe^{j\hat{\omega}(n-k)}u[n-k] \tag{6.3.5}$$

By considering different values of n and the fact that $u[n - k] = 0$ for $k > n$, it follows that the sum in (6.3.5) can be expressed as

$$y[n] = \begin{cases} 0 & n < 0 \\ \left(\displaystyle\sum_{k=0}^{n} b_k e^{-j\hat{\omega}k}\right) Xe^{j\hat{\omega}n} & 0 \le n < M \\ \left(\displaystyle\sum_{k=0}^{M} b_k e^{-j\hat{\omega}k}\right) Xe^{j\hat{\omega}n} & M \le n \end{cases} \tag{6.3.6}$$

That is, when the complex exponential signal is suddenly applied, the output can be considered to be defined over three distinct regions. In the first region, $n < 0$, the input is zero, and therefore the corresponding output is zero, too. The second region is a transition region whose length is M samples (i.e., the *order* of the FIR system). In this region, the complex multiplier of $e^{j\hat{\omega}n}$ depends upon n. This region is often called the *transient* part of the output. In the third region, $M \leq n$, the output is identical to the output that would be obtained if the input were defined over the doubly infinite interval. That is,

$$y[n] = \mathcal{H}(\hat{\omega})Xe^{j\hat{\omega}n} \qquad M \leq n \qquad (6.3.7)$$

This part of the output is generally called the *steady-state* part. While we have specified that the steady-state part exists for all $n \geq M$, it should be clear that (6.3.7) holds only as long as the input remains equal to $Xe^{j\hat{\omega}n}$. If, at some time $n > M$, the input changes frequency or goes to zero, another transient region will occur.

Example 6.5   A simple example will illustrate the above discussion. Consider the system of Exercise 6.1, whose filter coefficients are the sequence $\{b_k\} = \{1, -2, 4, -2, 1\}$. The frequency response of this system is

$$\mathcal{H}(\hat{\omega}) = [4 - 4\cos(\hat{\omega}) + 2\cos(2\hat{\omega})]e^{-j2\hat{\omega}}$$

If the input is the suddenly applied cosine signal

$$x[n] = \cos(0.2\pi n - \pi)u[n]$$

we can represent it as the sum of two suddenly applied complex exponential signals. Therefore, the frequency response can be used as discussed in Section 6.2 to determine the corresponding steady-state output as

$$y[n] = [4 - 4\cos(0.2\pi) + 2\cos(0.4\pi)]\cos(0.2\pi n - 2(0.2\pi) - \pi)$$

$$= 1.382\cos(0.2\pi(n-2) - \pi) \qquad n \geq 4$$

The frequency response has allowed us to find a simple expression for the output everywhere in the steady-state region. If we desire the values of the output in the transient region, we could compute them using the difference equation for the system.

The input and output signals for this example are shown in Fig. 6.1. Since $M = 4$ for this system, the transient region is $0 \leq n \leq 3$ (indicated by the gray region), and the steady-state region is $n \geq 4$. Also note that, as predicted by the steady-state analysis above, the signal in the steady-state region is simply a scaled and shifted (by 2 samples) version of the input.   ◇

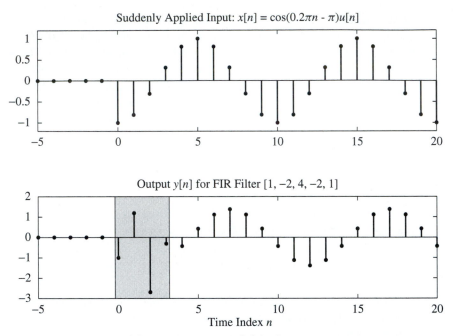

Figure 6.1   Input $x[n] = \cos(0.2\pi n - \pi)u[n]$ (top) and corresponding output $y[n]$ for FIR filter with coefficients $\{1, -2, 4, -2, 1\}$ (bottom). The transient region is the gray area in the lower panel. (Note the different vertical scales.)

6.4 PROPERTIES OF THE FREQUENCY RESPONSE

The frequency response function $\mathcal{H}(\hat{\omega})$ is a complex-valued function of the normalized frequency variable $\hat{\omega}$. This function has interesting properties that often can be used to simplify analysis.

6.4.1 Relation to Impulse Response and Difference Equation

$\mathcal{H}(\hat{\omega})$ can be calculated directly from the filter coefficients $\{b_k\}$. If (6.1.1) is compared to (6.1.4), we see that, given the difference equation, it is simple to write down an expression for $\mathcal{H}(\hat{\omega})$ by noting that each term $b_k x[n-k]$ in (6.1.1) corresponds to a term $b_k e^{-j\hat{\omega}k}$ or $h[k]e^{-j\hat{\omega}k}$ in (6.1.4), and vice versa. Likewise, $\mathcal{H}(\hat{\omega})$ can be determined directly from the impulse response since the impulse response of the FIR system consists of the sequence of filter coefficients; i.e., $h[k] = b_k$ for $k = 0, 1, \ldots, M$. To emphasize this point, we can write the correspondence

$$h[n] = \sum_{k=0}^{M} h[k]\delta[n-k] \quad\Longleftrightarrow\quad \mathcal{H}(\hat{\omega}) = \sum_{k=0}^{M} h[k]e^{-j\hat{\omega}k}$$

| *Time Domain* | $\Longleftrightarrow$ | *Frequency Domain* |

The process of going from the difference equation or impulse response to the frequency response is straightforward for the FIR filter. It is also simple to go from the frequency response to the difference equation or to the impulse response if we express $\mathcal{H}(\hat{\omega})$ in terms of powers of $e^{-j\hat{\omega}}$. These points are illustrated by the following examples.

Example 6.6   Consider the FIR filter defined by the impulse response

$$h[n] = -\delta[n] + 3\delta[n-1] - \delta[n-2]$$

By inspection, the filter coefficients are $\{b_k\} = \{-1, 3, -1\}$, so the difference equation corresponding to this impulse response is obviously

$$y[n] = -x[n] + 3x[n-1] - x[n-2]$$

and the frequency response of this system is

$$\mathcal{H}(\hat{\omega}) = -1 + 3e^{-j\hat{\omega}} - e^{-j2\hat{\omega}}$$

$\diamond$

Example 6.7   Suppose that the frequency response is given by the equation

$$\mathcal{H}(\hat{\omega}) = e^{-j\hat{\omega}}\left(3 - 2\cos\hat{\omega}\right)$$

Since $\cos\hat{\omega} = \frac{1}{2}(e^{j\hat{\omega}} + e^{-j\hat{\omega}})$, we can write

$$\mathcal{H}(\hat{\omega}) = e^{-j\hat{\omega}}\left[3 - 2\left(\frac{e^{j\hat{\omega}} + e^{-j\hat{\omega}}}{2}\right)\right] = -1 + 3e^{-j\hat{\omega}} - e^{-j\hat{\omega}2}$$

which corresponds to the following FIR difference equation:

$$y[n] = -x[n] + 3x[n-1] - x[n-2]$$

The impulse response, likewise, is easy to determine directly from $\mathcal{H}(\hat{\omega})$.

$\diamond$

Exercise 6.2. Use the inverse Euler formula for sines to find the impulse response and difference equation of $\mathcal{H}(\hat{\omega}) = 2j\sin(\hat{\omega}/2)e^{-j\hat{\omega}/2}$.

6.4.2 Periodicity of $\mathcal{H}(\hat{\omega})$

An important property of a discrete-time LTI system is that its frequency response $\mathcal{H}(\hat{\omega})$ is always a periodic function with period 2π. This can be seen by considering a frequency $\hat{\omega} + 2\pi$ where $\hat{\omega}$ is any frequency. Substituting into (6.1.3) gives

$$\mathcal{H}(\hat{\omega} + 2\pi) = \sum_{k=0}^{M} b_k e^{-j(\hat{\omega}+2\pi)k}$$

$$= \sum_{k=0}^{M} b_k e^{-j\hat{\omega}k} e^{-j2\pi k} = \mathcal{H}(\hat{\omega})$$

since $e^{-j2\pi k} = 1$ when k is an integer. It is not surprising that $\mathcal{H}(\hat{\omega})$ should have this property, since, as we have seen in Chapter 4, a change in the input frequency by 2π is not detectable; i.e.,

$$x[n] = Xe^{j(\hat{\omega}+2\pi)n} = Xe^{j\hat{\omega}n}e^{j2\pi n} = Xe^{j\hat{\omega}n}$$

In other words, two complex exponential signals with frequencies differing by 2π cannot be distinguished from their samples, so there is no reason to expect a discrete-time system to behave differently for two such frequencies. For this reason, it is always sufficient to specify the frequency response only over an interval of one period, e.g., $-\pi < \hat{\omega} \le \pi$.

6.4.3 Conjugate Symmetry

The frequency response $\mathcal{H}(\hat{\omega})$ is complex, but usually has a symmetry in its magnitude and phase that allows us to concentrate on just half of the period when plotting. This is the property of *conjugate symmetry*

$$\mathcal{H}(-\hat{\omega}) = \mathcal{H}^*(\hat{\omega}) \tag{6.4.1}$$

which is true whenever the filter coefficients are real so that $b_k = b_k^*$. We can prove this property for the FIR case as follows:

$$\mathcal{H}^*(\hat{\omega}) = \left(\sum_{k=0}^{M} b_k e^{-j\hat{\omega}k} \right)^*$$

$$= \sum_{k=0}^{M} b_k^* e^{+j\hat{\omega}k}$$

$$= \sum_{k=0}^{M} b_k e^{-j(-\hat{\omega})k} = \mathcal{H}(-\hat{\omega})$$

The conjugate symmetry property implies that the magnitude function is an even function of $\hat{\omega}$ and the phase is an odd function, i.e.,

$$|\mathcal{H}(-\hat{\omega})| = |\mathcal{H}(\hat{\omega})|$$

$$\angle \mathcal{H}(-\hat{\omega}) = -\angle \mathcal{H}(\hat{\omega})$$

Similarly, the real part is an even function of $\hat{\omega}$ and the imaginary part is an odd function, i.e.,

$$\Re e\{\mathcal{H}(-\hat{\omega})\} = \Re e\{\mathcal{H}(\hat{\omega})\}$$

$$\Im m\{\mathcal{H}(-\hat{\omega})\} = -\Im m\{\mathcal{H}(\hat{\omega})\}$$

As a result, plots of the frequency response are often shown only over half a period, $0 \le \hat{\omega} \le \pi$, because the negative frequency region can be constructed by symmetry. These symmetries are illustrated by the plots in Section 6.5.

> **Exercise 6.3.** Prove that the magnitude is an even function of $\hat{\omega}$ and the phase is an odd function of $\hat{\omega}$ for a conjugate-symmetric frequency response.

6.5 GRAPHICAL REPRESENTATION OF THE FREQUENCY RESPONSE

Two important points should be emphasized about the frequency response of an LTI system. The first is that for a given system, the frequency response usually varies with frequency, so that sinusoids of different frequencies are treated differently by the system. The second important point is that by appropriate choice of the coefficients, b_k, a wide variety of frequency response shapes can be realized. In order to visualize the variation of the frequency response with frequency, it is useful to plot $\mathcal{H}(\hat{\omega})$ versus $\hat{\omega}$. We will see that the plot tells us at a glance what the system does to complex exponential signals and sinusoids of different frequencies. Several examples are provided in this section to illustrate the value of plotting the frequency response.

6.5.1 Delay System

The delay system is a simple FIR filter given by the difference equation

$$y[n] = x[n - n_0]$$

It has only one nonzero filter coefficient, $b_{n_0} = 1$, so its frequency response is

$$\mathcal{H}(\hat{\omega}) = e^{-j\hat{\omega}n_0} \tag{6.5.1}$$

For this filter, a plot of the frequency response is easy to visualize; the magnitude response is one for all frequencies and the phase is linear with a slope equal to $-n_0$, as in Fig. 6.2. As a result, we can associate the property of linear phase with time delay in all filters. Since time delay affects only the time origin of the signal in a predictable way, we often think of linear phase as an ideal phase response.

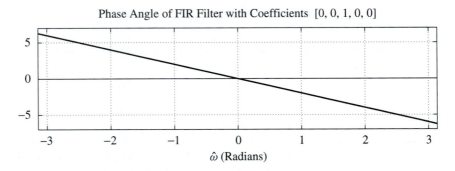

Figure 6.2   Phase response of pure delay ($n_0 = 2$) system, $\mathcal{H}(\hat{\omega}) = e^{-j2\hat{\omega}}$.

6.5.2 First Difference System

As another simple example, consider the first-difference system

$$y[n] = x[n] - x[n - 1]$$

The frequency response of this LTI system is

$$\mathcal{H}(\hat{\omega}) = 1 - e^{-j\hat{\omega}} = 1 - \cos\hat{\omega} + j\sin\hat{\omega}$$

The different parts of the complex representations are

$$\Re e\{\mathcal{H}(\hat{\omega})\} = (1 - \cos \hat{\omega})$$

$$\Im m\{\mathcal{H}(\hat{\omega})\} = \sin \hat{\omega}$$

$$|\mathcal{H}(\hat{\omega})| = [(1 - \cos \hat{\omega})^2 + \sin^2 \hat{\omega}]^{1/2} = [2(1 - \cos \hat{\omega})]^{1/2} = 2|\sin(\hat{\omega}/2)|$$

$$\angle \mathcal{H}(\hat{\omega}) = \arctan\left(\frac{\sin \hat{\omega}}{1 - \cos \hat{\omega}}\right)$$

The real and imaginary parts for this example are plotted in Fig. 6.3; the magnitude and phase are plotted in Fig. 6.4. All functions are plotted for $-3\pi < \hat{\omega} < 3\pi$, even though we would normally need to plot the frequency response of a discrete-time system only for $-\pi < \hat{\omega} < \pi$, or (because of conjugate symmetry) $0 \le \hat{\omega} < \pi$. These extended plots verify that $\mathcal{H}(\hat{\omega})$ is periodic with period 2π, and they verify the conjugate symmetry properties discussed in Section 6.4.3.

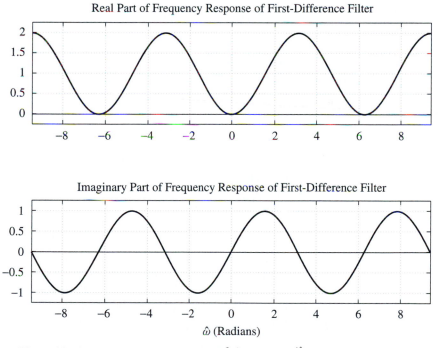

Figure 6.3   Real and imaginary parts for $\mathcal{H}(\hat{\omega}) = 1 - e^{-j\hat{\omega}}$ over three periods showing periodicity and conjugate symmetry of $\mathcal{H}(\hat{\omega})$.

The utility of the magnitude and phase plots of $\mathcal{H}(\hat{\omega})$ can be seen even for this simple example. In Fig. 6.4, $\mathcal{H}(0) = 0$, so we easily see that the system completely removes components with $\hat{\omega} = 0$ (i.e., DC). Furthermore, we can also see that the system emphasizes the higher frequencies (near $\hat{\omega} = \pi$) relative to the lower frequencies, so it would be called a *highpass filter*. This is another typical way of thinking about systems in the frequency domain.

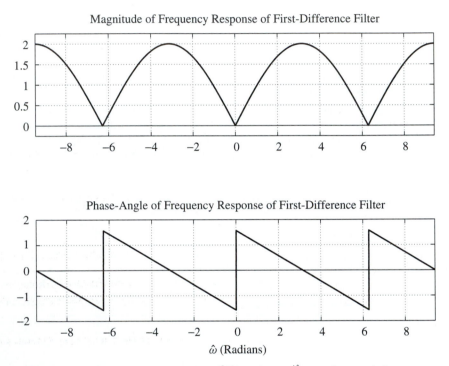

Figure 6.4   Magnitude and phase for $\mathcal{H}(\hat{\omega}) = 1 - e^{-j\hat{\omega}}$ over three periods showing periodicity and conjugate symmetry of $\mathcal{H}(\hat{\omega})$.

The real and imaginary parts and the magnitude and phase can always be determined as done above by standard manipulations of complex numbers. However, there is a simpler approach for getting the magnitude and phase when the sequence of coefficients is either symmetric or antisymmetric about a central point. The following algebraic manipulation of $\mathcal{H}(\hat{\omega})$ is possible because the $\{b_k\}$ coefficients satisfy the symmetry condition:

$$b_k = -b_{M-k}$$

The trick, which we have already used in Example 6.1, is to factor out an exponential whose phase is half of the filter order times $\hat{\omega}$, i.e.,

$$\mathcal{H}(\hat{\omega}) = 1 - e^{-j\hat{\omega}}$$

$$= e^{-j\hat{\omega}/2}\left(e^{j\hat{\omega}/2} - e^{-j\hat{\omega}/2}\right)$$

$$= 2je^{-j\hat{\omega}/2}\sin(\hat{\omega}/2)$$

$$= 2\sin(\hat{\omega}/2)e^{j(\pi/2-\hat{\omega}/2)}$$

The form derived for $\mathcal{H}(\hat{\omega})$ is almost a valid polar form, but since $\sin(\hat{\omega}/2)$ is negative for $-\pi < \hat{\omega} < 0$, we must write $|\mathcal{H}(\hat{\omega})| = 2|\sin(\hat{\omega}/2)|$ and absorb the algebraic sign[2] into the phase response for $-\pi < \hat{\omega} < 0$, i.e.,

$$\angle\mathcal{H}(\hat{\omega}) = \begin{cases} \pi/2 - \hat{\omega}/2 & 0 < \hat{\omega} < \pi \\ -\pi + \pi/2 - \hat{\omega}/2 & -\pi < \hat{\omega} < 0 \end{cases}$$

This formula for the phase is consistent with the phase plot in Fig. 6.4, which exhibits several linear segments. Notice also that the phase plot has discontinuities at $\hat{\omega} = 0$ and $\hat{\omega} = \pm 2\pi$. The size of these discontinuities is π, since they correspond to a sign change in $\mathcal{H}(\hat{\omega})$.

Example 6.8   Suppose that the input to a first-difference system is $x[n] = 4 + 2\cos(0.3\pi n - \pi/4)$. Since the output is related to the input by the equation $y[n] = x[n] - x[n-1]$, it follows that:

$$y[n] = 4 + 2\cos(0.3\pi n - \pi/4) - 4 - 2\cos(0.3\pi(n-1) - \pi/4)$$

$$= 2\cos(0.3\pi n - \pi/4) - 2\cos(0.3\pi(n-1) - \pi/4)$$

From this result, we see that the first-difference system removes the constant value and leaves two cosine signals of the same frequency, which could be combined by phasor addition. However, the solution using the frequency-response function is simpler. Since the first-difference system has frequency response

$$\mathcal{H}(\hat{\omega}) = 2\sin(\hat{\omega}/2)e^{j(\pi/2-\hat{\omega}/2)}$$

the output of this system for the given input is

$$y[n] = 4\mathcal{H}(0) + 2|\mathcal{H}(0.3\pi)|\cos(0.3\pi n - \pi/4 + \angle\mathcal{H}(0.3\pi))$$

---

[2] Remember that $-1 = e^{\pm j\pi}$, so we can either add or subtract π radians from the phase for $-\pi < \hat{\omega} < 0$. In this case, we subtract π so that the resulting phase curve remains between $-\pi$ and $+\pi$ radians for all $\hat{\omega}$.

Therefore, since $\mathcal{H}(0) = 0$, the output will be

$$y[n] = (2)(2)\sin(0.3\pi/2)\cos(0.3\pi n - \pi/4 + \pi/2 - 0.3\pi/2)$$

$$= 1.816\cos(0.3\pi n + 0.1\pi)$$

$\diamond$

6.5.3 A Simple Lowpass Filter

In Examples 6.1, 6.3, and 6.4, the system had frequency response

$$\mathcal{H}(\hat{\omega}) = 1 + 2e^{-j\hat{\omega}} + e^{-j2\hat{\omega}} = (2 + 2\cos\hat{\omega})e^{-j\hat{\omega}}$$

Since the factor $(2 + 2\cos\hat{\omega}) \geq 0$ for all $\hat{\omega}$, it follows that

$$|\mathcal{H}(\hat{\omega})| = (2 + 2\cos\hat{\omega})$$

and

$$\angle\mathcal{H}(\hat{\omega}) = -\hat{\omega}$$

These functions are plotted in Fig. 6.5 for $-\pi < \hat{\omega} < \pi$. Figure 6.5 shows at a glance that the system has a delay of 1 sample and that it tends to favor the low frequencies (close to $\hat{\omega} = 0$) with high gain, while it tends to suppress high frequencies (close to $\hat{\omega} = \pi$). In this case, there is a gradual decrease in gain from $\hat{\omega} = 0$ to $\hat{\omega} = \pi$, so that the midrange frequencies receive more gain than the high frequencies, but less than the low frequencies. Filters with magnitude responses that suppress the high frequencies of the input are called *lowpass filters*.

Example 6.9   If the input of Example 6.4

$$x[n] = 4 + 3\cos\left(\frac{\pi}{3}n - \frac{\pi}{2}\right) + 3\cos\left(\frac{20\pi}{21}n\right) \tag{6.5.2}$$

is the input to the lowpass filter whose frequency response is shown in Fig. 6.5, then we must evaluate $\mathcal{H}(\hat{\omega})$ at frequencies $0, \pi/3$, and $20\pi/21$ giving

$$\mathcal{H}(0) = 4$$

$$\mathcal{H}(\pi/3) = 3e^{-j\pi/3}$$

$$\mathcal{H}(20\pi/21) = 0.0223e^{-j20\pi/21}$$

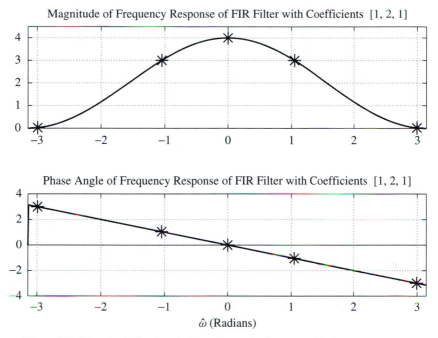

Figure 6.5 Magnitude (top) and phase (bottom) of system with frequency response $\mathcal{H}(\hat{\omega}) = (2 + 2\cos\hat{\omega})e^{-j\hat{\omega}}$. Stars indicate points where the frequency response is evaluated to calculate the sinusoidal response in Example 6.9.

These values are the points indicated with stars ($*$) on the graphs of Fig. 6.5. As in Example 6.4, the output is

$$
y[n] = 4 \cdot 4 + 3 \cdot 3\cos\left(\frac{\pi}{3}n - \frac{\pi}{3} - \frac{\pi}{2}\right) + 0.0223 \cdot 3\cos\left(\frac{20\pi}{21}n - \frac{20\pi}{21}\right)
$$

$$
= 16 + 9\cos\left(\frac{\pi}{3}(n-1) - \frac{\pi}{2}\right) + 0.067\cos\left(\frac{20\pi}{21}(n-1)\right)
$$

The plot of the frequency response clearly shows that all frequencies around $\hat{\omega} = \pi$ are greatly attenuated by the system. Also, the linear phase plot with slope of -1 indicates that all frequencies experience a time delay of 1 sample.

The effect of the simple lowpass filter on the time waveform of the signal can be seen in Fig. 6.6, which shows in the upper plot a segment of the signal $x[n]$ given by (6.5.2) and in the lower plot the corresponding segment of the output. Note that the DC component is indicated in both parts of the figure as a dotted horizontal line. We can see that the output consists a constant of 16 plus a cosine that has amplitude 9 and is periodic with period 6; i.e.,the output shows that the component at frequency $\hat{\omega} = 20\pi/21$ has been removed, leaving only the DC component and the component with frequency $\hat{\omega} = \pi/3$. $\diamond$

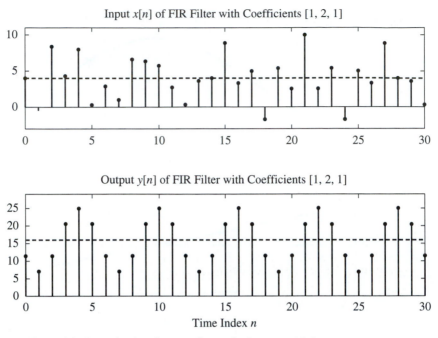

Figure 6.6   Input (top) and output (bottom) of system with frequency response $\mathcal{H}(\hat{\omega}) = (2 + 2\cos\hat{\omega})e^{-j\hat{\omega}}$.

6.6 CASCADED LTI SYSTEMS

CASCADING
LTI SYSTEM

In Chapter 5, we showed that if two LTI systems are connected in cascade (output of the first is input to the second), then the overall impulse response is the convolution of the two individual impulse responses, and therefore the cascade system is equivalent to a single system whose impulse response is the convolution of the two individual impulse responses. In this section, we will show that the frequency response of a cascade connection of two LTI systems is simply the product of the individual frequency responses.

The upper part of Fig. 6.7 shows two LTI systems in cascade. To find the frequency response of the overall system (from input $x[n]$ to output $y_2[n]$), we let

$$x[n] = e^{j\hat{\omega}n}$$

Then, the output of the first LTI system is

$$w[n] = \mathcal{H}_1(\hat{\omega})e^{j\hat{\omega}n}$$

and the output of the second system is

$$y_2[n] = \mathcal{H}_2(\hat{\omega}) \left(\mathcal{H}_1(\hat{\omega}) e^{j\hat{\omega}n} \right) = \mathcal{H}_2(\hat{\omega}) \mathcal{H}_1(\hat{\omega}) e^{j\hat{\omega}n}$$

From a similar analysis of the middle diagram of Fig. 6.7, it follows that

$$y_1[n] = \mathcal{H}_1(\hat{\omega}) \left(\mathcal{H}_2(\hat{\omega}) e^{j\hat{\omega}n} \right) = \mathcal{H}_1(\hat{\omega}) \mathcal{H}_2(\hat{\omega}) e^{j\hat{\omega}n}$$

Since $\mathcal{H}_2(\hat{\omega}) \mathcal{H}_1(\hat{\omega}) = \mathcal{H}_1(\hat{\omega}) \mathcal{H}_2(\hat{\omega})$ from the commutative property of multiplication, it follows that $y_1[n] = y_2[n]$; i.e., the two cascade systems are equivalent for the same complex exponential input, and both of them are equivalent to a single LTI system with frequency response

$$\mathcal{H}(\hat{\omega}) = \mathcal{H}_2(\hat{\omega}) \mathcal{H}_1(\hat{\omega}) = \mathcal{H}_1(\hat{\omega}) \mathcal{H}_2(\hat{\omega}) \tag{6.6.1}$$

The output $y[n]$ of any system with frequency response $\mathcal{H}(\hat{\omega})$ will be the same as either $y_1[n]$ or $y_2[n]$. This is depicted in the bottom part of Fig. 6.7.

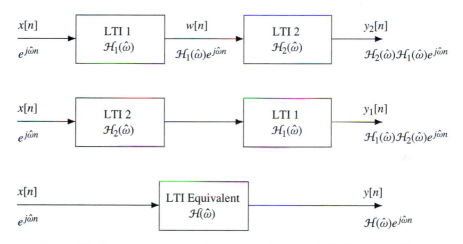

Figure 6.7   Frequency response of three equivalent cascaded LTI systems. All three systems have the same frequency response so that $y[n] = y_1[n] = y_2[n]$ for the same input $x[n]$

Recall from Chapter 5 that the overall impulse response is $h_1[n] * h_2[n]$. We can summarize this by the correspondence

$$
\begin{array}{ccc}
Convolution & \Longleftrightarrow & Multiplication \\
h_1[n] * h_2[n] & \Longleftrightarrow & \mathcal{H}_1(\hat{\omega}) \mathcal{H}_2(\hat{\omega})
\end{array}
\tag{6.6.2}
$$

That is, convolution of impulse responses corresponds to multiplication of the frequency responses of cascaded systems. The correspondence shown in (6.6.2) is useful because it provides another way of representing and manipulating LTI systems. This is illustrated by the following example.

Example 6.10   Suppose that the first system in a cascade of two systems is defined by the set of coefficients $\{2, 4, 6, 4, 2\}$ and the second system is defined by the coefficients $\{1, -2, 2, -1\}$. The frequency responses of the individual systems are

$$\mathcal{H}_1(\hat{\omega}) = 2 + 4e^{-j\hat{\omega}} + 6e^{-j\hat{\omega}2} + 4e^{-j\hat{\omega}3} + 2e^{-j\hat{\omega}4}$$

and

$$\mathcal{H}_2(\hat{\omega}) = 1 - 2e^{-j\hat{\omega}} + 2e^{-j\hat{\omega}2} - e^{-j\hat{\omega}3}$$

The overall frequency response is

$$\mathcal{H}(\hat{\omega}) = \mathcal{H}_1(\hat{\omega})\mathcal{H}_2(\hat{\omega})$$

$$= \left(2 + 4e^{-j\hat{\omega}} + 6e^{-j\hat{\omega}2} + 4e^{-j\hat{\omega}3} + 2e^{-j\hat{\omega}4}\right)\left(1 - 2e^{-j\hat{\omega}} + 2e^{-j\hat{\omega}2} - e^{-j\hat{\omega}3}\right)$$

$$= 2 + 0e^{-j\hat{\omega}} + 2e^{-j\hat{\omega}2} - 2e^{-j\hat{\omega}3} + 2e^{-j\hat{\omega}4} - 2e^{-j\hat{\omega}5} + 0e^{-j\hat{\omega}6} - 2e^{-j\hat{\omega}7}$$

Thus, the overall equivalent impulse response is

$$h[n] = 2\delta[n] + 2\delta[n - 2] - 2\delta[n - 3] + 2\delta[n - 4] - 2\delta[n - 5] - 2\delta[n - 7]$$

◇

This example illustrates that convolution of two impulse responses is equivalent to multiplying their corresponding frequency responses. Notice that, for FIR systems, the frequency response is just a polynomial in the variable $e^{-j\hat{\omega}}$. Thus, multiplying two frequency responses requires polynomial multiplication. This result provides the theoretical basis for the "synthetic" polynomial multiplication algorithm discussed in Chapter 5.

Exercise 6.4. Suppose that two systems are cascaded. The first system is defined by the set of coefficients $\{1, 2, 3, 4\}$, and the second system is defined by the coefficients $\{-1, 1, -1\}$. Determine the frequency response and the impulse response of the overall cascade system.

6.7   RUNNING-AVERAGE FILTERING

A simple linear time-invariant system is defined by the equation

$$y[n] = \frac{1}{L}\sum_{k=0}^{L-1} x[n-k] \tag{6.7.1}$$

$$= \frac{1}{L}\left(x[n] + x[n-1] + \ldots + x[n-L+1]\right)$$

This system (6.7.1) is called an *L-point running averager,* because the output at time n is computed as the average of $x[n]$ and the $L-1$ previous samples of the input. The system defined by (6.7.1) can be implemented in MATLAB for $L = 11$ by the statements:

RUNNING
AVG
LOWPASS
FILTERING of
SPEECH

```
bb = ones(11,1)/11;
yy = conv(bb, xx);
```

where xx is a vector containing the samples of the input. The vector bb contains the 11 filter coefficients, which are all the same size, in this case.

The frequency response of the L-point running averager is

$$\mathcal{H}(\hat{\omega}) = \frac{1}{L}\sum_{k=0}^{L-1} e^{-j\hat{\omega}k} \tag{6.7.2}$$

We can derive a simple formula for the magnitude and phase of the averager by making use of the formula for the sum of the first L terms of a geometric series,

$$\sum_{k=0}^{L-1} \alpha^k = \frac{1-\alpha^L}{1-\alpha} \tag{6.7.3}$$

First of all, we identify $e^{-j\hat{\omega}}$ as α, and then do the following steps:

$$\mathcal{H}(\hat{\omega}) = \frac{1}{L}\sum_{k=0}^{L-1} e^{-j\hat{\omega}k} = \frac{1}{L}\left(\frac{1-e^{-j\hat{\omega}L}}{1-e^{-j\hat{\omega}}}\right)$$

$$= \frac{1}{L}\left(\frac{e^{-j\hat{\omega}L/2}(e^{j\hat{\omega}L/2} - e^{-j\hat{\omega}L/2})}{e^{-j\hat{\omega}/2}(e^{j\hat{\omega}/2} - e^{-j\hat{\omega}/2})}\right)$$

$$= \left(\frac{\sin(\hat{\omega}L/2)}{L\sin(\hat{\omega}/2)}\right)e^{-j\hat{\omega}(L-1)/2} \tag{6.7.4}$$

The numerator and denominator are simplified by using the inverse Euler formula for sines. We will find it convenient to express (6.7.4) in the form

$$\mathcal{H}(\hat{\omega}) = \mathcal{D}_L(\hat{\omega})e^{-j\hat{\omega}(L-1)/2} \tag{6.7.5}$$

where

$$\mathcal{D}_L(\hat{\omega}) = \frac{\sin(\hat{\omega}L/2)}{L\sin(\hat{\omega}/2)} \tag{6.7.6}$$

The function $\mathcal{D}_L(\hat{\omega})$ is often called the Dirichlet function, and the subscript L indicates that it comes from an L-point averager. In MATLAB, it can be evaluated with the diric function.

6.7.1 Plotting the Frequency Response

The frequency response of an 11-point running-average filter is given by the equation

$$\mathcal{H}(\hat{\omega}) = \mathcal{D}_{11}(\hat{\omega})e^{-j\hat{\omega}5} \tag{6.7.7}$$

where, in this case, $\mathcal{D}_{11}(\hat{\omega})$ is the Dirichlet function defined by (6.7.6) with $L = 11$, i.e.,

$$\mathcal{D}_{11}(\hat{\omega}) = \frac{\sin(11\hat{\omega}/2)}{11\sin(\hat{\omega}/2)} \tag{6.7.8}$$

As Eq. (6.7.7) makes clear, the frequency-response function, $\mathcal{H}(\hat{\omega})$, can be expressed as the product of the real amplitude function $\mathcal{D}_{11}(\hat{\omega})$ and the complex exponential factor $e^{-j5\hat{\omega}}$. The latter has a magnitude of 1 and a phase angle $-5\hat{\omega}$. Figure 6.8 (top) shows a plot of the amplitude function, $\mathcal{D}_{11}(\hat{\omega})$; the phase function $-5\hat{\omega}$ is in the bottom part of the figure. We use the terminology amplitude rather than magnitude, because $\mathcal{D}_{11}(\hat{\omega})$ can be negative. We can obtain a plot of the magnitude $|\mathcal{H}(\hat{\omega})|$ by taking the absolute value of $\mathcal{D}_{11}(\hat{\omega})$. We shall consider the amplitude representation first, because it is simpler to examine the properties of the amplitude and phase functions. Figure 6.8 shows only one period, i.e., $-\pi < \hat{\omega} < \pi$. The frequency response is, of course, periodic with period 2π, so the plots in Fig. 6.8 would simply repeat with that period.

    In the case of the 11-point running averager, the phase factor is easy to plot, since it is just a straight line with slope of -5. The amplitude factor is somewhat more involved. First note that $\mathcal{D}_{11}(-\hat{\omega}) = \mathcal{D}_{11}(\hat{\omega})$; i.e., $\mathcal{D}_{11}(\hat{\omega})$ is an even function of $\hat{\omega}$ because it is the ratio of two odd functions. Since $\mathcal{D}_{11}(\hat{\omega})$ is even and periodic with period 2π, we need only to consider its values in the interval $0 \le \hat{\omega} \le \pi$. All others can be inferred from symmetry and periodicity. The numerator is $\sin(11\hat{\omega}/2)$, which, of course, oscillates between $+1$ and -1 and is zero whenever $11\hat{\omega}/2 = \pi k$, where k is an integer; solving for $\hat{\omega}$, $\mathcal{D}_{11}(\hat{\omega})$ is zero at frequencies $\hat{\omega}_k = 2\pi k/11$ where k is

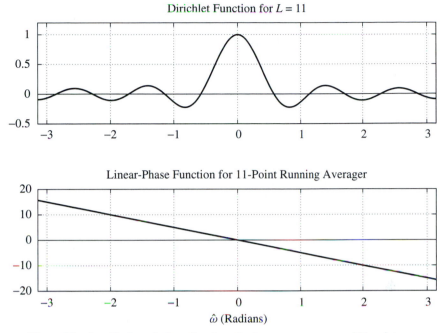

Figure 6.8   Amplitude and phase functions for frequency response of 11-point running-average filter.

a nonzero integer. The denominator of $\mathcal{D}_{11}(\hat{\omega})$ is $11 \sin(\hat{\omega}/2)$, which is zero at $\hat{\omega} = 0$ and increases to a maximum of one at $\hat{\omega} = \pi$. Therefore, $\mathcal{D}_{11}(\hat{\omega})$ is large around $\hat{\omega} = 0$, where the denominator is small, and it oscillates with decreasing amplitude as $\hat{\omega}$ increases to π. The behavior for $\hat{\omega} = 0$ is of particular interest because, at this frequency, $\mathcal{D}_{11}(\hat{\omega})$ is indeterminate, i.e.,

$$\mathcal{D}_{11}(0) \;=\; \frac{0}{0}$$

By l'Hôpital's rule, however, it is easily shown that $\lim\limits_{\hat{\omega}\to 0} \mathcal{D}_{11}(\hat{\omega}) = 1$. Thus, the function $\mathcal{D}_{11}(\hat{\omega})$ has the following properties:

1. $\mathcal{D}_{11}(\hat{\omega})$ is an even function of $\hat{\omega}$ that is periodic with period 2π.

2. $\mathcal{D}_{11}(\hat{\omega})$ has a maximum value of one at $\hat{\omega} = 0$.

3. $\mathcal{D}_{11}(\hat{\omega})$ decays as $\hat{\omega}$ increases, reaching its smallest nonzero amplitude at $\hat{\omega} = \pm\pi$.

4. $\mathcal{D}_{11}(\hat{\omega})$ has zeros at nonzero multiples of $2\pi/11$. (In general, the zeros of $\mathcal{D}_L(\hat{\omega})$ are at nonzero multiples of $2\pi/L$.)

Together, the amplitude and phase plots of Fig. 6.8 completely define the frequency response of the 11-point running-average filter. Normally, however, the frequency response is represented in the form

$$\mathcal{H}(\hat{\omega}) = |\mathcal{H}(\hat{\omega})|e^{j\angle\mathcal{H}(\hat{\omega})}$$

This would require plotting $|\mathcal{H}(\hat{\omega})|$ and $\angle\mathcal{H}(\hat{\omega})$ as functions of $\hat{\omega}$. It is easy to see from Eq. (6.7.7) that $|\mathcal{H}(\hat{\omega})| = |\mathcal{D}_{11}(\hat{\omega})|$. The top part of Fig. 6.9 shows $|\mathcal{H}(\hat{\omega})| = |\mathcal{D}_{11}(\hat{\omega})|$ for the 11-point running-average filter. On the other hand, the phase response, $\angle\mathcal{H}(\hat{\omega})$, is more complicated to plot than the linear function shown in the bottom plot of Fig. 6.8. There are two reasons for this:

1. The algebraic sign of $\mathcal{D}_{11}(\hat{\omega})$ must be represented in the phase function, since $|H(\hat{\omega})| = |\mathcal{D}_{11}(\hat{\omega})|$ discards the sign of $\mathcal{D}_{11}(\hat{\omega})$.

2. It is generally easiest to plot the *principal value* of the phase function.

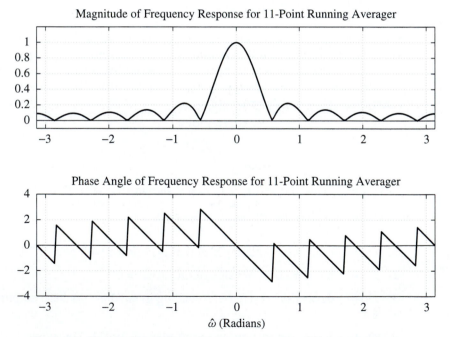

Figure 6.9 Magnitude and phase of frequency response of 11-point running-average filter. Compare to Fig. 6.8.

The sign of $\mathcal{D}_{11}(\hat{\omega})$ can be incorporated into the phase by noting that $\mathcal{D}_{11}(\hat{\omega}) = |\mathcal{D}_{11}(\hat{\omega})|e^{\pm j\pi}$ whenever $\mathcal{D}_{11}(\hat{\omega}) < 0$. The principal value of the angle of a complex number is defined to be the angle between $-\pi$ and $+\pi$ radians. Using the result

$$e^{j(\theta \pm 2\pi k)} = e^{j\theta}e^{\pm j2\pi k} = e^{j\theta}$$

where k is any integer, we see that we can add or subtract integer multiples of 2π from the angle of a complex number without changing the value of the complex number. We can always find a multiple of 2π, which, when added to or subtracted from θ, will produce a result in the range $-\pi < \theta < +\pi$.

This is called *reducing θ modulo 2π*. The principal value is generally what is computed when an arctangent function is evaluated in MATLAB or other computer languages. In Fig. 6.8, we were able to plot an angle whose values were outside the principal value range simply because we had an equation for the angle. Figure 6.9 was plotted using the following MATLAB code (not including labeling):

```
omega = -pi:(pi/500):pi;
bb = ones(1,11)/11;
H = freqz(bb,1,omega);
subplot(2,1,1),    plot(omega,abs(H))
subplot(2,1,2),    plot(omega,angle(H))
```

The MATLAB function angle( ) uses the arctangent to return the principal value of the angle determined by the real and imaginary parts of the elements of the vector H.

In Fig. 6.9, the phase curve is seen to have discontinuities that occur at the zeros of $\mathcal{D}_{11}(\hat{\omega})$. These discontinuities are due to the combination of multiples of π radians added to the phase due to the negative sign of $\mathcal{D}_{11}(\hat{\omega})$ in the intervals $2\pi/11 < |\hat{\omega}| < 4\pi/11$, $6\pi/11 < |\hat{\omega}| < 8\pi/11$, and $10\pi/11 < |\hat{\omega}| < \pi$ and multiples of 2π that are added implicitly in the computation of the principal value. The equation for the phase curve plotted in Fig. 6.9 is as follows for frequencies $0 \le \hat{\omega} \le \pi$:

$$\angle \mathcal{H}(\hat{\omega}) = \begin{cases} -5\hat{\omega} & 0 \le \hat{\omega} < 2\pi/11 \\ -5\hat{\omega} + \pi & 2\pi/11 < \hat{\omega} < 4\pi/11 \\ -5\hat{\omega} + 2\pi & 4\pi/11 < \hat{\omega} < 6\pi/11 \\ -5\hat{\omega} + \pi + 2\pi & 6\pi/11 < \hat{\omega} < 8\pi/11 \\ -5\hat{\omega} + 4\pi & 8\pi/11 < \hat{\omega} < 10\pi/11 \\ -5\hat{\omega} + \pi + 4\pi & 10\pi/11 < \hat{\omega} \le \pi \end{cases}$$

The values for $-\pi < \hat{\omega} < 0$ can be filled in using the fact that $\angle \mathcal{H}(-\hat{\omega}) = -\angle \mathcal{H}(\hat{\omega})$.

Exercise 6.5. Test yourself to see whether you understand why the principal value of $\angle \mathcal{H}(\hat{\omega})$ is as shown in Fig. 6.9 for the 11-point moving averager.

6.7.2 Cascade of Magnitude and Phase

It can be seen from (6.7.4) that $\mathcal{H}(\hat{\omega})$ is the product of two functions, i.e., $\mathcal{H}(\hat{\omega}) = \mathcal{H}_2(\hat{\omega})\mathcal{H}_1(\hat{\omega})$ where

$$\mathcal{H}_1(\hat{\omega}) = e^{-j\hat{\omega}(L-1)/2} \tag{6.7.9}$$

and

$$\mathcal{H}_2(\hat{\omega}) = \mathcal{D}_L(\hat{\omega}) = \frac{\sin(\hat{\omega}L/2)}{L\sin(\hat{\omega}/2)} \tag{6.7.10}$$

which is the Dirichlet function defined earlier. The component $\mathcal{H}_1(\hat{\omega})$ contributes only to the phase of $\mathcal{H}(\hat{\omega})$, and we see that this phase contribution is a linear function of $\hat{\omega}$. Earlier, we saw that a linear phase such as $\angle\mathcal{H}_1(\hat{\omega}) = -\hat{\omega}(L-1)/2$ corresponds to a time delay of $(L-1)/2$ samples. The linear-phase contribution (with slope of $-(L-1)/2 = -5$) is clearly in evidence in the lower part of Fig. 6.8. The frequency response of the second system $\mathcal{H}_2(\hat{\omega})$ is real. It contributes to the magnitude of $\mathcal{H}(\hat{\omega})$, and when it is negative, it also contributes $\pm\pi$ to the phase of $\mathcal{H}(\hat{\omega})$ causing the discontinuities at multiples of $2\pi/11$.

The product representation suggests the block diagram of Fig. 6.10, which shows that the running averager can be thought of as a cascade combination of a delay followed by a "lowpass filter" that accentuates low frequencies relative to high frequencies. The overall moving average system can only be implemented by (6.7.1). However, the block diagram is a useful convenience for thinking about the system. The system cannot be implemented by this cascade because $\mathcal{H}_2(\hat{\omega}) = \mathcal{D}_L(\hat{\omega})$ can never be implemented by itself. When $(L-1)/2$ is a integer, $w[n] = x[n-(L-1)/2]$ is a delay in Fig. 6.10. The case when $(L-1)/2$ is not an integer requires special interpretation, which will be provided in Section 6.8.2. For the present discussion, we will assume that L is an odd integer, so that $(L-1)/2$ is also an integer.

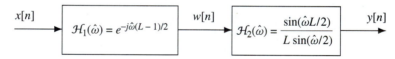

Figure 6.10   Representation of L-point running averager as the cascade of a delay and a real frequency response.

6.7.3 Experiment: Smoothing an Image

As a simple experiment to show the filtering effect of the running-average system, consider the image at the top of Fig. 6.11. The image is a two-dimensional discrete signal that can be represented as a two-dimensional array of samples $x_i[m, n]$. In an image, each sample value is called a *pixel*, which is shorthand for *picture element*. A single horizontal scan line (at $m = 40$) was extracted from the image yielding the one-dimensional signal $x[n] = x_i[40, n]$, plotted at the bottom of Fig. 6.11. The position in the image from which $x[n]$ was extracted is shown by the black line in the image. The values in the image signal are all positive integers in the range $0 \leq x_i[m, n] \leq 255$. These numbers can be represented by 8-bit binary numbers.[3] If you compare the one-dimensional plot to the gray levels in the region around the line, you will see that dark regions in the image have large values (near 255), and bright regions have low values (near zero). This is actually a "negative" image, but that is appropriate since it is a scan of a handwritten homework solution.

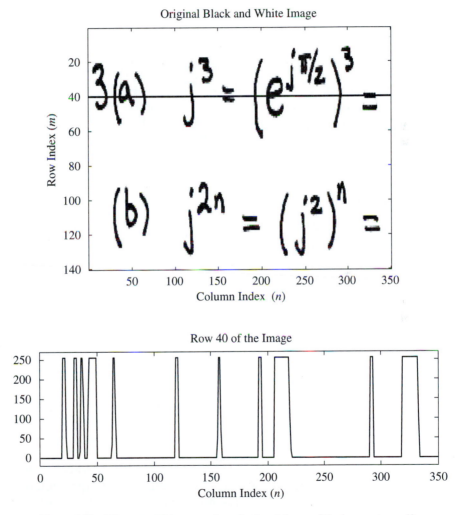

Figure 6.11   "Homework" image and one horizontal scan of the image at row 40.

---

[3] We often say that the total dynamic range of such image values is 8 bits.

An 11-point running averager was applied to $x[n]$, and the input and output were plotted on the same graph (Fig. 6.12). Notice that the output $y[n]$ appears to be a smoother version of $x[n]$ with a slight shift to the right. This shift is the 5-sample delay that we expect for an 11-point running averager. The smoothness is a result of the relative attenuation of the higher frequencies in the signal that correspond to the edges of the handwritten characters in the image. To verify the effect of the delay of the system, Fig. 6.13 shows a plot of $w[n] = x[n-5]$ and $y[n]$. Now we see that the output appears to be aligned with the input.

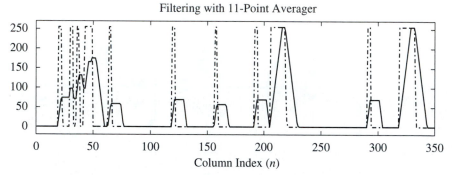

Figure 6.12   Input and output of 11-point running averager. The solid line is the output; the dot-dash is the input.

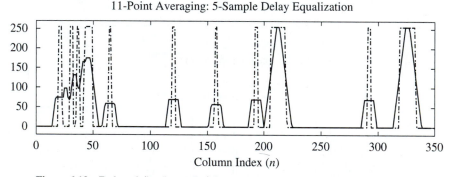

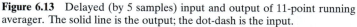

Figure 6.13   Delayed (by 5 samples) input and output of 11-point running averager. The solid line is the output; the dot-dash is the input.

The 11-point averager can be applied first over all the rows and then over all the columns of the image to get a visual assessment of the lowpass filtering operation.[4] Each row is filtered using the one-dimensional averager, then each column of that filtered image is processed. The result is shown in Fig. 6.14, where is it obvious that the lowpass filter has blurred the image. As we have seen, the filter attenuates the

---

[4] In MATLAB, the function `filter2( )` will do this two-dimensional filtering.

high-frequency components of the image. Thus, we can conclude that sharp edges in an image must be associated with high frequencies.

Row and Column Filtered Image

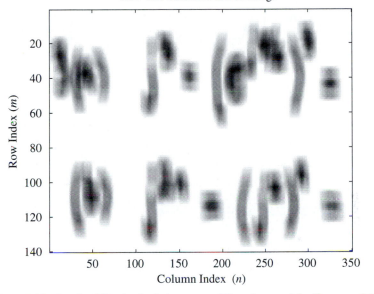

Figure 6.14   Result of filtering both the rows and the columns of the "homework" image with an 11-point running averager. The processed image had to be rescaled so that its values would occupy the entire gray scale range.

As another example of the effect of filtering, the image signal was distorted by adding a cosine signal to create a new input:

$$x_1[n] = x[n] + 100\cos(2\pi n/11)$$

This corrupted signal $x_1[n]$ was filtered with the 11-point running averager. The delayed input $x_1[n-5]$ and the corresponding output are shown in Fig. 6.15. By comparing Figs. 6.13 and 6.15, it is clear that the output is the same for $x[n]$ and $x_1[n]$.[5] The reason is clear: $2\pi n/11$ is one of the frequencies that is completely removed by the averaging filter, because $\mathcal{H}(2\pi/11) = 0$. Since the system is LTI and obeys superposition, the output due to $x_1[n]$ must be the same as the output due to $x[n]$ alone. If the cosine is added to each row of an image, it appears visible as vertical stripes (Fig. 6.16(a)). When each row is processed with an 11-point averaging filter, the cosine will be removed, but the image will be blurred horizontally (Fig. 6.16(b)).

---

[5] A careful comparison reveals that there is a small difference over the transient region $0 \le n \le 9$.

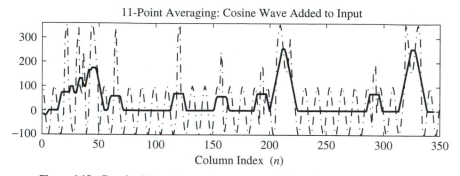

Figure 6.15 Result of 11-point running averager with $\cos(2\pi n/11)$ added to the image scan line. The solid line is the output; the dot-dash is the input.

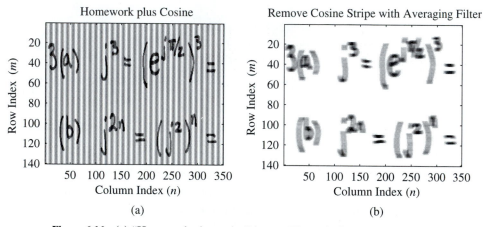

| (a) | (b) |

Figure 6.16 (a) "Homework plus cosine" image. The periodic nature of the cosine across each row causes a vertical striping. (b) After filtering the rows of the "homework plus cosine" image with an 11-point running averager, the processed image is blurred, but has no traces of the cosine stripes. (Both input and output were rescaled for 8-bit image display.)

6.8 FILTERING SAMPLED CONTINUOUS-TIME SIGNALS

Discrete-time filters can be used to filter continuous-time signals that have been sampled. In this section, we study the relationship between the frequency response of the discrete-time filter and the effective frequency response applied to the continuous-time signal. When the input to a discrete-time system is a sequence derived by sampling a continuous-time signal, we can use our knowledge of sampling and reconstruction to interpret the effect of the filter on the original continuous-time signal.

Consider the system depicted in Fig. 6.17, and assume that the input is the complex sinusoid

$$x(t) = Xe^{j\omega t}$$

with $X = Ae^{j\phi}$. After sampling, the input sequence to the discrete-time filter is

$$x[n] = x(nT_s) = Xe^{j\omega nT_s} = Xe^{j\hat{\omega}n}$$

The relationship between the discrete-time frequency $\hat{\omega}$ and the continuous-time frequency ω is

$$\hat{\omega} = \omega T_s$$

If the frequency of the continuous-time signal satisfies the condition of the sampling theorem, i.e., $|\omega| < \pi/T_s$, then there will be no aliasing, and the normalized discrete-time frequency is such that $|\hat{\omega}| < \pi$.

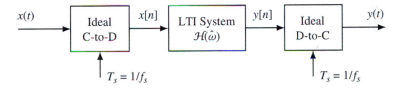

Figure 6.17   System for doing discrete-time filtering of continuous-time signals.

The frequency response of the discrete-time system gives us a quick way to calculate the output $y[n]$ in Fig 6.17.

$$y[n] = \mathcal{H}(\hat{\omega})Xe^{j\hat{\omega}n}$$

If we now make the substitution $\hat{\omega} = \omega T_s$, then we can write $y[n]$ in terms of the analog frequency ω.

$$y[n] = \mathcal{H}(\omega T_s)Xe^{j\omega T_s n}$$

Finally, since no aliasing occurred in the original sampling, the ideal D-to-C converter will reconstruct the original frequency, giving

$$y(t) = \mathcal{H}(\omega T_s)Xe^{j\omega t}$$

Remember that this formula for $y(t)$ is good only for frequencies such that $-\pi/T_s < \omega < \pi/T_s$, and recall that the ideal D-to-C converter reconstructs all digital-frequency components in the band $|\hat{\omega}| < \pi$ as analog frequencies in the band $|\omega| < \pi/T_s$. As long as there is no aliasing, the frequency band of the input signal $x(t)$ matches the frequency band of the output $y(t)$. Thus, the overall system of Fig. 6.17 behaves as though it is an LTI continuous-time system whose frequency response is $\mathcal{H}(\omega T_s)$.

It is very important to understand this analysis of the system of Fig. 6.17. We have just shown that the system of Fig. 6.17 can be used to implement LTI filtering operations on continuous-time signals. Furthermore, it is clear from this analysis that the block diagram of Fig. 6.17 represents an infinite number of systems. This is true in two ways. First, for any given discrete-time system, we can change the sampling period T_s and obtain a new system. Alternatively, if we fix the sampling period, we can change the discrete-time system to vary the overall response. In a specific case, all we have to do is select the sampling rate to avoid aliasing, and then design a discrete-time LTI filter whose frequency response $\mathcal{H}(\omega T_s)$ has the desired frequency-selective properties.

Exercise 6.6. In general, when we use the system of Fig. 6.17 to filter continuous-time signals, we would want to choose the sampling frequency $f_s = 1/T_s$ to be as low as possible. Why?

6.8.1 Example: Lowpass Averager

As an example, we use the 11-point moving averager

$$y[n] = \frac{1}{11} \sum_{k=0}^{10} x[n-k]$$

as the filter in Fig. 6.17. The frequency response of this discrete-time system is

$$\mathcal{H}(\hat{\omega}) = \frac{\sin(\hat{\omega}11/2)}{11\sin(\hat{\omega}/2)} e^{-j\hat{\omega}5}$$

The magnitude of this frequency response is shown in the top part of Fig. 6.18.

When this system is used as the discrete-time system in Fig. 6.17 with sampling frequency $f_s = 1000$, we want to answer two questions: What is the equivalent analog frequency response, and how would the signal

$$x(t) = \cos(2\pi(25)t) + \sin(2\pi(250)t)$$

be processed by the system?

The frequency-response question is easy. The equivalent analog-frequency response is

$$\mathcal{H}(\omega T_s) = \mathcal{H}(\omega/1000) = \mathcal{H}(2\pi f/1000)$$

where f is in Hz. A plot of the equivalent continuous-time frequency response versus f is shown in the bottom part of Fig. 6.18. Note that the frequency response of the overall system stops abruptly at $|f| = f_s/2 = 500$ Hz, since the ideal D-to-C converter does not reconstruct frequencies above $f_s/2$.

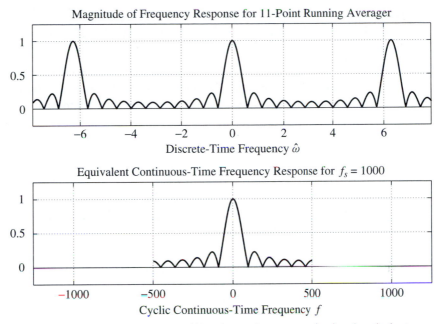

Figure 6.18   Frequency response of 11-point moving averager (top) and equivalent analog frequency response (bottom) when used to filter analog signals. The sampling frequency is $f_s = 1000$ Hz, so the maximum analog frequency that can be processed is 500 Hz.

The second question is also easy if we track the two frequencies of the input signal through the three systems of Fig. 6.17. The input $x(t)$ contains two frequencies at $\omega = 2\pi(25)$ and $\omega = 2\pi(250)$. Since $f_s = 1000 > 2(250) > 2(25)$, there is no aliasing,[6] so the same frequency components appear in the output signal $y(t)$. The magnitude and phase changes are found by evaluating the equivalent frequency response at 25 and 250 Hz.

$$\mathcal{H}(2\pi(25)/1000) = \frac{\sin(\pi(25)(11)/1000)}{11\sin(\pi(25)/1000)} e^{-j2\pi(25)(5)/1000} = 0.8811 e^{-j\pi/4}$$

$$\mathcal{H}(2\pi(250)/1000) = \frac{\sin(\pi(250)(11)/1000)}{11\sin(\pi(250)/1000)} e^{-j2\pi(250)(5)/1000} = 0.0909 e^{-j\pi/2}$$

These values can be checked against the plots in Fig. 6.18. Thus the final output is

$$y(t) = 0.8811 \cos(2\pi(25)t - \pi/4) + 0.0909 \sin(2\pi(250)t - \pi/2)$$

---

[6] When there is aliasing, this sort of problem is less straightforward because the output signal $y(t)$ will have different frequency components from the input.

The lowpass nature of the filter has greatly attenuated the 250-Hz component, while the 25-Hz component suffered only slight attenuation because it lies in the passband of the filter near 0 Hz.

> **Exercise 6.7.** Assuming the same input signal and the same discrete-time system, work the example of this section again, but use a sampling rate of $f_s = 500$ Hz.

6.8.2 Interpretation of Delay

We have seen that a frequency response of the form $\mathcal{H}(\hat{\omega}) = e^{-j\hat{\omega}n_0}$ implies a time delay of n_0 samples. For n_0, an integer, the interpretation of this is straightforward. If the input to the system is $x[n]$, the corresponding output is $y[n] = x[n - n_0]$. However, if n_0 is not an integer, the interpretation is less obvious. An example of where this can occur is the L-point running average system whose frequency response is

$$\mathcal{H}(\hat{\omega}) = \mathcal{D}_L(\hat{\omega})e^{-j\hat{\omega}(L-1)/2}$$

where $\mathcal{D}_L(\hat{\omega})$ is the real function

$$\mathcal{D}_L(\hat{\omega}) = \frac{\sin(\hat{\omega}L/2)}{L\sin(\hat{\omega}/2)}$$

Thus, the L-point running averager includes a delay of $\frac{1}{2}(L-1)$ samples. If L is an odd integer, this delay causes the output to be shifted $\frac{1}{2}(L-1)$ samples with respect to the input. However, if L is an even integer, then $\frac{1}{2}(L-1)$ is not an integer. The analyis of this section provides a useful interpretation of this delay factor.

Suppose that the input to the ideal C-to-D converter is

$$x(t) = Xe^{j\omega t}$$

and that there is no aliasing, so that the sampled input to the L-point running averager is

$$x[n] = Xe^{j\omega nT_s} = Xe^{j\hat{\omega}n}$$

where $\hat{\omega} = \omega T_s$. Now, the output of the L-point running average filter is

$$y[n] = \mathcal{H}(\hat{\omega})Xe^{j\hat{\omega}n} = \mathcal{D}_L(\hat{\omega})e^{-j\hat{\omega}(L-1)/2}Xe^{j\hat{\omega}n}$$

Finally, if $\omega < \pi/T_s$ (i.e., no aliasing occurred in the sampling operation), then the ideal D-to-C converter will reconstruct the complex exponential signal

$$y(t) = \mathcal{D}_L(\hat{\omega})Xe^{-j\hat{\omega}(L-1)/2}e^{j\omega t}$$
$$= \mathcal{D}_L(\omega T_s)Xe^{-j\omega T_s(L-1)/2}e^{j\omega t}$$
$$= \mathcal{D}_L(\omega T_s)Xe^{j\omega(t-T_s(L-1)/2)}$$

Thus, regardless of whether or not $\frac{1}{2}(L-1)$ is an integer, the delay factor $e^{-j\hat{\omega}(L-1)/2}$ corresponds to a delay of $\frac{1}{2}T_s(L-1)$ seconds with respect to continuous-time signals sampled with sampling period T_s.

Example 6.11   To illustrate the effect of non-integer delay with the running average filter, consider the cosine signal $x[n] = \cos(0.2\pi n)$, which could have resulted in Fig 6.17 from sampling the signal $x(t) = \cos(200\pi t)$ with sampling rate $f_s = 1000$ Hz. The top part of Fig. 6.19 shows $x(t)$ and $x[n]$. If $x[n]$ is the input to a 5-point running average filter, the steady-state part of the output is

$$y_5[n] = \frac{\sin(0.2\pi(5/2))}{5\sin(0.2\pi/2)}\cos(0.2\pi n - 0.2\pi(2)) = 0.6472\cos(0.2\pi(n-2))$$

For this filter output, the output of the D-to-C converter in Fig. 6.17 would be

$$y_5(t) = 0.6472\cos(200\pi(t-0.002))$$

The delay is 2 samples. On the other hand, if the same signal $x[n]$ is the input to a 4-point running average system, the steady-state part of the output is

$$y_4[n] = \frac{\sin(0.2\pi(4/2))}{4\sin(0.2\pi/2)}\cos(0.2\pi n - 0.2\pi(3/2)) = 0.7694\cos(0.2\pi(n - \tfrac{3}{2}))$$

Now the delay is 3/2 samples, so we cannot write $y_4[n]$ as an integer shift with respect to the input sequence. In this case, the "3/2 samples" delay introduced by the filter can be interpreted in terms of the corresponding output of the D-to-C converter in Fig. 6.17, which in this case would be

$$y_4(t) = 0.7694\cos(200\pi(t-0.0015)).$$

Figure 6.19 shows the input (top) and the corresponding outputs $y_5[n]$ and $y_5(t)$ (middle) and $y_4[n]$ and $y_4(t)$ (bottom). In all cases, the solid curve is the continuous-time cosine signal that would be reconstructed by the ideal D-to-C converter for the given discrete-time signal. The following specific points are made by this example:

1. Both outputs are smaller than the input. This is because $\mathcal{D}_L(0.2\pi) < 1$ for both cases.

2. The dashed vertical lines in the lower two panels show the peaks of the output cosine signals that correspond to the peak of the input at $n = 0$. Note that the delay is $(5-1)/2 = 2$ for the 5-point averager and $(4-1)/2 = 3/2$ for the 4-point averager.

3. The effect of the fractional delay is to implement an interpolation of the cosine signal at points halfway between the original samples.

$\Diamond$

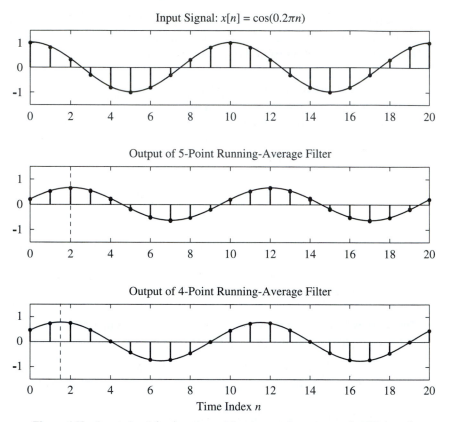

Figure 6.19   Input signal (top), output of 5-point running averager (middle), and output of 4-point running averager (bottom).

6.9   SUMMARY AND LINKS

This chapter introduced the concept of the frequency response for the class of FIR filters. The frequency response applies to any linear time-invariant system, as we will see in upcoming chapters.

LINKS

This chapter extends the discussion of Chapter 5, which introduced the basics of FIR filtering. The labs, in particular, require the student to be familiar with both chapters. Among the laboratory projects in Appendix C, we have provided four on the CD-ROM. Lab C.5 deals with FIR filtering of sinusoidal waveforms. The sinusoidal response leads naturally to a consideration of the frequency response. Lab C.6 deals with common filters such as the first-difference and the L-point averager. It also has exercises of the equivalence of different cascade forms. In Labs C.7 and C.8, FIR filters are used in practical systems such as a touch-tone decoder and image-smoothing, respectively. The labs are also found on the CD-ROM.

The CD-ROM also contains the following demonstrations of lowpass and high-pass filtering:

1. Filtering photographic images to show that lowpass filtering is blurring, and that highpass filtering enhances images.
2. Cascade processing of images to show that a highpass filtering can undo the blurring effects of a lowpass filter.
3. Filtering of sound signals to illustrate bass and treble emphasis.

Finally, the reader is reminded of the large number of solved homework problems available for review and practice on the CD-ROM.

PROBLEMS

6.1 Let the input signal to an FIR system be $x[n]$, where

$$x[n] = e^{j(0.4\pi n - 0.5\pi)}$$

If we define a new signal $y[n]$ to be the first difference $y[n] = x[n] - x[n-1]$, it is possible to express $y[n]$ in the form

$$y[n] = Ae^{j(\hat{\omega}_0 n + \phi)}$$

Determine the numerical values of A, ϕ, and $\hat{\omega}_0$.

6.2 Suppose that a discrete-time system is described by the input-output relation

$$y[n] = (x[n])^2$$

(a) Determine the output when the input is the complex exponential signal

$$x[n] = Ae^{j\phi}e^{j\hat{\omega}n}$$

(b) Is the output of the form

$$y[n] = \mathcal{H}(\hat{\omega})Ae^{j\phi}e^{j\hat{\omega}n}$$

If not, why not?

6.3 Suppose that a discrete-time system is described by the input-output relation

$$y[n] = x[-n]$$

(a) Determine the output when the input is the complex exponential signal

$$x[n] = Ae^{j\phi}e^{j\hat{\omega}n}$$

(b) Is the output of the form

$$y[n] = \mathcal{H}(\hat{\omega})Ae^{j\phi}e^{j\hat{\omega}n}$$

If not, why not?

6.4 A linear time-invariant system is described by the difference equation

$$y[n] = 2x[n] - 3x[n-1] + 2x[n-2]$$

(a) Find the frequency response $\mathcal{H}(\hat{\omega})$; then express it as a mathematical formula, in polar form (magnitude and phase).

(b) $\mathcal{H}(\hat{\omega})$ is a periodic function of $\hat{\omega}$; determine the period.

(c) Plot the magnitude and phase of $\mathcal{H}(\hat{\omega})$ as a function of $\hat{\omega}$ for $-\pi < \hat{\omega} < 3\pi$. Do this by hand, and then check with the MATLAB function freqz.

(d) Find all frequencies $\hat{\omega}$, for which the output response to the input $e^{j\hat{\omega}n}$ is zero.

(e) When the input to the system is $x[n] = \sin(\pi n/13)$, determine the output signal and express it in the form $y[n] = A\cos(\hat{\omega}_0 n + \phi)$.

6.5 A linear time-invariant filter is described by the difference equation

$$y[n] = x[n] + 2x[n-1] + x[n-2]$$

(a) Obtain an expression for the frequency response of this system.

(b) Sketch the frequency response (magnitude and phase) as a function of frequency.

(c) Determine the output when the input is $x[n] = 10 + 4\cos(0.5\pi n + \pi/4)$.

(d) Determine the output when the input is the unit-impulse sequence

$$x[n] = \delta[n] = \begin{cases} 1 & n = 0 \\ 0 & n \neq 0 \end{cases}$$

(e) Determine the output when the input is the unit-step sequence

$$x[n] = u[n] = \begin{cases} 0 & n < 0 \\ 1 & n \geq 0 \end{cases}$$

6.6 A linear time-invariant filter is described by the difference equation

$$y[n] = x[n] - x[n-2]$$

(a) Obtain an expression for the frequency response of this system.

(b) Sketch the frequency response (magnitude and phase) as a function of frequency.

(c) What is the output if the input is $x[n] = 4 + \cos(0.25\pi n - \pi/4)$?

(d) If the input is $x_1[n] = (4 + \cos(0.25\pi n - \pi/4))\, u[n]$, for what values of n will the corresponding output be equal to the output obtained in (c)?

6.7 The frequency response of a linear time-invariant filter is given by the formula

$$\mathcal{H}(\hat{\omega}) = (1 + e^{-j\hat{\omega}})(1 - e^{j2\pi/3}e^{-j\hat{\omega}})(1 - e^{-j2\pi/3}e^{-j\hat{\omega}})$$

(a) Write the difference equation that gives the relation between the input $x[n]$ and the output $y[n]$.

(b) What is the output if the input is $x[n] = \delta[n]$?

(c) If the input is of the form $x[n] = Ae^{j\phi}e^{j\hat{\omega}n}$, for what values of $-\pi \leq \hat{\omega} \leq \pi$ will $y[n] = 0$ for all n?

6.8 The frequency response of a linear time-invariant filter is given by the formula

$$H(\hat{\omega}) = (1 - e^{-j\hat{\omega}})(1 - 0.5e^{j\pi/6}e^{-j\hat{\omega}})(1 - 0.5e^{-j\pi/6}e^{-j\hat{\omega}})$$

(a) Write the difference equation that gives the relation between the input $x[n]$ and the output $y[n]$.

(b) What is the output if the input is $x[n] = \delta[n]$?

(c) If the input is of the form $x[n] = Ae^{j\phi}e^{j\hat{\omega}n}$, for what values of $-\pi \le \hat{\omega} \le \pi$ will $y[n] = 0$ for all n?

6.9 Suppose that S is a linear time-invariant system whose exact form is unknown. It is to be tested by observing the output signals corresponding to several different test inputs. Suppose that the following input-output pairs are the result of the tests:

$$x[n] = \delta[n] \longmapsto y[n] = \delta[n] - \delta[n-3]$$

$$x[n] = \cos(2\pi n/3) \longmapsto y[n] = 0$$

$$x[n] = \cos(\pi n/3 + \pi/2) \longmapsto y[n] = 2\cos(\pi n/3 + \pi/2)$$

(a) Make a plot of the signal $x[n] = 3\delta[n] - 2\delta[n-2] + \delta[n-3]$.

(b) What is the output of the system when the input is $x[n] = 3\delta[n] - 2\delta[n-2] + \delta[n-3]$?

(c) Determine the output when the input is $x[n] = \cos(\pi(n-3)/3)$.

(d) Is the following statement true or false? Explain.

$$H(\pi/2) = 0$$

6.10 The *Dirichlet* function is defined by

$$\mathcal{D}_L(\hat{\omega}) = \frac{\sin(L\hat{\omega}/2)}{L\sin(\hat{\omega}/2)}$$

For the case $L = 8$:

(a) Make a plot of $\mathcal{D}_L(\hat{\omega})$ over the range $-3\pi \le \hat{\omega} \le +3\pi$. Label all the zero crossings.

(b) Determine the period of $\mathcal{D}_8(\hat{\omega})$.

(c) Find the maximum value of the function.

6.11 A linear time-invariant filter is described by the difference equation

$$y[n] = x[n] - 3x[n-1] + 3x[n-2] - x[n-3]$$

(a) Obtain an expression for the frequency response of this system, and, using the fact that $(1-a)^3 = 1 - 3a + 3a^2 - a^3$, show that $H(\hat{\omega})$ can be expressed in the form

$$H(\hat{\omega}) = 8\sin^3(\hat{\omega}/2)e^{j(-\pi/2 - 3\hat{\omega}/2)}$$

(b) Sketch the frequency response (magnitude and phase) as a function of frequency.

(c) What is the output if the input is $x[n] = 10 + 4\cos(0.5\pi n + \pi/4)$?

(d) What is the output if the input is $x[n] = \delta[n]$?

(e) Use superposition to find the output when $x[n] = 10 + 4\cos(0.5\pi n + \pi/4) + 5\delta[n-3]$.

6.12 Suppose that three systems are connected in cascade; i.e., the output of S_1 is the input to S_2, and the output of S_2 is the input to S_3. The three systems are specified as follows:

$$S_1: \quad y_1[n] = x_1[n] - x_1[n-1]$$

$$S_2: \quad y_2[n] = x_2[n] + x_2[n-2]$$

$$S_3: \quad y_3[n] = x_3[n-1] + x_3[n-2]$$

where the output of S_i is $y_i[n]$ and its input is $x_i[n]$.

(a) Determine the equivalent system that is a single operation from the input $x[n]$ (into S_1) to the output $y[n]$, which is the output of S_3. Thus, $x[n]$ is $x_1[n]$ and $y[n]$ is $y_3[n]$.

(b) Use the frequency response to write one difference equation that defines the overall system in terms of $x[n]$ and $y[n]$ only.

6.13 An LTI filter is described by the difference equation

$$y[n] = \frac{1}{4}\left(x[n] + x[n-1] + x[n-2] + x[n-3]\right) = \frac{1}{4}\sum_{k=0}^{3} x[n-k]$$

(a) What is the impulse response $h[n]$ of this system?

(b) Obtain an expression for the frequency response of this system.

(c) Sketch the frequency response (magnitude and phase) as a function of frequency.

(d) Suppose that the input is

$$x[n] = 5 + 4\cos(0.2\pi n) + 3\cos(0.5\pi n + \pi/4) \qquad -\infty < n < \infty$$

Obtain an expression for the output in the form $y[n] = A + B\cos(\hat{\omega}_0 n + \phi_0)$.

(e) Suppose that the input is

$$x_1[n] = \left(5 + 4\cos(0.2\pi n) + 3\cos(0.5\pi n + \pi/4)\right)u[n]$$

where $u[n]$ is the unit-step sequence. For what values of n will the output $y_1[n]$ be equal to the output $y[n]$ in (d)?

6.14 A system for filtering continuous-time signals is shown in the following figure

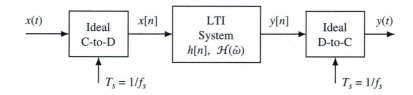

The input to the C-to-D converter in this system is

$$x(t) = 10 + 8\cos(200\pi t) + 6\cos(500\pi t + \pi/4) \qquad -\infty < t < \infty$$

The impulse response of the LTI system is

$$h[n] = \frac{1}{4}\sum_{k=0}^{3}\delta[n-k]$$

If $f_s = 1000$ samples/sec, determine an expression for $y(t)$, the output of the D-to-C converter. *Hint:* The results of Problem 6.13 can be applied here.

6.15 This diagram depicts a cascade connection of two linear time-invariant systems, where the first system is a 3-point moving averager and the second system is a first difference.

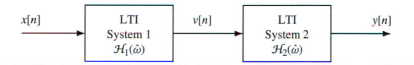

(a) If the input is of the form $x[n] = 10 + x_1[n]$, the output, $y[n]$, of the overall system will be of the form $y[n] = y_1[n]$, where $y_1[n]$ is the output due only to $x_1[n]$. Explain why this is true.

(b) Determine the frequency-response function of the overall cascade system.

(c) Sketch the frequency response (magnitude and phase) functions of the individual systems and the overall cascade system for $-\pi \le \hat{\omega} \le \pi$.

(d) Obtain a single difference equation that relates $y[n]$ to $x[n]$ for the overall cascade system.

6.16 A linear time-invariant system is described by the difference equation

$$y[n] = -x[n] + 2x[n-2] - x[n-4]$$

(a) Find the impulse response $h[n]$ and plot it.

(b) Determine an equation for the frequency response $\mathcal{H}(\hat{\omega})$ and express it in the form

$$\mathcal{H}(\hat{\omega}) = \mathcal{R}(\hat{\omega})e^{-j\hat{\omega}n_0}$$

where $\mathcal{R}(\hat{\omega})$ is a real function and n_0 is an integer.

(c) Carefully sketch and label a plot of $|\mathcal{H}(\hat{\omega})|$ for $-\pi < \hat{\omega} < \pi$.

(d) Carefully sketch and label a plot of the principal value of the $\angle\mathcal{H}(\hat{\omega})$ for $-\pi < \hat{\omega} < \pi$.

6.17 A linear time-invariant system has frequency response

$$\mathcal{H}(\hat{\omega}) = (1 - e^{j\pi/2}e^{-j\hat{\omega}})(1 - e^{-j\pi/2}e^{-j\hat{\omega}})(1 + e^{-j\hat{\omega}})$$

The input to the system is

$$x[n] = 5 + 20\cos\left(\frac{\pi}{2}n + \frac{\pi}{4}\right) + 10\delta[n - 3]$$

Use superposition to determine the corresponding output of the LTI system $y[n]$ for $-\infty < n < \infty$.

6.18 Consider this cascade system

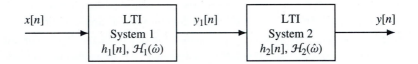

where

$$\mathcal{H}_1(\hat{\omega}) = 1 + 2e^{-j\hat{\omega}} + e^{-j\hat{\omega}2} \text{ and}$$

$$h_2[n] = \delta[n] - \delta[n - 1] + \delta[n - 2] - \delta[n - 3]$$

(a) Determine the frequency response $\mathcal{H}(\hat{\omega})$ for the overall cascade system (i.e., from input $x[n]$ to output $y[n]$). Simplify your answer as much as possible.

(b) Determine and plot the impulse response $h[n]$ of the overall cascade system.

(c) Write down the difference equation that relates $y[n]$ to $x[n]$.

6.19 The input to the C-to-D converter in this system is

$$x(t) = 10 + 20\cos(\omega_0 t + \pi/3) \qquad -\infty < t < \infty$$

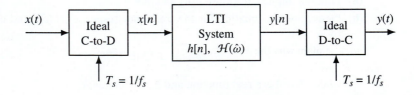

(a) Suppose that the impulse response of the LTI system is $h[n] = \delta[n]$. If $\omega_0 = 2\pi(500)$, for what values of $f_s = 1/T_s$ will it be true that $y(t) = x(t)$?

(b) Now suppose that the impulse response of the LTI system is changed to $h[n] = \delta[n - 10]$. Determine the sampling rate $f_s = 1/T_s$ and a range of values for ω_0 so that the output of the overall system is

$$y(t) = x(t - 0.001) = 10 + 20\cos(\omega_0(t - 0.001) + \pi/3) \qquad -\infty < t < \infty.$$

(c) Suppose that the LTI system is a 5-point moving averager whose frequency response is

$$\mathcal{H}(\hat{\omega}) = \frac{\sin(5\hat{\omega}/2)}{5\sin(\hat{\omega}/2)}e^{-j\hat{\omega}2}$$

If the sampling rate is $f_s = 2000$ samples/sec, determine all values of ω_0 such that the output is equal to a constant, i.e., $y(t) = A$ for $-\infty < t < \infty$. Also, determine the constant A in this case.

6.20 The frequency response of an LTI system is plotted as shown.

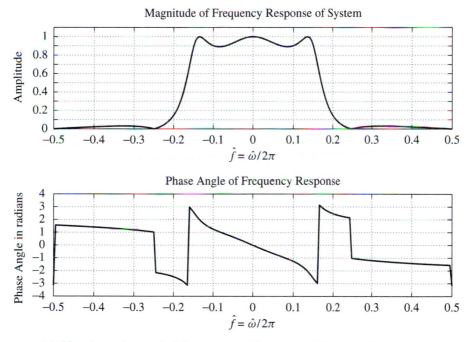

(a) Use these plots to find the output of the system if the input is

$$x[n] = 10 + 10\cos(0.2\pi n) + 10\cos(0.5\pi n)$$

(b) Explain why the phase-response curve is discontinuous at frequencies around $\hat{f} = \hat{\omega}/2\pi = 0.17$ and also around $\hat{f} = 0.25$.

7

z-Transforms

In this chapter we introduce the z-transform, which brings polynomials and rational functions into the analysis of linear discrete-time systems. We will show that convolution is equivalent to polynomial multiplication and that common algebraic operations, such as multiplying, dividing, and factoring polynomials, can be interpreted as combining or decomposing LTI systems. The most common z-transforms are rational functions, i.e., a numerator polynomial divided by a denominator polynomial. The roots of these polynomials are important, because most properties of digital filters can be restated in terms of the locations of these roots.

The z-transform method is introduced in this chapter for FIR filters and finite-length sequences in general. We will use the FIR case to introduce the important concept of "domains of representation" for discrete-time signals and systems. In this text, we consider three domains of representation of signals and systems: the *n-domain* or *time domain* (the domain of sequences, impulse responses and difference equations), the *$\hat{\omega}$-domain* or *frequency domain* (the domain of frequency responses and spectrum representations), and the *z-domain* (the domain of z-transforms, operators, and poles and zeros).[1] The value of having three different domains of representation is that a difficult analysis in one domain is often much easier in one of the other domains. Therefore, increased understanding will result from developing skills for moving from one representation to another. For example, the cascade combination of LTI systems, which in the n-domain seems to require the new (less familiar)

---

[1] Traditionally, signals and systems texts have identified just two domains: the time domain and the frequency domain. Many authors consider our $\hat{\omega}$-domain and z-domain together as the "frequency domain." This is mainly because, as we will see, the $\hat{\omega}$-domain can be viewed as a special case of the more general z-domain. However, we feel that because of the distinctly different character of the mathematical functions involved in the two domains, there is a distinct advantage in considering the $\hat{\omega}$- and z-domains as separate, but related, points of view.

technique of convolution, is converted in the z-domain into the more familiar algebraic operation of polynomial multiplication. It is important, however, to note that the "real" or "actual" domain is the n-domain where the signals are generated and processed, and where the implementation of filters takes place. The frequency domain has physical significance when analyzing sound, but is seldom used for implementation. The z-domain exists primarily for its convenience in mathematical analysis and synthesis.

7.1   DEFINITION OF THE z-TRANSFORM

A finite-length signal $x[n]$ can be represented by the relation

$$x[n] = \sum_{k=0}^{N} x[k]\delta[n - k] \qquad (7.1.1)$$

and the z-transform of such a signal is defined by the formula

$$X(z) = \sum_{k=0}^{N} x[k]z^{-k} \qquad (7.1.2)$$

where we will assume that z represents any complex number; i.e., z is the independent (complex) variable of the z-transform $X(z)$. Although (7.1.2) is the conventional definition of the z-transform,[2] it is instructive to note that $X(z)$ can be written in the form

$$X(z) = \sum_{k=0}^{N} x[k](z^{-1})^k$$

which emphasizes the fact that $X(z)$ is simply a polynomial of degree N in the variable z^{-1}.

When we use (7.1.2) to determine the z-transform of the signal $x[n]$, we *transform* $x[n]$ into a new representation $X(z)$. Indeed, it is often said that we "take the z-transform of $x[n]$." All that we have to do to obtain $X(z)$ is to construct a polynomial whose coefficients are the values of the sequence $x[n]$. Specifically, the kth sequence value is the coefficient of the kth power of z^{-1} in the polynomial $X(z)$. Obviously, it is just as easy to go from (7.1.2) back to (7.1.1). We can recover $x[n]$ from $X(z)$ simply by extracting the coefficient of the kth power of z^{-1} and placing that coefficient in the kth position in the sequence $x[n]$. This operation is sometimes called "taking the inverse z-transform." In order to emphasize this unique correspondence between a sequence $x[n]$ and its z-transform, we will use the notation

---

[2] Some authors use positive powers of z in the definition of the z-transform, but this convention is not common in signal processing.

$$n\text{-}Domain \qquad \Longleftrightarrow \qquad z\text{-}Domain$$

$$x[n] = \sum_{k=0}^{N} x[k]\delta[n - k] \qquad \Longleftrightarrow \qquad X(z) = \sum_{k=0}^{N} x[k]z^{-k} \quad (7.1.3)$$

In general, a "z-transform pair" is a sequence and its corresponding z-transform, which we will denote as

$$x[n] \Longleftrightarrow X(z)$$

Notice that n is the independent variable of the sequence $x[n]$. Thus, we say that (7.1.1) represents the signal in the *n-domain*. Since n is often an index that counts time in a sampled time waveform, we also refer to (7.1.1) as the *time–domain* representation of the signal. Similarly, note that z is the independent variable of the z-transform $X(z)$. Thus, we say that (7.1.2) represents the signal in the *z-domain*, and in "taking the z-transform" of a signal, we move from the time domain to the z-domain.

As a simple, but very important, example of a z-transform pair, suppose that $x[n] = \delta[n - n_0]$. Then, from the definition, (7.1.2), it follows that $X(z) = z^{-n_0}$. To emphasize this correspondence we use the notation

$$n\text{-}Domain \qquad \Longleftrightarrow \qquad z\text{-}Domain$$

$$x[n] = \delta[n - n_0] \qquad \Longleftrightarrow \qquad X(z) = z^{-n_0} \qquad (7.1.4)$$

When the sequence is defined with numerical values, we can take the z-transform and get a polynomial.

Example 7.1   Consider the sequence $x[n]$ given in the following table:

n	$n < -1$	-1	0	1	2	3	4	5	$n > 5$
$x[n]$	0	0	2	4	6	4	2	0	0

The z-transform of this sequence is

$$X(z) = 2 + 4z^{-1} + 6z^{-2} + 4z^{-3} + 2z^{-4}$$

$\diamond$

This example shows how to find the z-transform given the sequence. The following example illustrates the inverse z-transform operation, i.e., finding the sequence if we are given its z-transform.

Example 7.2   Consider the z-transform $X(z)$ given by the equation

$$X(z) = 1 - 2z^{-1} + 3z^{-3} - z^{-5}$$

We can give $x[n]$ in tabular form as in Example 7.1, or we can give an equation for the sequence values as a function of n in the form

$$x[n] = \begin{cases} 0 & n < 0 \\ 1 & n = 0 \\ -2 & n = 1 \\ 0 & n = 2 \\ 3 & n = 3 \\ 0 & n = 4 \\ -1 & n = 5 \\ 0 & n > 5 \end{cases}$$

Alternatively, using the representation (7.1.1) in terms of impulse sequences, the corresponding sequence $x[n]$ is

$$x[n] = \delta[n] - 2\delta[n-1] + 3\delta[n-3] - \delta[n-5]$$

$\diamond$

At this point, we have a definition of the z-transform, and we have seen how to find it for a given sequence and how to find the sequence given the z-transform, but why would we want to transform from the n-domain to the z-domain? This is the obvious question to ask at this point, and the remainder of this chapter will attempt to answer it.

7.2   THE z-TRANSFORM AND LINEAR SYSTEMS

z-transforms are indispensable in the design and analysis of LTI systems. The fundamental reason for this has to do with the way that LTI systems respond to the particular input signal z^n for $-\infty < n < \infty$.

7.2.1 The z-Transform of an FIR Filter

Recall that the general difference equation of an FIR filter is

$$y[n] = \sum_{k=0}^{M} b_k\, x[n-k] \tag{7.2.1}$$

An alternative representation of the input-output relation is the convolution sum

$$y[n] = x[n] * h[n]$$

where $h[n]$ is the impulse response of the FIR filter. Remember that the impulse response $h[n]$ is identical to the sequence of difference equation coefficients b_n, as shown in the following table:

n	$n < 0$	0	1	2	$\ldots$	M	$n > M$
$h[n]$	0	b_0	b_1	b_2	$\ldots$	b_M	0

which can be represented in the more compact notation,

$$h[n] = \sum_{k=0}^{M} b_k \, \delta[n - k] \tag{7.2.2}$$

To see why the z-transform is of interest to us for FIR filters, let the input to system in (7.2.1) be the signal

$$x[n] = z^n \qquad \text{for all } n$$

where z is any complex number. Recall that we have already considered such inputs in Chapter 6, where we used $z = e^{j\hat{\omega}}$. As with our discussion of the frequency response, the qualification "for all n" is an extremely important detail. Because we want to avoid any consideration of what might happen at a starting point such as $n = 0$, we think of this as having the input start at $n = -\infty$, and we assume that for finite values of n, the start-up effects have disappeared, i.e., we are concerned only with the "steady state" part of the output. For the more general complex exponential input z^n, the corresponding output signal is

$$y[n] = \sum_{k=0}^{M} b_k \, x[n - k] = \sum_{k=0}^{M} b_k z^{n-k} = \sum_{k=0}^{M} b_k z^n z^{-k} = \left(\sum_{k=0}^{M} b_k z^{-k} \right) z^n$$

The term inside the parentheses is a polynomial whose form depends on the coefficients of the FIR filter. It is called the *system function* of the FIR filter. From our previous definition of the z-transform, this polynomial is observed to be the z-transform of the impulse-response sequence. Using the notation introduced in the previous section, we define the *system function* of an FIR filter to be

$$H(z) = \sum_{k=0}^{M} b_k z^{-k} = \sum_{k=0}^{M} h[k] z^{-k} \tag{7.2.3}$$

Therefore, we have the following important result:

The system function $H(z)$ is the z-transform of the impulse response.

$$h[n] = \sum_{k=0}^{M} b_k \delta[n - k] \qquad \Longleftrightarrow \qquad H(z) = \sum_{k=0}^{M} b_k z^{-k} \qquad (7.2.4)$$

We have just shown that for FIR filters, if the input is z^n for $-\infty < n < \infty$, then the corresponding output is

$$y[n] = h[n] * z^n = H(z)z^n \qquad (7.2.5)$$

That is, the result of convolving the sequence $h[n]$ with the sequence z^n is $H(z)z^n$, where $H(z)$ is the z-transform of $h[n]$. This is a very general statement. In Chapter 8, it will be shown that it applies to any LTI system, not just FIR filters. Thus, the operation of convolution, which really is synonymous with the definition of an LTI system, appears to be closely linked to the z-transform.

Equation (7.2.3) is general enough to find the "z-transform representation" of any FIR filter, because the polynomial coefficients are exactly the same as the filter coefficients $\{b_k\}$ from the FIR filter difference equation (7.2.1), or, equivalently, the same as the impulse-response sequence from (7.2.2). Thus, the FIR filter difference equation can be transformed easily into a polynomial in the z-domain simply by replacing each "delay by k" (i.e., $x[n - k]$ in (7.2.1)) by z^{-k}.

The system function $H(z)$ is a function of the complex variable z. As we have already noted, $H(z)$ in (7.2.3) is the z-transform of the impulse response, and for the FIR case, it is an Mth-degree polynomial in the variable z^{-1}. Therefore $H(z)$ will have M zeros (i.e., M values z_0 such that $H(z_0) = 0$) that (according to the fundamental theorem of algebra) completely define the polynomial to within a multiplicative constant.

Example 7.3   Consider the FIR filter

$$y[n] = 6x[n] - 5x[n - 1] + x[n - 2]$$

The z-transform system function is

$$H(z) = 6 - 5z^{-1} + z^{-2} = (3 - z^{-1})(2 - z^{-1}) = 6\frac{(z - \frac{1}{3})(z - \frac{1}{2})}{z^2}$$

Thus, the zeros of $H(z)$ are $\frac{1}{3}$ and $\frac{1}{2}$. Note that the filter

$$w[n] = x[n] - \frac{5}{6}x[n - 1] + \frac{1}{6}x[n - 2]$$

has a system function with the same zeros, but the overall constant is 1 rather than 6. This simply means that $w[n] = y[n]/6$. $\diamond$

Exercise 7.1. Find the system function $H(z)$ of an FIR filter whose impulse response is

$$h[n] = \delta[n] - 7\delta[n-2] - 3\delta[n-3]$$

Exercise 7.2. Find the impulse response $h[n]$ of an FIR filter whose system function is

$$H(z) = 4(1 - z^{-1})(1 + z^{-1})(1 + 0.8z^{-1})$$

Hint: Multiply out the factors to get a polynomial and then determine the impulse response by "inverse z-transformation."

7.3 PROPERTIES OF THE z-TRANSFORM

In Section 7.1 we gave the general definition of the z-transform, and we showed that for finite-length sequences, it is possible to go uniquely back and forth between the sequence and its z-transform. In this sense, we have demonstrated that the z-transform is a unique representation of *any* finite-length sequence (including the impulse response of an FIR filter). In Section 7.2 we showed that the z-transform arises naturally out of the convolution of the impulse-response sequence with a sequence z^n. In this section, we will explore several properties of the z-transform representation and indicate how the z-transform can be extended to the infinite-length case.

7.3.1 The Superposition Property of the z-Transform

The z-transform is a linear transformation. This is easily seen by considering the sequence $x[n] = ax_1[n] + bx_2[n]$, where both $x_1[n]$ and $x_2[n]$ are assumed to be finite with length less than or equal to N. Using the definition of (7.1.2), we write

$$X(z) = \sum_{n=0}^{N} (ax_1[n] + bx_2[n]) \, z^{-n}$$

$$= a \sum_{n=0}^{N} x_1[n] z^{-n} + b \sum_{n=0}^{N} x_2[n] z^{-n}$$

$$= aX_1(z) + bX_2(z)$$

Thus, we have demonstrated the *superposition property* for the z-transform:

The z-transform is a linear transformation.

$$ax_1[n] + bx_2[n] \qquad \Longleftrightarrow \qquad aX_2(z) + bX_2(z) \qquad\qquad (7.3.1)$$

This property leads to another way to interpret the z-transform of a finite-length sequence, as illustrated in Example 7.4.

Example 7.4 Recall that any finite-length sequence $x[n]$ can be represented as a sum of scaled and shifted impulse sequences as in

$$x[n] = \sum_{k=0}^{N} x[k]\delta[n-k] \qquad\qquad (7.3.2)$$

Furthermore, recall from (7.1.4) that for a single shifted unit impulse sequence,

$$\delta[n-k] \Longleftrightarrow z^{-k} \qquad\qquad (7.3.3)$$

Thus, applying (7.3.3) to each impulse in (7.3.2) and then adding the individual z-transforms according to (7.3.1), we obtain as before

$$X(z) = \sum_{k=0}^{N} x[k]z^{-k}$$

$\diamond$

7.3.2 The Time-Delay Property of the z-Transform

Another important property of the z-transform is that the quantity z^{-1} in the z-domain corresponds to a time shift of 1 in the n-domain. We will illustrate this property with a numerical example. Consider the length-6 signal $x[n]$ defined by the following table of values:

n	$n < 0$	0	1	2	3	4	5	$n > 5$
$x[n]$	0	3	1	4	1	5	9	0

The z-transform of $x[n]$ is the polynomial (in z^{-1})

$$X(z) = 3 + z^{-1} + 4z^{-2} + z^{-3} + 5z^{-4} + 9z^{-5}$$

Recall that the signal values $x[n]$ are the coefficients of the polynomial $X(z)$ and that the exponents correspond to the time locations of the values. For example, the term $4z^{-2}$ indicates that the signal value at $n = 2$ is 4, i.e., $x[2] = 4$.

Now consider the effect of multiplying the polynomial by z^{-1}:

$$Y(z) = z^{-1}X(z)$$

$$= z^{-1}(3 + z^{-1} + 4z^{-2} + z^{-3} + 5z^{-4} + 9z^{-5})$$

$$= 0z^0 + 3z^{-1} + z^{-2} + 4z^{-3} + z^{-4} + 5z^{-5} + 9z^{-6}$$

The resulting polynomial $Y(z)$ is the z-transform representation of a signal $y[n]$, which is found by using the polynomial coefficients and exponents in $Y(z)$ to determine the values of $y[n]$ at all time positions. The result is the following table of values for $y[n]$:

n	$n < 0$	0	1	2	3	4	5	6	$n > 6$
$y[n]$	0	0	3	1	4	1	5	9	0

Each of the signal samples has moved over one position in the table; i.e., $y[n] = x[n-1]$. In general, for any finite-length sequence, multiplication of the z-transform polynomial by z^{-1} simply subtracts one from each exponent in the polynomial, thereby creating a delay of one. Thus, we have the following fundamental relation:

A delay of one sample multiplies the z-transform by z^{-1}.

$$x[n-1] \qquad \Longleftrightarrow \qquad z^{-1}X(z) \tag{7.3.4}$$

---

which we will refer to as the *unit-delay property* of the z-transform.

The unit-delay property can be generalized for the case of shifting by more than one sample by simply applying (7.3.4) n_0 times. The general result is

---

Time delay of n_0 samples multiplies the z-transform by z^{-n_0}

$$x[n-n_0] \qquad \Longleftrightarrow \qquad z^{-n_0}X(z) \tag{7.3.5}$$

7.3.3 A General z-Transform Formula

So far, we have defined the z-transform only for finite-length signals.

$$X(z) = \sum_{n=0}^{N} x[n]z^{-n} \tag{7.3.6}$$

Our definition assumes that the sequence is nonzero only in the interval $0 \le n \le N$. It is possible to extend the definition to signals of infinite length by simply extending the upper or lower limits to $+\infty$ and $-\infty$ respectively, i.e.,

$$X(z) = \sum_{n=-\infty}^{\infty} x[n]z^{-n} \tag{7.3.7}$$

However, infinite sums may cause serious mathematical difficulties and require special attention. Summing an infinite number of complex numbers could result in an infinite result. In mathematical terms, the sum might not converge. Although we will consider the infinite-length case in Chapter 8, the careful mathematical development of a complete z-transform theory for signal and system analysis is better left to another, more advanced, course.

7.4   THE z-TRANSFORM AS AN OPERATOR

The delay property stated in Section 7.3.2 suggests that the quantity z^{-1} is in some sense equivalent to a delay or time shift. This point of view leads to a useful, but potentially confusing, interpretation of the z-transform as an *operator*. To see how this interpretation comes about, we will consider the system function of the unit-delay system.

7.4.1 Unit-Delay Operator

The unit-delay system is one of the basic building blocks for the FIR difference equation. In the time domain, the unit-delay operator $\mathcal{D}$ is defined by

$$y[n] = \mathcal{D}\{x[n]\} = x[n-1] \tag{7.4.1}$$

It is instructive to find the z-transform representation of this system by letting the input to the unit-delay system be the signal

$$x[n] = z^n \qquad \text{for all } n$$

where z is a complex number. With the z^n input signal, the output of the unit delay is simply

$$
\begin{aligned}
y[n] &= \mathcal{D}\{x[n]\} = \mathcal{D}\{z^n\} \\
&= z^{n-1} = z^{-1}z^n = z^{-1}x[n] \tag{7.4.2}
\end{aligned}
$$

In other words, the input signal is multiplied by z^{-1}, *in the particular case where* $x[n] = z^n$.

Strictly speaking, the expression $z^{-1}x[n]$ in (7.4.2) is misleading, because we must remember that it holds only for $x[n] = z^n$. However, it is common to use the quantity z^{-1} interchangeably with the unit-delay operator symbol $\mathcal{D}$, so that we can say that for *any* input $x[n]$ the action of the unit-delay system is *represented* by the operator z^{-1}; i.e.,

$$y[n] = z^{-1}\{x[n]\} = x[n-1]$$

The brackets enclose the signal operated on by z^{-1} just as in (7.4.1). Thus, if we are careful in our interpretation, we can use the symbol z^{-1} to stand for the delay operator, and many authors use $\mathcal{D}$ and z^{-1} interchangeably.

We know from the delay property of Section 7.3.2 that if $y[n] = x[n-1]$, then $Y(z) = z^{-1}X(z)$; i.e., for *any* finite-length sequence, z^{-1} multiplies $X(z)$ to produce $Y(z)$. This is the precise way in which z^{-1} represents a unit delay; i.e., it is *not* appropriate to write $z^{-1}x[n]$ without the brackets around $x[n]$, since this mixes the z-domain and the n-domain.

7.4.2 Operator Notation

Consider a system that calculates the *first difference* of two successive signal values:

$$y[n] = x[n] - x[n-1]$$

The z-transform operator that represents the first-difference system is $(1 - z^{-1})$, because we can write the "operator" equation

$$y[n] = (1 - z^{-1})\{x[n]\} = x[n] - x[n-1] \tag{7.4.3}$$

This equation (7.4.3) has the following interpretation: The operator "1" leaves $x[n]$ unchanged, and the operator z^{-1} delays $x[n]$ before subtracting it from $x[n]$.

Another simple example would be a system that delays by more than one sample, e.g., by n_d samples:

$$y[n] = x[n - n_d] \qquad n_d \text{ is an integer}$$

In this case, the system function is $H(z) = z^{-n_d}$ and the operator is z^{-n_d}, an obvious generalization of the unit-delay case.

Exercise 7.3. Derive the z-transform operator for the first-difference system by working the input $x[n] = z^n$ through the system. Write $y[n]$ as $y[n] = H(z)\{x[n]\}$.

7.4.3 Operator Notation in Block Diagrams

The delay operator concept is particularly useful in block diagrams of LTI systems. In a block diagram representation of the FIR filter, the z-transform works as follows: All the unit delays become z^{-1} operators in the transform domain, and, owing to the superposition property of the z-transform, the scalar multipliers and adders are the same as in the time-domain representation. Figure 7.1a shows the n-domain and z-domain representations of a unit delay, and Figure 7.1b shows a block diagram representation of a two-point FIR filter using a z^{-1} operator to represent the delay operator.

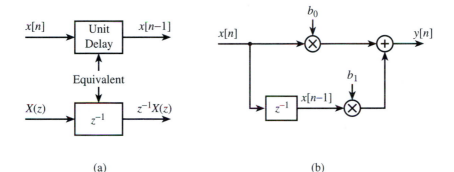

(a)   (b)

Figure 7.1   Computational structure for a first-order FIR filter: (a) the equivalence between z^{-1} and the unit delay; (b) block diagram for the first-order filter whose difference equation is $y[n] = b_0 x[n] + b_1 x[n-1]$.

Exercise 7.4. Draw a block diagram similar to Fig. 7.1(b) for the first difference system: $y[n] = \left(1 - z^{-1}\right)\{x[n]\}$

7.5 CONVOLUTION AND THE z-TRANSFORM

In Section 7.3.2, we observed that a unit delay of a signal in the n-domain is equivalent to multiplication by z^{-1} of the corresponding z-transform in the z-domain. The impulse response of the unit-delay system is

$$h[n] = \delta[n - 1]$$

so a delay by one sample is equivalent to the convolution

$$y[n] = x[n] * \delta[n - 1] = x[n - 1]$$

The system function of the unit-delay system is the z-transform of its impulse response so

$$H(z) = z^{-1}$$

Furthermore, the unit-delay property (7.3.4) states that delay by one sample multiplies the z-transform by z^{-1}; i.e.,

$$Y(z) = z^{-1}X(z)$$

Therefore, we observe that in the case of the unit delay, the z-transform of the output is equal to the z-transform of the input multiplied by the system function of the LTI system, i.e.,

$$Y(z) = H(z)X(z) \qquad (7.5.1)$$

More importantly, this result (7.5.1) is true for any LTI system.

To show that convolution is converted into a product of z-transforms (7.5.1), recall that the discrete convolution of two finite-length sequences $x[n]$ and $h[n]$ is given by the formula:

$$y[n] = x[n] * h[n] = \sum_{k=0}^{M} h[k]x[n - k] \qquad (7.5.2)$$

where M is the order of the FIR filter. To prove the desired result, we can apply the superposition property (7.3.1) and the general delay property (7.3.5) to find the z-transform of $y[n]$ as given by (7.5.2). This leads to

$$Y(z) = \sum_{k=0}^{M} h[k]\left(z^{-k}X(z)\right) = \left(\sum_{k=0}^{M} h[k]z^{-k}\right)X(z) = H(z)X(z). \qquad (7.5.3)$$

If $x[n]$ is a finite-length sequence, $X(z)$ is a polynomial, so (7.5.3) proves that convolution is equivalent to polynomial multiplication. This result is illustrated in the following example.

Example 7.5   Suppose that

$$x[n] = \delta[n - 1] - \delta[n - 2] + \delta[n - 3] - \delta[n - 4]$$

$$\text{and} \quad h[n] = \delta[n] + 2\delta[n - 1] + 3\delta[n - 2] + 4\delta[n - 3]$$

The z-transforms of the sequences $x[n]$ and $h[n]$ are:

$$X(z) = 0 + 1z^{-1} - 1z^{-2} + 1z^{-3} - 1z^{-4}$$

$$\text{and} \quad H(z) = 1 + 2z^{-1} + 3z^{-2} + 4z^{-3}$$

Both $X(z)$ and $H(z)$ are polynomials in z^{-1}, so we can compute the z-transform of the convolution by multiplying these two polynomials, i.e.,

$$
\begin{aligned}
Y(z) &= H(z)X(z) \\
&= (1 + 2z^{-1} + 3z^{-2} + 4z^{-3})(z^{-1} - z^{-2} + z^{-3} - z^{-4}) \\
&= z^{-1} + (-1 + 2)z^{-2} + (1 - 2 + 3)z^{-3} + (-1 + 2 - 3 + 4)z^{-4} \\
&\quad + (-2 + 3 - 4)z^{-5} + (-3 + 4)z^{-6} + (-4)z^{-7} \\
&= z^{-1} + z^{-2} + 2z^{-3} + 2z^{-4} - 3z^{-5} + z^{-6} - 4z^{-7}
\end{aligned}
$$

Since the coefficients of any z-polynomial are just the sequence values, with their position in the sequence being indicated by the power of (z^{-1}), we can "inverse transform" $Y(z)$ to obtain

$$
\begin{aligned}
y[n] &= \delta[n-1] + \delta[n-2] + 2\delta[n-3] + 2\delta[n-4] \\
&\quad - 3\delta[n-5] + \delta[n-6] - 4\delta[n-7]
\end{aligned}
$$

Now we look at the convolution sum for computing the output. If we write out a few terms, we can detect a pattern that is similar to the z-transform polynomial multiplication.

$$
y[0] = h[0]x[0] = 1(0) = 0
$$

$$
y[1] = h[0]x[1] + h[1]x[0] = 1(1) + 2(0) = 1
$$

$$
y[2] = h[0]x[2] + h[1]x[1] + h[2]x[0] = 1(-1) + 2(1) + 3(0) = 1
$$

$$
y[3] = h[0]x[3] + h[1]x[2] + h[2]x[1] + h[3]x[0] = 1(1) + 2(-1) + 3(1) = 2
$$

$$
y[4] = h[0]x[4] + h[1]x[3] + h[2]x[2] + h[3]x[1] = 1(-1) + 2(1) + 3(-1) + 4(1) = 2
$$

$$
\vdots = \vdots
$$

Notice how the index of $h[k]$ and the index of $x[n-k]$ sum to the same value (i.e., n) for all products that contribute to $y[n]$. The same thing happens in polynomial multiplication because exponents add.

In Chapter 5 we demonstrated a synthetic multiplication tableau for evaluating the convolution of $x[n]$ with $h[n]$. Now we see that this is also a process for multiplying the polynomials $X(z)$ and $H(z)$. The procedure is repeated below for the numerical example of this section.

x[n], X(z)	0	+1	-1	+1	-1			
h[n], H(z)	1	2	3	4				

0	+1	-1	+1	-1			
	0	+2	-2	+2	-2		
		0	+3	-3	+3	-3	
			0	+4	-4	+4	-4

y[n], Y(z)	0	+1	+1	+2	+2	-3	+1	-4

In the z-transforms $X(z)$, $H(z)$, and $Y(z)$, the power of z^{-1} is implied by the horizontal position of the coefficient in the tableau. Each row is produced by multiplying the $x[n]$ row by one of the $h[n]$ values and shifting the result right by the implied power of z^{-1}. The final answer is obtained by summing down the columns. The final row is the sequence of values of $y[n] = x[n] * h[n]$ or, equivalently, the coefficients of the polynomial $Y(z)$. ◇

In this section we have established that convolution and polynomial multiplication are essentially the same thing.[3] Indeed, the most important result of z-transform theory is:

Convolution in the n-domain corresponds to multiplication in the z-domain.

$$y[n] = h[n] * x[n] \qquad \Longleftrightarrow \qquad Y(z) = H(z)X(z)$$

This result will be seen to have many implications far beyond its use as a basis for understanding and implementing convolution.

Exercise 7.5. Use the z-transform of

$$x[n] = \delta[n - 1] - \delta[n - 2] + \delta[n - 3] - \delta[n - 4]$$

and the system function $H(z) = 1 - z^{-1}$ to find the output of a first-difference filter when $x[n]$ is the input. Compute your answer by using polynomial multiplication and also by using the difference equation:

$$y[n] = x[n] - x[n - 1]$$

What is the degree of the output z-transform polynomial that represents $y[n]$?

---

[3] In MATLAB, there is no special function for multiplying polynomials. Instead, you simply use the convolution function conv to multiply polynomials since polynomial multiplication is identical to discrete convolution of the sequences of coefficients.

7.5.1 Cascading Systems

One of the main applications of the z-transform in system design is its use in creating alternative filters that have exactly the same input–output behavior. An important example is the cascade connection of two or more LTI systems. In block diagram form, the cascade is drawn with the output of the first system connected to the input of the second. The input signal is $x[n]$ and the overall output is $y[n]$. The sequence $w[n]$ is an intermediate signal that can be thought of as temporary storage.

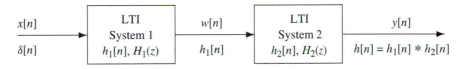

Figure 7.2   Cascade of two LTI systems.

As we have already seen in Chapter 5, if $h_1[n]$ and $h_2[n]$ are the respective impulse responses of the first and second systems, then, as shown in Fig. 7.2, the overall impulse response from input $x[n]$ to output $y[n]$ is $h[n] = h_1[n] * h_2[n]$. Therefore, the z-transform of the overall impulse response of the cascade of the two systems is the product of the individual z-transforms of the two impulse responses. That is,

> *The system function for a cascade of two LTI systems*
> *is the product of the individual system functions.*
>
> $$h[n] = h_1[n] * h_2[n] \qquad \Longleftrightarrow \qquad H(z) = H_1(z)H_2(z) \qquad (7.5.4)$$

An important consequence of this result follows easily from the fact that multiplication is commutative; i.e., $H(z) = H_1(z)H_2(z) = H_2(z)H_1(z)$. This implies that convolution must also be a commutative operation and that the two systems can be cascaded in either order to obtain the same overall system response.

Example 7.6   To give a simple example of this idea, consider a system described by the difference equations

$$w[n] = 3x[n] - x[n-1] \qquad (7.5.5)$$

$$y[n] = 2w[n] - w[n-1] \qquad (7.5.6)$$

which represent a cascade of two first-order systems as in Fig. 7.2. The output $w[n]$ of the first system is the input to the second system, and the overall output is the output of the second system. The intermediate signal $w[n]$ in (7.5.5) must be computed prior to being used in (7.5.6). We can combine the two filters into a single difference equation by substituting $w[n]$

from the first system into the second, which gives

$$y[n] = 2w[n] - w[n-1]$$
$$= 2(3x[n] - x[n-1]) - (3x[n-1] - x[n-2])$$
$$= 6x[n] - 5x[n-1] + x[n-2] \tag{7.5.7}$$

Thus we have proved that the cascade of the two first-order systems is equivalent to a single second-order system. It is important to notice that the difference equation (7.5.7) defines an algorithm for computing $y[n]$ that is different from the algorithm specified by (7.5.5) and (7.5.6) together. However, the above analysis shows that with perfectly accurate computation, the outputs of the two different implementations would be exactly the same.

Working out the details of the overall difference equation as we have just done would be extremely tedious if the systems were higher-order. The z-transform simplifies these operations into the multiplication of polynomials. The first-order systems have system functions:

$$H_1(z) = 3 - z^{-1} \qquad \text{and} \qquad H_2(z) = 2 - z^{-1}.$$

Therefore, the overall system function is

$$H(z) = (3 - z^{-1})(2 - z^{-1}) = 6 - 5z^{-1} + z^{-2},$$

which matches the difference equation in (7.5.7). Note that, even in this simple example, the z-domain solution is more straightforward than the n-domain solution.   ◇

Exercise 7.6. Use z-transforms to combine the following cascaded systems

$$w[n] = x[n] + x[n-1]$$
$$y[n] = w[n] - w[n-1] + w[n-2]$$

into a single difference equation for $y[n]$ in terms of $x[n]$.

7.5.2 Factoring z-Polynomials

If we can multiply z-transforms to get higher-order systems, we can also factor z-transform polynomials to break down a large system into smaller modules. Since cascading systems is equivalent to multiplying their system functions, the factors of a high-order polynomial $H(z)$ would represent component systems that make up $H(z)$ in a cascade connection.

Example 7.7   Consider the following example

$$H(z) = 1 - 2z^{-1} + 2z^{-2} - z^{-3}$$

One of the roots of $H(z)$ is $z = 1$, so $H_1(z) = (1 - z^{-1})$ is a factor of $H(z)$. The other factor can be obtained by division

$$H_2(z) = \frac{H(z)}{H_1(z)} = \frac{H(z)}{1 - z^{-1}} = 1 - z^{-1} + z^{-2}$$

The factorization of $H(z)$ as $H(z) = (1 - z^{-1})(1 - z^{-1} + z^{-2})$ gives the cascade shown in the block diagram of Fig. 7.3. The resulting difference equations for the cascade are

$$w[n] = x[n] - x[n - 1]$$

$$y[n] = w[n] - w[n - 1] + w[n - 2]$$

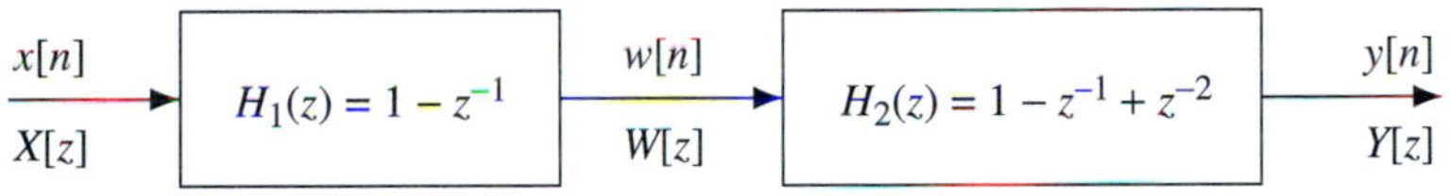

Figure 7.3   Factoring $H(z) = 1 - 2z^{-1} + 2z^{-2} - z^{-3}$ into the product of a first-order system and a second-order system.

◇

7.5.3 Deconvolution

The cascading property leads to an interesting question that has practical application. Can we use the second filter in a cascade to undo the effect of the first filter? What we would like is for the output of the second filter to be equal to the input to the first. Stated more precisely, suppose that we have the cascade of two filters $H_1(z)$ and $H_2(z)$, and $H_1(z)$ is known. Is it possible to find $H_2(z)$ so that the overall system has its output equal to the input? If so, the z-transform analysis tells us that its system function would have to be $H(z) = 1$, so that

$$Y(z) = H_1(z)H_2(z)X(z) = H(z)X(z) = X(z)$$

Since the first system processes the input via convolution, the second filter tries to undo convolution, so the process is called *deconvolution*. Another term for this is *inverse filtering*, and if $H_1(z)H_2(z) = 1$, then $H_2(z)$ is said to be the *inverse* of $H_1(z)$ (and vice versa).

Example 7.8   If we take a specific example, we can generate a solution in terms of z-transforms. Suppose that $H_1(z) = 1 - z^{-1} + \frac{1}{2}z^{-2}$. We want

$$H(z) = 1 \qquad \Longrightarrow \qquad H_1(z)H_2(z) = 1$$

Since $H_1(z)$ is known we can solve for $H_2(z)$ to get

$$H_2(z) = \frac{1}{H_1(z)} = \frac{1}{1 - z^{-1} + \frac{1}{2}z^{-2}}$$

$\diamond$

What are we to make of this example? It seems that the deconvolver for an FIR filter must have a system function that is not a polynomial, but a rational function (ratio of two polynomials) instead. This means that the inverse filter for an FIR filter cannot be also an FIR filter, and deconvolution suddenly is not as simple as it appeared. Since we have not yet considered the possibility of anything but polynomial system functions, we cannot give the solution in the form of a difference equation. However, in Chapter 8 we will see that other types of LTI systems exist that do have rational system functions. We will therefore return to the inverse filtering problem in Chapter 8.

7.6   RELATIONSHIP BETWEEN THE z-DOMAIN AND THE $\hat{\omega}$-DOMAIN

The system function $H(z)$ has a functional form that is identical to the form of the frequency response formula $\mathcal{H}(\hat{\omega})$. This is quite easy to see for the FIR filter if we repeat the formula for the frequency response (6.1.3) alongside the formula for the system function (7.2.3):

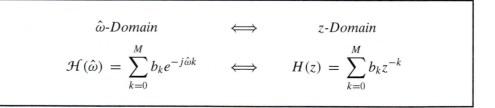

$$\hat{\omega}\text{-Domain} \qquad \Longleftrightarrow \qquad z\text{-Domain}$$

$$\mathcal{H}(\hat{\omega}) = \sum_{k=0}^{M} b_k e^{-j\hat{\omega}k} \qquad \Longleftrightarrow \qquad H(z) = \sum_{k=0}^{M} b_k z^{-k}$$

MATLAB GUI:
PeZ

There is a clear correspondence between the z- and $\hat{\omega}$-domains if we make the substitution $z = e^{j\hat{\omega}}$ in $H(z)$. Specifically, it is exceedingly important to note that the connection between $\mathcal{H}(\hat{\omega})$ and the z-transform $H(z)$ is

$$\mathcal{H}(\hat{\omega}) = H(e^{j\hat{\omega}}) = H(z)\big|_{z=e^{j\hat{\omega}}} \qquad (7.6.1)$$

The relationship between the z-domain and the $\hat{\omega}$-domain hinges on the important formula

$$z = e^{j\hat{\omega}} \qquad\qquad (7.6.2)$$

To see why this relationship is the key, we need only recall that if the signal z^n is the input to an LTI filter, the resulting output is $y[n] = H(z)z^n$. If the value of z is $z = e^{j\hat{\omega}}$, then

$$y[n] = H(e^{j\hat{\omega}})e^{j\hat{\omega}n}$$

where $H(e^{j\hat{\omega}})$ is obviously the same as what we have called the frequency response and which, previously, we denoted $\mathcal{H}(\hat{\omega})$.

7.6.1 The z-Plane and the Unit Circle

From now on, we will use $H(e^{j\hat{\omega}})$ in place of $\mathcal{H}(\hat{\omega})$ to emphasize the connection between the $\hat{\omega}$-domain and the z-domain. When we use this notation, we are indicating explicitly that the frequency response $H(e^{j\hat{\omega}})$ is obtained from the system function $H(z)$ by evaluating $H(z)$ for a specific set of values of z. Recall that since the frequency response is periodic with period 2π, we need only evaluate it over one period, such as $-\pi < \hat{\omega} < \pi$. If we substitute these values of $\hat{\omega}$ into (7.6.2), we see that the corresponding values of z all have unit magnitude and that the angle $\hat{\omega}$ varies from $-\pi$ to $+\pi$. In other words, the values of $z = e^{j\hat{\omega}}$ lie on a circle of radius 1 and range from the point $z = -1$ all the way around the circle and back to the point $z = -1$. Quite naturally, the contour on which all the values of $z = e^{j\hat{\omega}}$ lie is called the *unit circle*. This is illustrated in Fig. 7.4, which shows the unit circle and a typical point $z = e^{j\hat{\omega}}$, which is at a distance 1 from the origin and at an angle of $\hat{\omega}$ with respect to the real axis of the z-plane.

The graphical representation of Fig. 7.4 gives us a convenient way of visualizing the relationship between the $\hat{\omega}$-domain and the z-domain. Because the $\hat{\omega}$-domain lies on a special part of the z-domain — the unit circle — many properties of the frequency response are evident from plots of system function properties in the z-plane. For example, the periodicity of the frequency response is obvious from Fig. 7.4, which shows that evaluating the system function at all points on the unit circle requires moving through an angle of 2π radians. Since frequency $\hat{\omega}$ is equivalent to angle in the z-plane, 2π radians in the z-plane correspond to an interval of 2π radians of frequency. Continuing around the unit circle more times simply cycles through more periods of the frequency response.

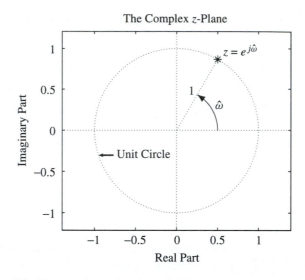

Figure 7.4   The complex z-plane including the unit circle, where $z = e^{j\hat{\omega}}$.

7.6.2 The Zeros and Poles of $H(z)$

We have already seen that the system function for an FIR system is essentially determined by its zeros. This is illustrated by the following example.

Example 7.9   Consider the system function $H(z) = 1 - 2z^{-1} + 2z^{-2} - z^{-3}$, which can be expressed in the following different forms:

$$H(z) = 1 - 2z^{-1} + 2z^{-2} - z^{-3} \tag{7.6.3}$$

$$= (1 - z^{-1})(1 - e^{j\pi/3}z^{-1})(1 - e^{-j\pi/3}z^{-1}) \tag{7.6.4}$$

or, if we multiply $H(z)$ by z^3/z^3, we obtain the following two equivalent forms:

$$H(z) = \frac{z^3 - 2z^2 + 2z - 1}{z^3} \tag{7.6.5}$$

$$= \frac{(z - 1)(z - e^{j\pi/3})(z - e^{-j\pi/3})}{z^3} \tag{7.6.6}$$

Equations (7.6.3)–(7.6.6) give four different equivalent forms of $H(z)$. The factored form in (7.6.6) shows clearly that the zeros of $H(z)$ are at locations $z_1 = 1$, $z_2 = e^{j\pi/3}$, and $z_3 = e^{-j\pi/3} = z_2^*$ in the z-plane. Equation (7.6.6) also shows that $H(z) \to \infty$ for $z \to 0$. Values of z for which $H(z)$ is undefined (infinite) are called *poles* of $H(z)$. In this case, we say that the term z^3 represents three poles at $z = 0$ or that $H(z)$ has a third-order pole at $z = 0$.

We have stated that the poles and zeros determine the system function to within a constant. As an illustration, note that the polynomial $\frac{1}{2}H(z) = 0.5 - z^{-1} + z^{-2} - 0.5z^{-3}$ has exactly the same poles and zeros as $H(z)$ in (7.6.3). $\diamond$

Although it is perhaps less obvious, the locations of both the poles and the zeros are also clear when $H(z)$ is written in the form (7.6.4), since each factor of the form $(1 - az^{-1})$ always can be expressed as

$$(1 - az^{-1}) = \frac{(z - a)}{z}$$

which shows that each factor of the form $(1 - az^{-1})$ represents a zero at $z = a$ and a pole at $z = 0$. When $H(z)$ contains only negative powers of z, it is usually most convenient to use the representations of the form of (7.6.3) and (7.6.4) since the negative powers of z have a direct correspondence to the difference equation and the impulse response.

It is useful to display the zeros and poles of $H(z)$ as points in the complex z-plane. The plot in Fig. 7.5 shows the three zeros and the three poles for Example 7.9. Such a plot is called a *pole-zero plot*. This plot was generated in MATLAB using the zplane function. Each zero location is denoted by a small circle, and the three poles at $z = 0$ are indicated by a single $\times$ with a numeral 3 beside it. In general, when all the poles are not concentrated at $z = 0$, the $\times$ symbol will mark the location of each pole. Since the unit circle is where $H(z)$ is evaluated to obtain the frequency response of the LTI system whose system function is $H(z)$, it is also shown for reference as a dotted circle in Fig. 7.5.

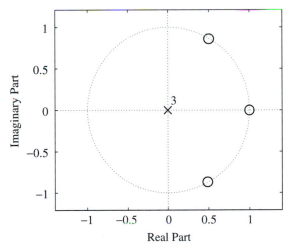

Figure 7.5   Zeros and poles of $H(z)$ marked in a z-plane that includes the unit circle.

7.6.3 Significance of the Zeros of $H(z)$

In Section 7.6.2 we showed that the zeros of a polynomial system function are sufficient to determine $H(z)$ except for a constant multiplier. The system function determines the difference equation of the filter because the polynomial coefficients of $H(z)$ are the coefficients of the difference equation. The difference equation is the direct link between an input $x[n]$ and the corresponding output $y[n]$. However, there are some inputs where knowledge of the zero locations is sufficient to make a precise statement about the output without actually computing it using the difference equation. Such signals are of the form $x[n] = z_0^n$ for all n, where the subscript signifies that z_0 is a particular complex number. In this case, the output is

$$y[n] = H(z_0)z_0^n$$

The quantity $H(z_0)$ is a complex constant, which, through complex multiplication, causes a magnitude and phase change of the input signal z_0^n. In particular, if z_0 is one of the zeros of $H(z)$, then $H(z_0) = 0$ so the output will be zero.

Example 7.10   For example, when $H(z) = 1 - 2z^{-1} + 2z^{-2} - z^{-3}$, the roots are

$$z_1 = 1$$

$$z_2 = \tfrac{1}{2} + j\tfrac{1}{2}\sqrt{3} = 1e^{j\pi/3}$$

$$z_3 = \tfrac{1}{2} - j\tfrac{1}{2}\sqrt{3} = 1e^{-j\pi/3}$$

As shown in Fig. 7.5, these zeros are all on the unit circle, so complex sinusoids with frequencies 0, $\pi/3$, and $-\pi/3$ will be set to zero by the system. That is, the output resulting from each of the following three signals will be zero:

$$x_1[n] = (z_1)^n = 1$$

$$x_2[n] = (z_2)^n = e^{j\pi n/3}$$

$$x_3[n] = (z_3)^n = e^{-j\pi n/3}$$

$\diamond$

As illustrated by this example, the zeros of the system function that lie on the unit circle correspond to frequencies at which the gain of the system is zero. Thus, complex sinusoids at those frequencies are blocked, or "nulled" by the system.

Exercise 7.7. Double-check the fact that the inputs $x_1[n]$, $x_2[n]$, and $x_3[n]$ determined in Example 7.10 produce outputs that are zero everywhere by substituting these signals into the difference equation $y[n] = x[n] - 2x[n-1] + 2x[n-2] - x[n-3]$ to show that the complex phasors cancel out for all values of n. Also show that since the filter is linear, it will also null out signals such as $2\cos(\pi n/3)$, which is the sum of $x_2[n]$ and $x_3[n]$.

7.6.4 Nulling Filters

We have just shown that if the zeros of $H(z)$ lie on the unit circle, then certain sinusoidal input signals are removed or nulled by the filter. Therefore, it should be possible to use this result in designing an FIR filter that can null a particular sinusoidal input. Such capability is often needed to eliminate jamming signals in a radar or communications system. Similarly, 60-Hz interference from a power line could be eliminated by placing a null at the correct frequency.

Zeros in the z-plane can remove only signals that have the special form $x[n] = z_0^n$. If we want to eliminate a sinusoidal input signal, then we actually have to remove two signals of the form $z_1^n + z_2^n$; i.e.,

$$x[n] = \cos(\hat{\omega}_0 n) = \tfrac{1}{2}e^{j\hat{\omega}_0 n} + \tfrac{1}{2}e^{-j\hat{\omega}_0 n}$$

Each complex exponential can be removed with a first-order FIR filter, and then the two filters would be cascaded to form the second-order nulling filter that removes the cosine. The second-order FIR filter will have two zeros at $z_1 = e^{j\hat{\omega}_0}$ and $z_2 = e^{-j\hat{\omega}_0}$. The signal z_1^n will be nulled by a filter with system function

$$H_1(z) = 1 - z_1 z^{-1}$$

because $H_1(z_1) = 0$ at $z = z_1$, i.e.,

$$H_1(z_1) = 1 - z_1(z_1)^{-1} = 1 - 1 = 0$$

Similarly, $H_2(z) = 1 - z_2 z^{-1}$ will remove z_2^n. Thus the second-order nulling filter will be the product

$$
\begin{aligned}
H(z) &= H_1(z)H_2(z) \\[4pt]
&= (1 - z_1 z^{-1})(1 - z_2 z^{-1}) \\[4pt]
&= 1 - (z_1 + z_2)z^{-1} + (z_1 z_2)z^{-2} \\[4pt]
&= 1 - (e^{j\hat{\omega}_0} + e^{-j\hat{\omega}_0})z^{-1} + (e^{j\hat{\omega}_0}e^{-j\hat{\omega}_0})z^{-2} \\[4pt]
&= 1 - 2\cos(\hat{\omega}_0)z^{-1} + z^{-2}
\end{aligned}
$$

Figure 7.6 shows the two zeros needed to remove components at $z = e^{\pm j\pi/4}$. For the example depicted in Fig. 7.6, the numerical values for the coefficients of $H(z)$ are

$$H(z) = 1 - 2\cos(\pi/4)z^{-1} + z^{-2} = 1 - \sqrt{2}z^{-1} + z^{-2}$$

Thus the nulling filter that will remove the signal $\cos(0.25\pi n)$ from its input is the FIR filter whose difference equation is

$$y[n] = x[n] - \sqrt{2}x[n-1] + x[n-2] \tag{7.6.7}$$

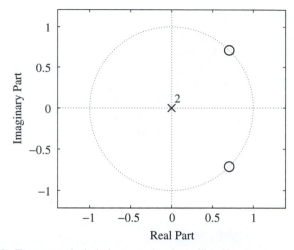

Figure 7.6 Zeros on unit circle for second-order nulling filter to remove sinusoidal components at $\hat{\omega}_0 = \pm\pi/4$. There are two poles at the origin.

7.6.5 Graphical Relation Between z and $\hat{\omega}$

The equation $z = e^{j\hat{\omega}}$ provides the link between the z-domain and the $\hat{\omega}$-domain. As we have shown in (7.6.1), the frequency response is obtained by evaluating the system function on the unit circle of the z-plane. This correspondence can be given a useful graphical interpretation. By considering the pole-zero plot of the system function, we can visualize how the frequency response plot of $\mathcal{H}(\hat{\omega}) = H(e^{j\hat{\omega}})$ results from evaluating $H(z)$ on the unit circle, and also how it depends on the poles and zeros of $H(z)$. As an example, we show in Fig. 7.7 a plot obtained by evaluating the z-transform magnitude $|H(z)|$ over a region of the z-plane that includes the unit circle as well as values both inside and outside the unit circle. The system in this case is an 11-point running sum; i.e., it is an FIR filter where the coefficients are all equal to one.[4] The system function for this filter is

$$H(z) = \sum_{k=0}^{10} z^{-k} \tag{7.6.8}$$

In Section 7.7 we will show that the zeros of the system function of this filter are on the unit circle at angles $\hat{\omega} = 2\pi k/11$, for $k = 1, 2, \ldots, 10$. This means that the polynomial in (7.6.8) can be represented in the form

$$H(z) = (1 - e^{j2\pi/11}z^{-1})(1 - e^{j4\pi/11}z^{-1})\cdots(1 - e^{j20\pi/11}z^{-1}) \tag{7.6.9}$$

---

[4] This is the same system as the 11-point running average filter that we discussed in detail in Section 6.7, except that we have omitted the gain constant 1/11.

Recall that each factor of the form $(1 - e^{j2\pi k/11}z^{-1})$ represents a zero at $z = e^{j2\pi k/11}$ and a pole at $z = 0$. Thus, (7.6.9) displays the 10 zeros of $H(z)$ at $z = e^{j2\pi k/11}$, for $k = 1, 2, \ldots, 10$ and the 10 poles at $z = 0$.

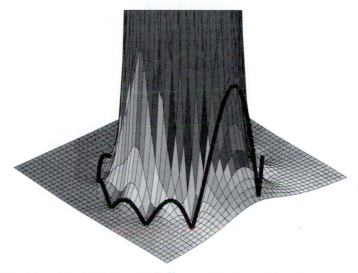

Figure 7.7 z-transform for an FIR filter evaluated over the region $[-1.4 \le \Re e\{z\} \le 1.4] \times [-1.4 \le \Im m\{z\} \le 1.4]$ of the z-plane that includes the unit circle. Values along the unit circle are shown as a dark line where the frequency response (magnitude) is evaluated. The view is from the fourth quadrant, so the point $z = 1$ is in the foreground on the right.

Z-to-FREQ
FLYING
THRU
Z-PLANE

In the magnitude plot of Fig. 7.7, we observe that the zeros pin down the three-dimensional plot around the unit circle. Inside the unit circle, the values of $H(z)$ become very large owing to the poles at $z = 0$. The frequency response $\mathcal{H}(\hat{\omega}) = H(e^{j\hat{\omega}})$ is obtained by selecting the values of the z-transform along the unit circle in Fig. 7.7. A plot of $|H(e^{j\hat{\omega}})|$ versus $\hat{\omega}/2\pi$ is given in Fig. 7.8. The shape of the frequency response can be explained in terms of the zero locations shown in Fig. 7.9 by recognizing that the poles at $z = 0$ push $H(e^{j\hat{\omega}})$ up, but the zeros along the unit circle make $H(e^{j\hat{\omega}}) = 0$ at regular intervals except for the region near $\hat{\omega} = 0$ (i.e., around $z = 1$). The unit circle values follow the ups and downs of $H(z)$ as $\hat{\omega}$ goes from $-\pi$ to $+\pi$ with $|z| = 1$.

This example illustrates that an intuitive picture of the frequency response of an LTI system can be visualized from a plot of the poles and zeros of the system function $H(z)$. We simply need to remember that a pole will "push up" the frequency response and a zero will "pull it down." Furthermore, a zero on the unit circle will force the frequency response to be zero at the frequency corresponding to the angular position of the zero.

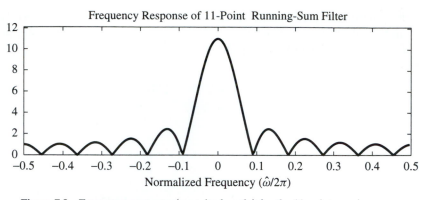

Figure 7.8   Frequency response (magnitude only) for the 11-point running sum. These are the values along the unit circle in the z-plane. There are 10 zeros spread out uniformly along the frequency axis.

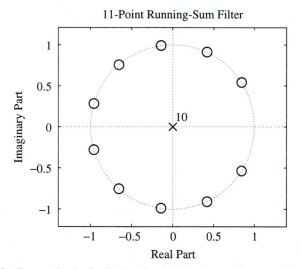

Figure 7.9   Zero and pole distribution for the 11-point running sum. There are 10 zeros spread out uniformly along the unit circle and 10 poles at the origin.

7.7   USEFUL FILTERS

Now that we understand the tie between the z and $\hat{\omega}$ domains, we can exploit that knowledge to design filters with desirable characteristics. In this section, we will look at a special class of bandpass filters (BPFs) that are all close relatives of the running-sum filter.

7.7.1 The L-Point Running Sum Filter

Generalizing from the previous section, the L-point running sum filter

$$y[n] = \sum_{k=0}^{L-1} x[n-k]$$

has system function

$$H(z) = \sum_{k=0}^{L-1} z^{-k}$$

Recalling the formula (6.7.3) for the sum of L terms of a geometric series, $H(z)$ can be represented in the forms

$$H(z) = \sum_{k=0}^{L-1} z^{-k} = \frac{1-z^{-L}}{1-z^{-1}} = \frac{z^L - 1}{z^{L-1}(z-1)} \tag{7.7.1}$$

The final form for $H(z)$ is a rational function where the numerator polynomial is $z^L - 1$ and the denominator is $z^{L-1}(z-1)$. The zeros of $H(z)$ will be determined by the roots of the numerator polynomial, i.e., the values of z such that

$$z^L - 1 = 0 \qquad \Longrightarrow \quad z^L = 1 \tag{7.7.2}$$

Since $e^{j2\pi k} = 1$ for k, an integer, it is easy to see by substitution that each of the values

$$z = e^{j2\pi k/L} \quad \text{for } k = 0, 1, 2, \ldots, L-1 \tag{7.7.3}$$

satisfy (7.7.2) and therefore these L numbers are the roots of the Lth-order equation in (7.7.2). Because the values in (7.7.3) satisfy the equation $z^L = 1$, they are called *the Lth roots of unity*. The zeros of the denominator in (7.7.1), which are either $z = 0$ (of order $L - 1$) or $z = 1$, would normally be the poles of $H(z)$. However, since one of the Lth roots of unity is $z = 1$, i.e., $k = 0$ in (7.7.3), that zero of the numerator cancels the corresponding zero of the denominator, so that only the term z^{L-1} really causes a pole of $H(z)$. Therefore, it follows that $H(z)$ can be expressed in the factored form

$$H(z) = \sum_{k=0}^{L-1} z^{-k} = \prod_{k=1}^{L-1} (1 - e^{j2\pi k/L} z^{-1}) \qquad (7.7.4)$$

Example 7.11   For a 10-point running-sum filter $(L = 10)$, the system function is

$$H(z) = \sum_{k=0}^{9} z^{-k} = \frac{1 - z^{-10}}{1 - z^{-1}} = \frac{z^{10} - 1}{z^9(z - 1)} \qquad (7.7.5)$$

A pole-zero diagram for this case is shown in Fig. 7.10, and the corresponding frequency response for the running-sum filter is shown in Fig. 7.11. The factors of the numerator are the tenth roots of unity, and the zero at $z = 1$ is canceled by the corresponding term in the denominator. This explains why we have only nine zeros around the unit circle with the gap at $z = 1$. The nine zeros around the unit circle in Fig 7.10 show up as zeros along the $\hat{\omega}$ axis in Fig. 7.11 at $\hat{\omega} = 2\pi k/10$, and it is the gap at $z = 1$ that allows the frequency response to be larger at $\hat{\omega} = 0$. The other zeros around the unit circle keep $H(e^{j\hat{\omega}})$ small, thereby creating the "lowpass" filter frequency response shown in Fig. 7.11. (Note that $H(e^{j\hat{\omega}})$ is plotted versus $\hat{\omega}/2\pi$ in Fig. 7.11.)

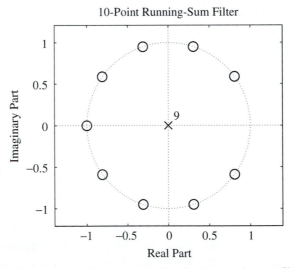

Figure 7.10   Zero and pole distribution for the 10-point running-sum filter. There are nine zeros spread out uniformly along the unit circle, and nine poles at the origin.

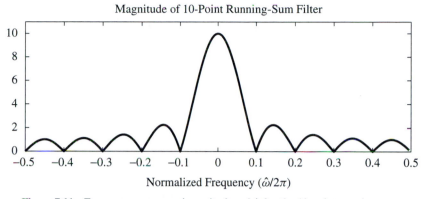

Magnitude of 10-Point Running-Sum Filter

Normalized Frequency ($\hat{\omega}/2\pi$)

Figure 7.11   Frequency response (magnitude only) for the 10-point running-sum filter. These are the values along the unit circle in the z-plane. There are nine zeros spread out uniformly along the frequency axis.

$\diamond$

7.7.2 A Complex Bandpass Filter

Now we have a new insight that tells us how to control the frequency response of an FIR filter by placing its zeros on the unit circle. This viewpoint makes it easy to create other FIR filters where we control the location of the passband. If we move the passband to a frequency away from $\hat{\omega} = 0$, then we have a filter that passes a small band of frequency components—a bandpass filter, or BPF.

The obvious way to move the passband is to use all but one of the roots of unity as the zeros of an FIR filter. A formula for this new filter is

$$H(z) = \prod_{\substack{k=0 \\ k \neq k_0}}^{L-1} (1 - e^{j2\pi k/L} z^{-1}) \tag{7.7.6}$$

where the index k_0 denotes the one omitted root at $z = e^{j2\pi k_0/L}$. An example is shown in Fig. 7.12 for $k_0 = 2$ and $L = 10$. In the general case, the passband of the filter whose system function is given by (7.7.6) should move from the interval around $\hat{\omega} = 0$ to a like interval around

$$\hat{\omega} = \frac{2\pi k_0}{L}$$

because the zero is missing at that frequency. Figure 7.13 confirms our intuition, because with $k_0 = 2$ the normalized frequency of the peak is $\hat{\omega}/2\pi = k_0/L = 2/10$. This filter is a *bandpass filter*, since frequencies outside the narrow band around $\hat{\omega} = 0.4\pi$ are given much less gain than those in the passband of the filter.

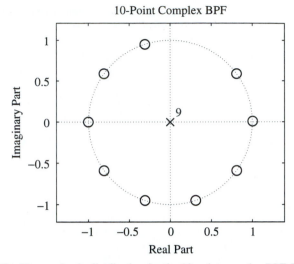

Figure 7.12 Zero and pole distribution for the 10-point complex BPF. The zero at angle $2\pi k_0/L = 2\pi(0.2)$ is the one missing from the tenth roots of unity. The other nine zeros are spread out at equal angles $(2\pi k/10)$ around the unit circle; there are nine poles at the origin.

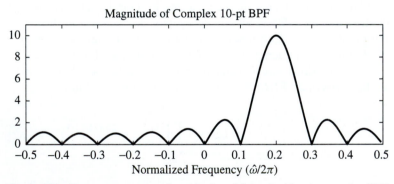

Figure 7.13 Frequency response (magnitude only) for the 10-point complex BPF. These are the values along the unit circle in the z-plane. There are nine zeros spread out uniformly along the frequency axis.

The formula in (7.7.6) is ideal for seeing how to make the frequency response of a BPF, but it is not very useful for calculating the filter coefficients of the bandpass filter. If one attempts a direct multiplication of the factors in (7.7.6), nine complex terms must be combined. When all the algebra is finished, the resulting filter coefficients will turn out to be complex-valued. This fact is obvious if we realize that the zeros in Fig. 7.12 *cannot* all be grouped as complex-conjugate pairs.

Another strategy is needed to get the filter coefficients. One idea is to view the zero distribution in Fig. 7.12 as a rotation of the zeros of the running-sum filter in Fig. 7.10. Note that rotation of the z-plane representation will have the corresponding effect of shifting the frequency response along the $\hat{\omega}$-axis by the amount of the rotation. The desired rotation in this case is by the angle $2\pi k_0/L$. So the question is how to move the roots of a polynomial through a rotation. The answer is that we must multiply the kth filter coefficient b_k by $e^{-jk\theta}$ where θ is the desired angle of rotation.

Consider the following general operation on a polynomial $G(z)$:

$$H(z) = G(z/r)$$

Every occurrence of the variable z in the polynomial $G(z)$ is replaced by z/r. The effect of this substitution on the roots of $G(z)$ is to multiply them by r and make these the roots of $H(z)$. For the simple example $G(z) = z^2 - 3z + 2 = (z-2)(z-1)$,

$$H(z) = G(z/r) = (z/r)^2 - 3(z/r) + 2 = \frac{z^2 - 3rz + 2r^2}{r^2} = \frac{(z - 2r)(z - r)}{r^2}$$

The two roots of $H(z)$ are now $z = 2r$ and $z = r$.

In the case of the complex bandpass filter, $G(z)$ is the running-sum system function

$$G(z) = \sum_{k=0}^{L-1} z^{-k}$$

and the parameter r is a complex exponential $r = e^{j2\pi k_0/L}$. Remember that multiplication by a complex exponential will rotate a complex number through the angle $2\pi k_0/L$. Now it is easy to get the new filter coefficients

$$H(z) = G(z/r) = G(ze^{-j2\pi k_0/L}) = \sum_{k=0}^{L-1} z^{-k} e^{j2\pi k_0 k/L}$$

Thus the filter coefficients of the complex bandpass filter are

$$b_k = e^{j2\pi k_0 k/L} \qquad \text{for } k = 0, 1, 2, \ldots, L - 1 \qquad (7.7.7)$$

Another way to determine the frequency response of the complex bandpass filter is to compute it directly, as in

$$H(e^{j\hat{\omega}}) = \sum_{k=0}^{L-1} e^{j2\pi k_0 k/L} e^{-j\hat{\omega}k} = \sum_{k=0}^{L-1} e^{-j(\hat{\omega}-j2\pi k_0/L)k} = G(e^{j(\hat{\omega}-j2\pi k_0/L)}) \quad (7.7.8)$$

This equation shows that the frequency response of the system whose filter coefficients are given by (7.7.7) is a shifted (by $2\pi k_0/L$) version of the frequency response of the L-point running-sum filter.

7.7.3 A Bandpass Filter with Real Coefficients

The obvious disadvantage of the previous strategy is that the resulting filter coefficients (7.7.8) are complex. We can modify the strategy slightly to get a bandpass filter with real coefficients if we just take the real part of the complex BPF coefficients. Thus the kth filter coefficient is now

$$b_k = \cos(2\pi k_0 k/L) \quad \text{for } k = 0, 1, 2, \ldots, L-1$$

With these real filter coefficients, the new BPF can be written as the sum of two complex BPFs. By expanding the coefficients of z^{-k} in terms of complex exponentials we obtain

$$H(z) = \sum_{k=0}^{L-1} (\cos(2\pi k_0 k/L)) z^{-k}$$

$$= \sum_{k=0}^{L-1} z^{-k} \left(\tfrac{1}{2} e^{j2\pi k_0 k/L} + \tfrac{1}{2} e^{-j2\pi k_0 k/L} \right)$$

$$= \tfrac{1}{2} \sum_{k=0}^{L-1} z^{-k} e^{j2\pi k_0 k/L} + \tfrac{1}{2} \sum_{k=0}^{L-1} z^{-k} e^{-j2\pi k_0 k/L}$$

$$= H_1(z) + H_2(z)$$

where $H_1(z)$ is a complex bandpass filter centered on frequency $2\pi k_0/L$ and $H_2(z)$ is a complex bandpass filter centered on frequency $-2\pi k_0/L$. Figure 7.14 shows the frequency response for $L = 10$ and $k_0 = 2$. There are zeros of the frequency response at some of the frequencies $\hat{\omega} = 2\pi k/L$ because both component filters have zeros at all these frequencies except for $\pm 2\pi k_0/L$. As is the case with any real-valued filter, the magnitude of the frequency response exhibits a symmetry about $\hat{\omega} = 0$.

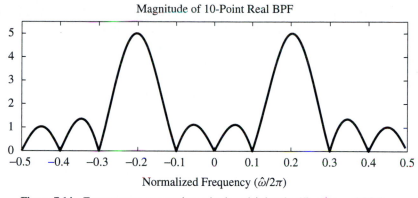

Figure 7.14   Frequency response (magnitude only) for the 10-point real BPF. Notice the two passbands at $\hat{\omega}/2\pi = \pm 2/10$—one at positive frequency, the other at negative frequency.

The frequency response in Fig. 7.14 has two peaks at $\hat{\omega} = \pm 4\pi/10$, so there must be two missing zeros on the unit circle at angles $\pm 4\pi/10$. In Fig. 7.15 we see that the two zeros at $z = e^{j4\pi/10} = e^{j2\pi(2)/10}$ and $z = e^{-j4\pi/10} = e^{j2\pi(8)/10}$ have been replaced by a single real zero. Thus, there are eight zeros on the unit circle and one on the real axis for a total of nine zeros, which is the order of the z-transform polynomial. The location of this new zero appears to be at $z = \cos(2\pi k_0/L) = \cos(0.4\pi) = 0.309$, which is the real part of the missing unit-circle zeros.

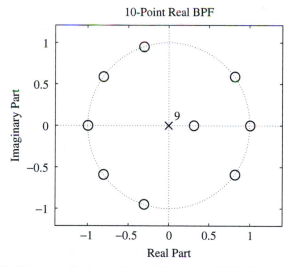

Figure 7.15   Pole-zero distribution for the 10-point real BPF. Of the original 10 roots of unity, two have been dropped off the unit circle at the angles $\pm 4\pi/10$, but a new one appears on the real axis. There are nine poles at the origin.

An algebraic manipulation will uncover the exact location of the new zero. We use the numerator-denominator representation and combine the two terms over a common denominator. To make the notation simpler, let $p = e^{j2\pi k_0/L}$, so that the conjugate is $p^* = e^{-j2\pi k_0/L}$. Then,

$$H(z) = H_1(z) + H_2(z)$$

$$= \frac{1}{2} \frac{z^L - 1}{z^{L-1}(z - p)} + \frac{1}{2} \frac{z^L - 1}{z^{L-1}(z - p^*)}$$

$$= \frac{1}{2} \frac{(z^L - 1)(z - p^*) + (z^L - 1)(z - p)}{z^{L-1}(z - p)(z - p^*)}$$

$$= \frac{(z^L - 1)(z - \frac{1}{2}(p + p^*))}{z^{L-1}(z - p)(z - p^*)}$$

The two factors $(z - p)(z - p^*)$ in the denominator cancel corresponding factors in the numerator polynomial $z^L - 1$, leaving $L - 2$ (in this case $L - 2 = 8$) of the Lth roots of unity. The term $(z - \frac{1}{2}(p + p^*))$ is the new zero at

$$z = \frac{1}{2}(p + p^*) = \frac{1}{2}\left(e^{j2\pi k_0/L} + e^{-j2\pi k_0/L}\right) = \cos(2\pi k_0/L)$$

which is the real part of the canceled zeros.

7.8 PRACTICAL BANDPASS FILTER DESIGN

Although much better filters can be designed by more sophisticated methods, the example of bandpass filter design discussed in Section 7.7 is a useful illustration of the power of the z-transform to simplify the analysis of such problems. Its characterization of a filter by the zeros (and poles) of $H(z)$ is used continually in filter design and implementation. The underlying reason is that the z-transform converts difficult problems involving convolution and frequency response into simple algebraic ideas based on multiplying and factoring polynomials. Thus, skills with basic algebra become essential everyday tools in design.

As a final example of FIR filters, we present a high-order FIR filter that has been designed by a computer-aided filter design program. Most digital filter design is carried out by software tools that permit much more sophisticated control of the passband and stopband characteristics than we were able to achieve with the simple analysis of Section 7.7. Design programs such as `remez` and `fir1` can be found in the MATLAB software. Although it is not our intention to discuss any of these methods or even how they are used, it is interesting to examine the output from the program `fir1` to get a notion of what can be achieved with a sophisticated design method.

The software allows us to specify the frequency range for the passband, and then computes a good approximation to an ideal filter that has unity gain over the passband and zero gain in the stopband. An example is shown in Figs. 7.16, 7.17, and 7.18 for a length–24 FIR bandpass filter.

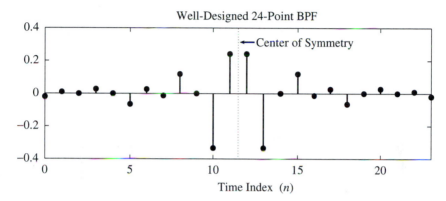

Figure 7.16   Impulse response for the 24-point BPF designed with MATLAB's `fir1` function. These are the FIR filter coefficients $\{b_k\}$, $k = 0, 1, 2, \ldots, 23$, which are needed in the difference equation implementation.

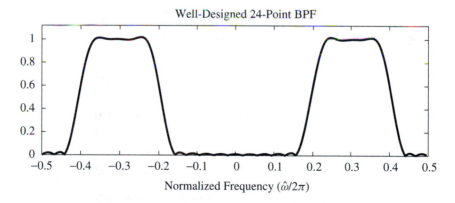

Figure 7.17   Frequency response (magnitude only) for the 24-point BPF.

The impulse response of the 24-point FIR bandpass filter is given in Fig. 7.16. Notice that the coefficients would be labeled $\{b_k\}$, from $k = 0, 1, 2, \ldots, 23$. The plot shows an obvious symmetry about a midpoint at $n = 23/2$. Indeed, for the impulse response and filter coefficients, we have

$$b_0 = -0.0193 = b_{23}$$
$$b_1 = 0.0099 = b_{22}$$
$$b_2 = -0.0003 = b_{21}$$
$$b_3 = 0.0276 = b_{20}$$
$$b_4 = 0.0000 = b_{19}$$
$$b_5 = -0.0649 = b_{18}$$
$$b_6 = 0.0264 = b_{17}$$
$$b_7 = -0.0126 = b_{16}$$
$$b_8 = 0.1188 = b_{15}$$
$$b_9 = 0.0000 = b_{14}$$
$$b_{10} = -0.3331 = b_{13}$$
$$b_{11} = 0.2422 = b_{12}$$

These are the coefficients used in the difference equation

$$y[n] = \sum_{k=0}^{23} b_k x[n - k]$$

to implement the filter whose frequency response is shown in Fig. 7.17. Notice the wide passbands in the frequency ranges corresponding to $2\pi(0.24) < |\hat{\omega}| < 2\pi(0.36)$. In these intervals, the gain of the filter deviates only slightly from one, so the amplitudes of sinusoidal signals with these frequencies will not be affected by the filter. We will see in Section 7.9 that this filter has linear phase, so these frequencies will only experience delay. Also note that in the regions $|\hat{\omega}| < 2\pi(0.16)$ and $2\pi(0.44) < |\hat{\omega}| < 2\pi(0.5)$ the gain is very nearly equal to zero. These regions are the "stopbands" of the filter, since the amplitudes of sinusoids with these frequencies will be multiplied by a very small gain and thus blocked from appearing in the output. Observe that the frequency response tapers smoothly from the passbands to the stopbands. In these transition regions, sinusoids will be reduced in amplitude according to the gain shown in Fig. 7.17. In many applications, we would want such transition regions to be very narrow. Ideally, we might even want them to have zero width. While this is theoretically achievable, it comes with a high price. It turns out that for an FIR frequency selective (lowpass, bandpass, highpass) filter, the width of the transition region is inversely proportional to M, the order of the system function polynomial. The higher the order, the narrower the transition regions can be, and as $M \to \infty$, the transition regions shrink to zero. Unfortunately, increasing M also increases the amount of computation required to compute each sample of the output, so a trade-off is always required in any practical application.

Figure 7.18 shows the pole-zero plot of the FIR filter. Notice the distinctive pattern of locations of the zeros. In particular, note how the zeros off the unit circle seem to be grouped into groups of four zeros. Indeed, for each zero that is not on the unit circle, there are also zeros at the conjugate location, at the reciprocal location, and at the conjugate reciprocal location. These groups of four zeros are strategically placed by the design process to form the passband of the filter. Similarly, the design process places zeros on the unit circle to ensure that the gain of the filter is low in the stopband regions of the frequency axis. Also, note that all complex zeros appear in conjugate pairs, and since the system function is a twenty-third-order polynomial, there are 23 poles at $z = 0$. In the Section 7.9, we will show that these properties of the pole-zero distribution are the direct result of the symmetry of the filter coefficients.

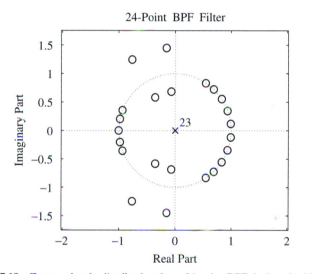

Figure 7.18   Zero and pole distribution for a 24-point BPF designed with MATLAB's `fir1` function.

7.9   PROPERTIES OF LINEAR-PHASE FILTERS

The filter that was discussed in Section 7.8 is an example of a class of systems where the sequence of coefficients (impulse response) has symmetry of the form $b_k = b_{M-k}$, for $k = 0, 1, \ldots, M$. Such systems have a number of interesting properties that are easy to show in the z-domain representation.

7.9.1   The Linear-Phase Condition

FIR systems that have symmetric filter coefficients (and, therefore, symmetric impulse responses) have frequency responses with linear phase. An example that we have already studied is the L-point moving averager, whose coefficients are all the same and therefore clearly satisfy the condition $b_k = b_{M-k}$, for $k = 0, 1, \ldots, M$. The example of Section 7.8 also has linear phase because it satisfies the same symmetry condition. To see why linear phase results from this symmetry, let us consider a simple example where the system function is of the form

$$H(z) = b_0 + b_1 z^{-1} + b_2 z^{-2} + b_1 z^{-3} + b_0 z^{-4} \qquad (7.9.1)$$

Thus, $M = 4$ and the length of the sequence is $L = M + 1 = 5$ samples. After working out the frequency response for this special case, the generalization will be obvious. First, observe that we can write $H(z)$ as

$$H(z) = [b_0(z^2 + z^{-2}) + b_1(z^1 + z^{-1}) + b_2]z^{-2}$$

If M were greater, we would simply have more groups of factors of the form $(z^k + z^{-k})$. Now when we substitute $z = e^{j\hat{\omega}}$, each of these factors will become a cosine term, i.e.,

$$H(e^{j\hat{\omega}}) = [2b_0 \cos(2\hat{\omega}) + 2b_1 \cos(\hat{\omega}) + b_2]e^{-j\hat{\omega}M/2}$$

In our specific example, we have shown that $H(e^{j\hat{\omega}})$ is of the form

$$H(e^{j\hat{\omega}}) = R(e^{j\hat{\omega}})e^{-j\hat{\omega}M/2} \qquad (7.9.2)$$

where, in this case, $M = 4$ and $R(e^{j\hat{\omega}})$ is the real function

$$R(e^{j\hat{\omega}}) = [2b_0 \cos(2\hat{\omega}) + 2b_1 \cos(\hat{\omega}) + b_2]$$

By following this style of analysis for the general case, it is easy to show that (7.9.2) holds whenever $b_k = b_{M-k}$, for $k = 0, 1, \ldots, M$. In the general case, it can be shown that

$$R(e^{j\hat{\omega}}) = \begin{cases} b_{\frac{M}{2}} + \sum_{k=0}^{\frac{M-2}{2}} 2b_k \cos[(\frac{M}{2} - k)\hat{\omega}] & \text{when } M \text{ is an even integer} \\\\ \sum_{k=0}^{\frac{M-1}{2}} 2b_k \cos[(\frac{M}{2} - k)\hat{\omega}] & \text{when } M \text{ is an odd integer} \end{cases} \qquad (7.9.3)$$

Equation (7.9.2) shows that the frequency response of any symmetric filter has the form of a real amplitude function $R(e^{j\hat{\omega}})$ times a linear-phase factor $e^{-j\hat{\omega}M/2}$. The latter factor, as we have seen in Chapter 6, corresponds to a delay of $M/2$ samples. Thus, the analysis presented in Section 6.7 for the running-average filter is typical of what happens in the general symmetric FIR case. The main difference is that by choosing the filter coefficients carefully, as in Section 7.8, we can shape the function $R(e^{j\hat{\omega}})$ to have a much more selective frequency response.

7.9.2 Locations of the Zeros of FIR Linear-Phase Systems

If the filter coefficients satisfy the condition $b_k = b_{M-k}$, for $k = 0, 1, \ldots, M$, it follows that

$$H(1/z) = z^M H(z) \qquad (7.9.4)$$

To demonstrate this "reciprocal property" of linear-phase filters, let us again consider a 5-point system of the form of (7.9.1). In this case,

$$
\begin{aligned}
H(1/z) &= b_0 + b_1(1/z)^{-1} + b_2(1/z)^{-2} + b_1(1/z)^{-3} + b_0(1/z)^{-4} \\
&= b_0 + b_1 z^1 + b_2 z^2 + b_1 z^3 + b_0 z^4 \\
&= z^4 (b_0 + b_1 z^{-1} + b_2 z^{-2} + b_1 z^{-3} + b_0 z^{-4}) \\
&= z^4 H(z)
\end{aligned}
$$

Following the same style for a general M, (7.9.4) is easily proved for the general case.

The reciprocal property of linear-phase filters is responsible for the distinctive pattern of zeros in the pole-zero plot of Fig. 7.18. The zeros that are not on the unit circle occur as quadruples. Furthermore, these quadruples are responsible for creating the passband of the BPF. Zeros on the unit circle occur in pairs because the complex conjugate partner must be present, and these zeros are mainly responsible for creating the stopband of the filter.

These properties can be shown to be true in general for linear-phase filters. Specifically,

> When $b_k = b_{M-k}$, for $k = 0, 1, \ldots, M$, then if z_0 is a zero of $H(z)$, so are its conjugate, its inverse, and its conjugate inverse; i.e., $\{\, z_0,\ z_0^*,\ 1/z_0,\ 1/z_0^* \,\}$ are all zeros of $H(z)$.

The conjugate zeros are included because the filter coefficients, which are also the coefficients of $H(z)$, are real. Therefore, all of the zeros of $H(z)$ must occur in complex-conjugate pairs. The inverse property is true because the filter coefficients are symmetric. Using (7.9.4), and assuming that z_0 is a zero of $H(z)$, we get

$$H(1/z_0) = z_0^M H(z_0) = 0$$

Since $H(z_0) = 0$, then we must have $H(1/z_0) = 0$ also. Most FIR filters are designed with a symmetry property, so the zero pattern of Fig. 7.18 is typical.

7.10 SUMMARY AND LINKS

The z-transform method was introduced in this chapter for FIR filters and finite-length sequences in general. The z-transform reduces the manipulation of LTI systems into simple operations on polynomials and rational functions. Roots of these z-transform polynomials are quite important because filter properties such as the frequency response can be inferred directly from the root locations.

We also introduced the important concept of "domains of representation" for discrete-time signals and systems. There are three domains: the n-domain or time domain, the $\hat{\omega}$-domain or frequency domain, and the z-domain. With three different domains at our disposal, even the most difficult problems generally can be simplified by switching to one of the other domains.

LINKS

Among the laboratory projects in Appendix C, we have already provided four on the topic of FIR filtering in Chapters 5 and 6. Lab C.5 deals with FIR filtering of sinusoidal waveforms. Lab C.6 deals with common filters such as the first difference and the L-point averager. In Labs C.7 and C.8, FIR filters will be used in practical systems, such as a touch-tone decoder in Lab C.7 and image smoothing in Lab C.8. Each of these labs should be easier to understand and simpler to carry out with the newly acquired background on z-transforms.

The CD-ROM also contains some demonstrations of the relationship between the z-plane and the frequency domain and time domain.

Z-Transform Demos

1. A three-domain movie that shows how the frequency response and the impulse response of an FIR filter change as a zero location is moved. Several different filters are demonstrated.

2. A movie that animates the relationship between the z-plane and the unit circle where the frequency response lies.

3. The MATLAB program pez, which was written by Craig Ulmer, to facilitate exploring the three domains. The M-files for pez can be copied and run under MATLAB.

HW Problems and Solutions

As in previous chapters, the reader is reminded of the large number of solved homework problems on the CD-ROM that are available for review and practice.

PROBLEMS

7.1   Use the superposition and time-delay properties to find the z-transforms of the following signals

$$x_1[n] = \delta[n]$$

$$x_2[n] = \delta[n-1]$$

$$x_3[n] = \delta[n-7]$$

$$x_4[n] = 2\delta[n] - 3\delta[n-1] + 4\delta[n-3]$$

7.2 Use the superposition and time-delay properties of (7.3.1) and (7.3.4) to determine the z-transform $Y(z)$ in terms of $X(z)$ if

$$y[n] = x[n] - x[n-1]$$

and in the process show that for the first difference system, $H(z) = 1 - z^{-1}$.

7.3 Suppose that an LTI system has a system function

$$H(z) = 1 + 5z^{-1} - 3z^{-2} + 2.5z^{-3} + 4z^{-8}$$

   (a) Determine the difference equation that relates the output $y[n]$ of the system to the input $x[n]$.

   (b) Determine and plot the output sequence $y[n]$ when the input is $x[n] = \delta[n]$.

7.4 An LTI system is described by the difference equation

$$y[n] = \frac{1}{3}(x[n] + x[n-1] + x[n-2])$$

   (a) Determine the system function $H(z)$ for this system.

   (b) Plot the poles and zeros of $H(z)$ in the z-plane.

   (c) From $H(z)$, obtain an expression for $H(e^{j\hat{\omega}})$, the frequency response of this system.

   (d) Sketch the frequency response (magnitude and phase) as a function of frequency for $-\pi \leq \hat{\omega} \leq \pi$.

   (e) What is the output if the input is

$$x[n] = 4 + \cos[0.25\pi(n-1)] - 3\cos[(2\pi/3)n]$$

7.5 Consider the following LTI system:

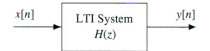

The system function of the LTI system is given by the formula

$$H(z) = (1 - z^{-1})(1 - e^{j\pi/2}z^{-1})(1 - e^{-j\pi/2}z^{-1})(1 - 0.9e^{j\pi/3}z^{-1})(1 - 0.9e^{-j\pi/3}z^{-1})$$

   (a) Write the difference equation that gives the relation between the input $x[n]$ and the output $y[n]$. (*Hint:* Multiply out the factors of $H(z)$.)

   (b) Plot the poles and zeros of $H(z)$ in the complex z-plane.

   (c) If the input is of the form $x[n] = Ae^{j\phi}e^{j\hat{\omega}n}$, for what values of $-\pi \leq \hat{\omega} \leq \pi$ will $y[n] = 0$?

7.6 The diagram in Fig. 7.19 depicts a cascade connection of two LTI systems; i.e., the output of the first system is the input to the second system, and the overall output is the output of the second system.

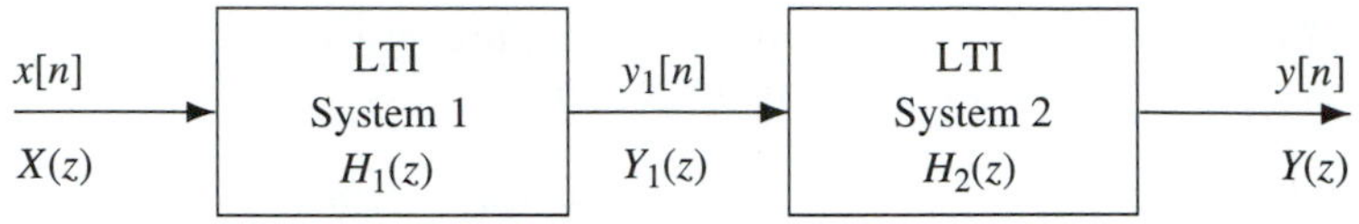

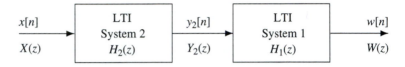

Figure 7.19   Cascade connection of two LTI systems.

(a) Use z-transforms to show that the system function for the overall system (from $x[n]$ to $y[n]$) is $H(z) = H_2(z)H_1(z)$, where $Y(z) = H(z)X(z)$.

(b) Use the result of (a) to show that the order of the systems is not important; i.e., show that for the same input $x[n]$ into the systems of Figs. 7.19 and 7.20, the overall outputs are the same ($w[n] = y[n]$).

$x[n]$ $\rightarrow$	LTI System 2 $H_2(z)$	$y_2[n]$ $\rightarrow$	LTI System 1 $H_1(z)$	$w[n]$ $\rightarrow$
$X(z)$		$Y_2(z)$		$W(z)$

Figure 7.20   Equivalent cascade system.

(c) Suppose that System 1 is a 3-point averager described by the difference equation $y_1[n] = \frac{1}{3}(x[n] + x[n-1] + x[n-2])$ and System 2 is described by the system function $H_2(z) = \frac{1}{3}(1 + z^{-1} + z^{-2})$. Determine the system function of the overall cascade system.

(d) Obtain a single difference equation that relates $y[n]$ to $x[n]$ in Fig. 7.19. Is the cascade of two 3-point averagers the same as a 6-point averager? Why would a better term be "weighted averager"?

(e) Plot the poles and zeros of $H(z)$ in the complex z-plane.

(f) From $H(z)$ obtain an expression for the frequency response $H(e^{j\hat{\omega}})$ and sketch the magnitude of the frequency response of the overall cascade system as a function of frequency for $-\pi \le \hat{\omega} \le \pi$.

7.7   Factor the following polynomial and plot its roots in the complex plane.

$$P(z) = 1 + \tfrac{1}{2}z^{-1} + \tfrac{1}{2}z^{-2} + z^{-3}$$

In MATLAB, see the functions called `roots` and `zplane`.

7.8   An LTI filter is described by the difference equation

$$y[n] = \frac{1}{4}\{x[n] + x[n-1] + x[n-2] + x[n-3]\} = \frac{1}{4}\sum_{k=0}^{3} x[n-k]$$

(a) What is $h[n]$, the impulse response of this system?

(b) Determine the system function $H(z)$ for this system.

(c) Plot the poles and zeros of $H(z)$ in the complex z-plane. *Hint:* Remember the Lth roots of unity.

(d) From $H(z)$, obtain an expression for the frequency response $H(e^{j\hat{\omega}})$ of this system.

(e) Sketch the frequency response (magnitude and phase) as a function of frequency (or plot it using `freqz( )` in MATLAB).

(f) Suppose that the input is

$$x[n] = 5 + 4\cos(0.2\pi n) + 3\cos(0.5\pi n + \pi/4) \quad \text{for } -\infty < n < \infty$$

Obtain an expression for the output in the form $y[n] = A + B\cos(\hat{\omega}_0 n + \phi_0)$.

7.9 The diagram in the following figure depicts a cascade connection of two LTI systems; i.e., the output of the first system is the input to the second system, and the overall output is the output of the second system.

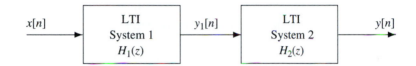

In this diagram, both systems are 4-point moving averagers.

(a) Determine the system function $H(z) = H_1(z)H_2(z)$ for the overall system.

(b) Plot the poles and zeros of $H(z)$ in the z-plane. *Hint:* The poles and zeros of $H(z)$ are the combined poles and zeros of $H_1(z)$ and $H_2(z)$.

(c) From $H(z)$, obtain an expression for the frequency response $H(e^{j\hat{\omega}})$ of the overall cascade system.

(d) Sketch the frequency response (magnitude and phase) functions of the overall cascade system for $-\pi \leq \hat{\omega} \leq \pi$.

(e) Use multiplication of z-transform polynomials to determine the impulse response $h[n]$ of the overall cascade system.

7.10 Suppose that an LTI system has system function equal to

$$H(z) = 1 - 3z^{-2} + 2z^{-3} + 4z^{-6}$$

The input to the system is the sequence

$$x[n] = 2\delta[n] + \delta[n-1] - 2\delta[n-2] + 4\delta[n-4]$$

(a) Without actually computing the output, determine from the above information the values of N_1 and N_2 so that the following is true:

$$y[n] = 0 \quad \text{for } n < N_1 \text{ and } n > N_2$$

(b) Use z-transforms and polynomial multiplication to find the sequence $y[n] = x[n] * h[n]$.

7.11 The intention of the following MATLAB program is to filter a sinusoid using the conv function.

```
omega = pi/6;
nn = [ 0:29 ];
xn = cos(omega*nn - pi/4);
bb = [ 1  0  0  1 ];
yn = conv( bb, xn );
```

(a) Determine $H(z)$ and the zeros of the FIR filter.

(b) Determine a formula for $y[n]$, the signal contained in the vector yn. Ignore the first few points, so your formula must be correct for $n \geq 3$. This formula should give numerical values for the amplitude, phase and frequency of $y[n]$.

(c) Give a value of omega such that the output is guaranteed to be zero, for $n \geq 3$.

7.12 Suppose that a system is defined by the operator

$$H(z) = (1 - z^{-1})(1 + z^{-2})(1 + z^{-1})$$

(a) Write the time–domain description of this system in the form of a difference equation.

(b) Write a formula for the frequency response of the system.

(c) Derive simple formulas for the magnitude response versus $\hat{\omega}$ and the phase response versus $\hat{\omega}$. These formulas must contain no complex terms and no square roots.

(d) This system can "null" certain input signals. For which input frequencies $\hat{\omega}_0$ is the response to $x[n] = \cos(\hat{\omega}_0 n)$ equal to zero?

(e) When the input to the system is $x[n] = \cos(\pi n/3)$ determine the output signal $y[n]$ in the form:

$$A \cos(\hat{\omega}_0 n + \phi)$$

Give numerical values for the constants A, $\hat{\omega}_0$ and ϕ.

7.13 Show that the system defined by the difference equation (7.6.7) will null *any* sinusoid of the form $A \cos(0.25\pi n + \phi)$ independent of the specific values of A and ϕ.

7.14 An LTI system has system function

$$H(z) = (1 + z^{-2})(1 - 4z^{-2}) = 1 - 3z^{-2} - 4z^{-4}$$

The input to this system is

$$x[n] = 20 - 20\delta[n] + 20\cos(0.5\pi n + \pi/4) \qquad \text{for } -\infty < n < \infty$$

Determine the output of the system $y[n]$ corresponding to the above input $x[n]$. Give an equation for $y[n]$ that is valid for all n. (*Note:* This is an easy problem if you approach it correctly!)

7.15 The input to the C-to-D converter in Fig. 7.21 is

$$x(t) = 4 + \cos(250\pi t - \pi/4) - 3\cos[(2000\pi/3)t]$$

The system function for the LTI system is

$$H(z) = \frac{1}{3}(1 + z^{-1} + z^{-2})$$

If $f_s = 1000$ samples/sec, determine an expression for $y(t)$, the output of the D-to-C converter.

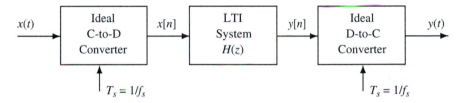

$$T_s = 1/f_s \qquad\qquad\qquad T_s = 1/f_s$$

Figure 7.21   Continuous-time filter implemented using a discrete-time filter.

7.16 Consider the following cascade system:

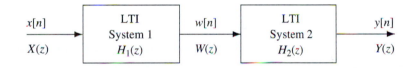

It is known that the system function of the overall system is

$$H(z) = (1 - z^{-2})(1 - 0.8e^{j\pi/4}z^{-1})(1 - 0.8e^{-j\pi/4}z^{-1})(1 + z^{-2})$$

   (a) Determine the poles and zeros of $H(z)$ and plot them in the complex z-plane.

   (b) It is possible to determine two system functions $H_1(z)$ and $H_2(z)$ so that: (1) the overall cascade system has the given system function $H(z)$, and (2) $w[n] = x[n] - x[n-4]$. Find $H_1(z)$ and $H_2(z)$.

   (c) Determine the difference equation that relates $y[n]$ to $w[n]$ for your answer in (b).

7.17 In Section 7.9 we showed that symmetric FIR filters have special properties. In this problem, we consider the antisymmetric case where $b_k = -b_{M-k}$, for $k = 0, 1, \ldots, M$. Consider the specific example

$$H(z) = b_0 + b_1 z^{-1} - b_1 z^{-3} - b_0 z^{-4}$$

where in this case, $b_2 = -b_2 = 0$.

(a) Show that, for this example,

$$H(e^{j\hat{\omega}}) = [2b_0 \sin(2\hat{\omega}) + 2b_1 \sin(\hat{\omega})]e^{j(\pi/2 - j\hat{\omega}2)}$$

(b) Show that, for this example,

$$H(1/z) = -z^4 H(z)$$

(c) Generalize these results for any M, i.e., both even and odd.

7.18 This diagram depicts a cascade connection of two LTI systems.

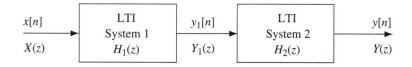

(a) Suppose that System 1 and System 2 both have a "square pulse" impulse response of the form

$$h_1[n] = h_2[n] = \delta[n] + \delta[n-1] + \delta[n-2] + \delta[n-3]$$

Determine the system functions $H_1(z)$ and $H_2(z)$ for the two systems.

(b) Use z-transforms to determine the system function $H(z)$ of the overall system.

(c) Use polynomial multiplication to determine the corresponding impulse response $h[n]$ for the overall system.

(d) Obtain a single difference equation that relates $y[n]$ to $x[n]$ in the cascade system.

(e) Can you see how this result could be used to do linear interpolation of a signal that had been subsampled by four? State a complete procedure based on your results in (a)–(d).

(f) Show that $H_1(z) = H_2(z)$ can be expressed as

$$H_1(z) = H_2(z) = \frac{1 - z^{-4}}{1 - z^{-1}}$$

(g) Find $H(z)$ and plot its poles and zeros in the complex z-plane.

(h) Show that the frequency responses of the two systems are

$$H_1(e^{j\hat{\omega}}) = H_2(e^{j\hat{\omega}}) = \frac{\sin(2\hat{\omega})}{\sin(\hat{\omega}/2)}e^{-j3\hat{\omega}/2}$$

(i) From $H(z)$ in (g) obtain an expression for the frequency response $H(e^{j\hat{\omega}})$ and sketch the magnitude of the frequency response of the overall cascade system as a function of frequency for $-\pi \le \hat{\omega} \le \pi$. (*Hint:* Note that $H(e^{j\hat{\omega}}) = H_2(e^{j\hat{\omega}})H_1(e^{j\hat{\omega}}) = [H_1(e^{j\hat{\omega}})]^2 = [H_2(e^{j\hat{\omega}})]^2$.)

8

IIR Filters

This chapter introduces a new class of LTI systems that have infinite duration impulse responses. For this reason, systems of this class are often called infinite-impulse-response (IIR) systems or also IIR filters. In contrast to FIR filters, IIR digital filters involve previously computed values of the output signal as well as values of the input signal in the computation of the present output. Since the output is "fed back" to be combined with the input, these systems are examples of the general class of *feedback systems*. From a computational point of view, since output samples are computed in terms of previously computed values of the output, the term *recursive filter* is also used for these filters.

The z-transform system functions for IIR filters are rational functions that have both poles and zeros at nonzero locations in the z-plane. Just as for the FIR case, we will show that many insights into the important properties of IIR filters can be obtained directly from the pole-zero representation.

We begin this chapter with the first-order IIR system, which is the simplest case because it involves feedback of only the previous output sample. We show by construction that the impulse response of this system has an infinite duration. Then the frequency response and the z-transform are developed for this class of filters. After showing the relationship among the three domains of representation for this simple case, we consider second-order filters. These filters are particularly important because they can be used to model "resonances" such as would occur in a speech synthesizer, as well as many other natural phenomena that exhibit vibratory behavior. The frequency response for the second-order case can exhibit a narrowband character that leads to the definition of bandwidth and center frequency, both of which can be controlled by appropriate choice of the feedback coefficients of the filter. The analysis and insights developed for the first- and second-order cases are readily generalized to higher-order systems.

8.1 THE GENERAL IIR DIFFERENCE EQUATION

RECURSIVE
FILTERS

FIR filters are extremely useful and have many nice properties, but that class of filters is not the most general class of LTI systems. This is because the output $y[n]$ is formed solely from a finite segment of the input signal $x[n]$. The most general class of digital filters that can be implemented with a finite amount of computation is obtained when the output is formed not only from the input, but also from previously computed outputs. The defining equation for this class of digital filters is the difference equation

$$y[n] = \sum_{\ell=1}^{N} a_\ell y[n - \ell] + \sum_{k=0}^{M} b_k x[n - k] \qquad (8.1.1)$$

The filter coefficients consist of two sets: $\{b_k\}$ and $\{a_\ell\}$. For reasons that will become obvious in the following simple example, the coefficients $\{a_\ell\}$ are called the *feedback* coefficients, and the $\{b_k\}$ are called the *feed-forward* coefficients. In all, $N + M + 1$ coefficients are needed to define the recursive difference equation (8.1.1).

Notice that if the coefficients $\{a_\ell\}$ are all zero, the difference equation (8.1.1) reduces to the difference equation of an FIR system. Indeed, we have asserted that (8.1.1) defines the most general class of LTI systems that can be implemented with finite computation, so FIR systems must be a special case. When discussing FIR systems we have referred to M as the "order" of the system. In this case, M is the number of delay terms in the difference equation and the degree or order of the polynomial system function. For IIR systems, we have both M and N as measures of the number of delay terms, and we will see that the system function of an IIR system is the ratio of an Mth-order polynomial to an Nth-order polynomial. Thus, there can be some ambiguity as to the order of an IIR system. In general, we will define N, the number of feedback terms, to be the order of an IIR system.

Example 8.1 Rather than tackle the general form given in (8.1.1), consider the first-order case where $M = N = 1$, i.e.,

$$y[n] = a_1 y[n - 1] + b_0 x[n] + b_1 x[n - 1] \qquad (8.1.2)$$

The block diagram representation of this difference equation, which is shown in Fig. 8.1, is constructed by noting that the signal $v[n] = b_0 x[n] + b_1 x[n - 1]$ is computed by the left half of the diagram, and we "close the loop" by computing $a_1 y[n - 1]$ from the delayed output and adding it to $v[n]$ to produce the output $y[n]$. This diagram clearly shows that the terms feed-forward and feedback describe the direction of signal flow in the block diagram.

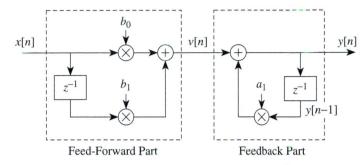

Feed-Forward Part   Feedback Part

Figure 8.1   First-order IIR system showing one feedback coefficient a_1 and two feed-forward coefficients b_0 and b_1.

$\diamond$

We will begin by studying a simplified version of the system defined by (8.1.2) and depicted in Fig. 8.1. This will involve characterizing the filter in each of the three domains: time domain, frequency domain, and z-domain. Since the filter is defined by a time–domain difference equation (8.1.2), we begin by studying how the difference equation is used to compute the output from the input, and we will illustrate how feedback results in an impulse response of infinite duration.

8.2   TIME–DOMAIN RESPONSE

To illustrate how the difference equation can be used to implement an IIR system, we will begin with a numerical example. Assume that the filter coefficients in (8.1.2) are $a_1 = 0.8$, $b_0 = 5$, and $b_1 = 0$, so that

$$y[n] = 0.8y[n-1] + 5x[n] \tag{8.2.1}$$

and assume that the input signal is

$$x[n] = 2\delta[n] - 3\delta[n-1] + 2\delta[n-3] \tag{8.2.2}$$

In other words, the total duration of the input is four samples, as shown in Fig. 8.2.

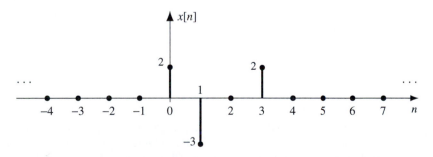

Figure 8.2   Input signal to recursive difference equation.

Although it is not a requirement, it is logical that the output sequence values should be computed in normal order; i.e., from left to right in a plot of the sequence. Furthermore, since the input is zero before $n = 0$, it would be natural to assume that $n = 0$ is the starting time of the output. Thus, we will consider computing the output from the difference equation (8.2.1) in the order $n = 0, 1, 2, 3, \ldots$. For example, the value of $x[0]$ is 2, so we can evaluate (8.2.1) at $n = 0$ obtaining

$$y[0] = 0.8y[0 - 1] + 5x[0] = 0.8y[-1] + 5(2) \tag{8.2.3}$$

Immediately we run into a problem: The value of $y[n]$ at $n = -1$ is not known. This is a serious problem, because no matter where we start computing the output, we will always have the same problem; at any point along the n-axis, we will need to know the output at the previous time $n - 1$. If we know the value $y[n - 1]$, however, we can use the difference equation to compute the next value of the output signal at time n. Once the process is started, it can be continued indefinitely by iteration of the difference equation. The solution requires the following two assumptions, which together are called the *initial rest conditions.*

INITIAL REST CONDITIONS

1. The input must be assumed to be zero prior to some starting time n_0, i.e., $x[n] = 0$ for $n < n_0$. We say that such inputs are *suddenly applied.*

2. The output is likewise assumed to be zero prior to the starting time of the signal, i.e., $y[n] = 0$ for $n < n_0$. We say that the system is *initially at rest* if its output is zero prior to the application of a suddenly applied input.

These conditions are not particularly restrictive, especially in the case of a real-time system, where a new output must be computed as each new sample of the input is taken. Real-time systems must, of course, be *causal* in the sense that the computation of the present output sample must not involve future samples of the input or output. Furthermore, any practical device would have a time at which it first begins to operate. All that is needed is for the memory containing the delayed output samples to be set initially to zero.[1]

With the initial rest assumption, we let $y[n] = 0$ for $n < 0$, so now we can evaluate $y[0]$ as

$$y[0] = 0.8y[-1] + 5(2) = 0.8(0) + 5(2) = 10$$

---

[1] In the case of a digital filter applied to sampled data stored in computer memory, the causality condition is not required, but, generally, the output is computed in the same order as the natural order of the input samples. The difference equation could be "recursed" backwards through the sequence, but this would require a different definition of "initial conditions."

Once we have started the recursion, the rest of the values follow easily, since the input signal and previous outputs are known.

$$y[1] = 0.8y[0] + 5x[1] = 0.8(10) + 5(-3) = -7$$

$$y[2] = 0.8y[1] + 5x[2] = 0.8(-7) + 5(0) = -5.6$$

$$y[3] = 0.8y[2] + 5x[3] = 0.8(-5.6) + 5(2) = 5.52$$

$$y[4] = 0.8y[3] + 5x[4] = 0.8(5.52) + 5(0) = 4.416$$

$$y[5] = 0.8y[4] + 5x[5] = 0.8(4.416) + 5(0) = 3.5328$$

$$y[6] = 0.8y[5] + 5x[6] = 0.8(3.5328) + 5(0) = 2.8262$$

$$\vdots \qquad \vdots$$

This output sequence is plotted in Fig. 8.3 up to $n = 7$.

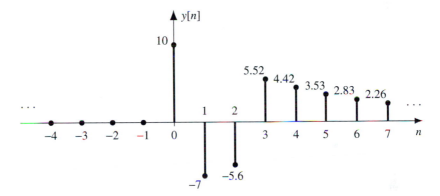

Figure 8.3   Output signal from recursive difference equation of (8.2.3) for the input of Fig. 8.2. For $n > 3$, the sequence is proportional to $(0.8)^n$, because the input signal ends at $n = 3$.

One key feature to notice in Fig. 8.3 is the structure of the output signal after the input turns off ($n > 3$). For this range of n, the difference equation becomes

$$y[n] = 0.8y[n-1] \qquad \text{for } n > 3$$

Thus the ratio between successive terms is a constant, and the output signal decays exponentially with a rate determined by $a_1 = 0.8$. Therefore, we can write the closed form expression

$$y[n] = y[3](0.8)^{n-3} \qquad \text{for } n \geq 3$$

for the rest of the sequence $y[n]$ once the value for $y[3]$ is known.

8.2.1 Linearity and Time Invariance of IIR Filters

When applied to the general IIR difference equation of (8.1.1), the condition of initial rest is sufficient to guarantee that the system implemented by iterating the difference equation is both linear and time-invariant. Although the feedback terms make the proof more complicated than the FIR case (see Section 5.5.3), we can show that, for suddenly applied inputs and initial rest conditions, the principle of superposition will hold because the difference equation involves only linear combinations of the input and output samples. Furthermore, since the initial rest condition is always applied just before the beginning of a suddenly applied input, time invariance also holds.

> **Exercise 8.1.** Assume that the input to the difference equation (8.2.1) is $x_1[n] = 10x[n-4]$, where $x[n]$ is given by (8.2.2) and Fig. 8.2. Use iteration to compute the corresponding output $y_1[n]$ for $n = 0, 1, \ldots, 11$ using the assumption of initial rest. Compare your result to the output plotted in Fig. 8.3 and verify that the system behaves as if it is both linear and time-invariant.

8.2.2 Impulse Response of a First-Order IIR System

In Chapter 5 we showed that the response to a unit impulse sequence characterizes a linear time-invariant system completely. Recall that when $x[n] = \delta[n]$, the resulting output signal, denoted by $h[n]$, is by definition the *impulse response*. Since all other input signals can be written as a superposition of weighted and delayed impulses, the corresponding output for all other signals can be constructed from weighted and shifted versions of the impulse response, $h[n]$. That is, since the recursive difference equation with initial rest conditions is an LTI system, its output can always be represented as the convolution sum

$$y[n] = \sum_{k=-\infty}^{\infty} x[k]h[n-k] \qquad (8.2.4)$$

Therefore, it is of interest to characterize the recursive difference equation by its impulse response.

To illustrate the nature of the impulse response of an IIR system, consider the first-order recursive difference equation with $b_1 = 0$,

$$y[n] = a_1 y[n-1] + b_0 x[n]. \qquad (8.2.5)$$

By definition, the difference equation

$$h[n] = a_1 h[n-1] + b_0 \delta[n] \qquad (8.2.6)$$

must be satisfied by the impulse response $h[n]$ for all values of n. We can construct a general formula for the impulse response in terms of the parameters a_1 and b_0 by simply constructing a table of a few values and then writing down the general formula by inspection. The following table shows the sequences involved in the computation:

n	$n < 0$	0	1	2	3	4	$\cdots$
$\delta[n]$	0	1	0	0	0	0	$\cdots$
$h[n-1]$	0	0	b_0	$b_0(a_1)$	$b_0(a_1)^2$	$b_0(a_1)^3$	$\cdots$
$h[n]$	0	b_0	$b_0(a_1)$	$b_0(a_1)^2$	$b_0(a_1)^3$	$b_0(a_1)^4$	$\cdots$

From this table it is obvious that the general formula is

$$h[n] = \begin{cases} b_0(a_1)^n & \text{for } n \geq 0 \\ 0 & \text{for } n < 0 \end{cases} \tag{8.2.7}$$

If we recall the definition of the unit step sequence

$$u[n] = \begin{cases} 1 & \text{for } n \geq 0 \\ 0 & \text{for } n < 0 \end{cases} \tag{8.2.8}$$

(8.2.7) can be expressed in the form

$$h[n] = b_0(a_1)^n u[n] \tag{8.2.9}$$

where multiplication of $(a_1)^n$ by $u[n]$ provides a compact representation of the conditions $n < 0$ and $n \geq 0$.

Example 8.2   For the example of (8.2.1) with $a_1 = 0.8$ and $b_0 = 5$, the impulse response is

$$h[n] = 5(0.8)^n u[n] = \begin{cases} 5(0.8)^n & \text{for } n \geq 0 \\ 0 & \text{for } n < 0 \end{cases} \tag{8.2.10}$$

This is the impulse response of the system in (8.2.1).   ◇

Exercise 8.2.   Substitute the solution (8.2.7) into the difference equation (8.2.6) and verify that the difference equation is satisfied for all values of n.

Exercise 8.3.   Find the impulse response of the following first-order system:

$$y[n] = 0.5y[n-1] + 5x[n-7]$$

Assume that the system is "at rest" for $n < 0$. Plot the resulting signal $h[n]$ as a function of n. Pay careful attention to where the nonzero portion of the impulse response begins.

A slightly more general problem would be to find the impulse response of the first-order system when a shifted version of the input signal is also included in the difference equation, i.e.,

$$y[n] = a_1 y[n-1] + b_0 x[n] + b_1 x[n-1]$$

Because this system is linear and time-invariant, it follows that its impulse response can be thought of as a sum of two terms as in

$$h[n] = b_0(a_1)^n u[n] + b_1(a_1)^{n-1} u[n-1] = \begin{cases} 0 & n < 0 \\ b_0 & n = 0 \\ (b_0 + b_1 a_1^{-1})(a_1)^n & n \geq 1 \end{cases}$$

Notice that the impulse response still decays exponentially with rate dependent only on a_1.

Exercise 8.4. Find the impulse response of the following first-order system:

$$y[n] = -0.5y[n-1] - 4x[n] + 5x[n-1].$$

Plot the resulting impulse response $h[n]$ as a function of n.

8.2.3 Response to Finite-Length Inputs

For finite-length inputs, the convolution sum is easy to evaluate for either FIR or IIR systems. Suppose that the finite-length input sequence is

$$x[n] = \sum_{k=N_1}^{N_2} x[k]\delta[n-k]$$

so that $x[n] = 0$ for $n < N_1$ and $n > N_2$. Then it follows from (8.2.4) that the corresponding output satisfies

$$y[n] = \sum_{k=N_1}^{N_2} x[k]h[n-k]$$

Example 8.3   As an example, consider again the LTI system defined by the difference equation (8.2.1), whose impulse response was shown in Example 8.2 to be $h[n] = 5(0.8)^n u[n]$. For the input of (8.2.2) and Fig. 8.2,

$$x[n] = 2\delta[n] - 3\delta[n-1] + 2\delta[n-3]$$

it is easily seen that

$$y[n] = 2h[n] - 3h[n-1] + 2h[n-3]$$

$$= 10(0.8)^n u[n] - 15(0.8)^{n-1}u[n-1] + 10(0.8)^{n-3}u[n-3]$$

To evaluate this expression for a specific time index, we need to take into account the different regions over which the individual terms are nonzero. If we do, we obtain

$$y[n] = \begin{cases} 0 & n < 0 \\ 10 & n = 0 \\ 10(0.8) - 15 = 7 & n = 1 \\ 10(0.8)^2 - 15(0.8) = -5.6 & n = 2 \\ 10(0.8)^3 - 15(0.8)^2 + 10 = 5.52 & n = 3 \\ 10(0.8)^n - 15(0.8)^{n-1} + 10(0.8)^{n-3} = 5.52(0.8)^{n-3} & n > 3 \end{cases}$$

A comparison to the output obtained by iterating the difference equation (see p. 253) shows that we have obtained the same values of the output sequence by superposition of scaled and shifted impulse responses.   ◇

Example 8.3 illustrates two important points about IIR systems.

1. The initial rest condition guarantees that the output sequence does not begin until the input sequence begins (or later).
2. Because of the feedback, the impulse response is infinite in extent, and the output due to a finite-length input sequence, being a superposition of scaled and shifted impulse responses, is therefore also infinite in extent. This is in contrast to the FIR case, where a finite-length input always produces a finite-length output sequence.

Exercise 8.5. Find the impulse response of the first-order system

$$y[n] = -0.5y[n-1] + 5x[n]$$

and use it to find the output due to the input signal

$$x[n] = \begin{cases} 0 & n < 1 \\ 3 & n = 1 \\ -2 & n = 2 \\ 0 & n = 3 \\ 3 & n = 4 \\ -1 & n = 5 \\ 0 & n > 5 \end{cases}$$

Write a formula that is the sum of four terms, each of which is a shifted impulse response. Assume the "at rest" initial condition. Plot the resulting signal $y[n]$ as a function of n for $0 \le n \le 10$.

8.2.4 Step Response of a First-Order Recursive System

When the input signal is infinitely long, the computation of the output of an IIR system using the difference equation is no different than for an FIR system; we simply continue to iterate the difference equation as long as samples of the output are desired. In the FIR case, the difference equation and the convolution sum are the same thing. This is not true in the IIR case, and computing the output using convolution is practical only in cases where simple formulas exist for both the input and the impulse response. Thus, in general, IIR filters must be implemented by iterating the difference equation. The computation of the response of a first-order IIR system to a unit step input provides a relatively simple illustration of the issues involved.

Again, assume that the system is defined by

$$y[n] = a_1 y[n-1] + b_0 x[n]$$

and assume that the input is the unit step sequence given by

$$u[n] = \begin{cases} 1 & \text{for } n \geq 0 \\ 0 & \text{for } n < 0 \end{cases} \tag{8.2.11}$$

As before, the difference equation can be iterated to produce the output sequence one sample at a time. The first few values are tabulated here. Work through the table to be sure that you understand the computation.

n	$x[n]$	$y[n]$
$n < 0$	0	0
0	1	b_0
1	1	$b_0 + b_0(a_1)$
2	1	$b_0 + b_0(a_1) + b_0(a_1)^2$
3	1	$b_0(1 + a_1 + a_1^2 + a_1^3)$
4	1	$b_0(1 + a_1 + a_1^2 + a_1^3 + a_1^4)$
⋮	1	⋮

From the tabulated values, it is not difficult to see that a general formula for $y[n]$ is

$$y[n] = b_0(1 + a_1 + a_1^2 + \ldots + a_1^n) = b_0 \sum_{k=0}^{n} a_1^k$$

With a bit of manipulation, we can get a simple closed-form expression for the general term in the sequence $y[n]$. For this we need to recall the formula

$$\sum_{k=0}^{L} r^k = \begin{cases} \dfrac{1 - r^{L+1}}{1 - r} & r \neq 1 \\ L + 1 & r = 1 \end{cases} \qquad (8.2.12)$$

which is the formula for summing the first $L+1$ terms of a geometric series. Armed with this fact, the formula for $y[n]$ is easily seen to be

$$y[n] = b_0 \frac{1 - a_1^{n+1}}{1 - a_1} \qquad \text{for } n \geq 0, \qquad \text{if } a_1 \neq 1 \qquad (8.2.13)$$

Three cases can be identified: $|a_1| > 1$, $|a_1| < 1$, and $|a_1| = 1$. Further investigation of these cases reveals two types of behavior.

1. When $|a_1| > 1$, the term a_1^{n+1} in the numerator will dominate and the values for $y[n]$ will get larger and larger without bound. This is called an *unstable* condition and is usually a situation to avoid. We will say more about the issue of stability later in Sections 8.4.2 and 8.8.

2. When $|a_1| < 1$, the term a_1^{n+1} will decay to zero as $n \to \infty$. In this case, the system is *stable*. Therefore, we can find a limiting value for $y[n]$ as $n \to \infty$

$$\lim_{n \to \infty} y[n] = \lim_{n \to \infty} b_0 \frac{1 - a_1^{n+1}}{1 - a_1} = \frac{b_0}{1 - a_1}$$

3. When $|a_1| = 1$, we might have an unbounded output, but not always. For example, when $a_1 = 1$, we get $y[n] = (n+1)b_0$ for $n \geq 0$, and the output $y[n]$ grows as $n \to \infty$. On the other hand, for $a_1 = -1$, the output alternates; it is $y[n] = b_0$ for n even, but $y[n] = 0$ for n odd.

The MATLAB plot in Fig. 8.4 shows the step response for the filter

$$y[n] = 0.8y[n-1] + 5x[n]$$

Notice that the limiting value is 25, which can be calculated from the filter coefficients

$$\lim_{n \to \infty} y[n] = \frac{b_0}{1 - a_1} = \frac{5}{1 - 0.8} = 25$$

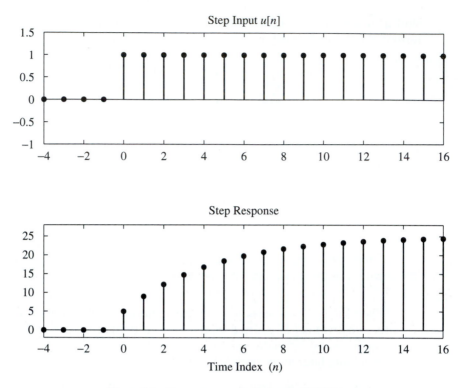

Figure 8.4 Step response of a first-order IIR filter.

Now suppose that we try to compute the step response by the convolution sum

$$y[n] = x[n] * h[n] = \sum_{k=-\infty}^{\infty} x[k]h[n-k] \qquad (8.2.14)$$

Since both the input and the impulse response have infinite durations, we might have difficulty in carrying out the computation. However, the fact that the input and output are given by the formulas $x[n] = u[n]$ and $h[n] = b_0(a_1)^n u[n]$ makes it possible to obtain a result. Substituting these formulas into (8.2.14) gives

$$y[n] = \sum_{k=-\infty}^{\infty} u[k]b_0(a_1)^{n-k}u[n-k]$$

The $u[k]$ and $u[n-k]$ terms inside the sum will change the limits on the sum because $u[k] = 0$ for $k < 0$ and $u[n-k] = 0$ for $n-k < 0$ (or $n < k$). The final result is

$$y[n] = \begin{cases} 0 & \text{for } n < 0 \\ \sum_{k=0}^{n} b_0(a_1)^{n-k} & \text{for } n \geq 0 \end{cases}$$

Using (8.2.12) we can write the step response for $n \geq 0$ as

$$y[n] = \sum_{k=0}^{n} b_0 (a_1)^{n-k} = b_0 (a_1)^n \sum_{k=0}^{n} (a_1)^{-k}$$

$$= b_0 (a_1)^n \frac{1 - (1/a_1)^{n+1}}{1 - (1/a_1)} = b_0 \frac{1 - (a_1)^{n+1}}{1 - a_1} \qquad (8.2.15)$$

which is identical to (8.2.13), the step response computed by iterating the difference equation. Notice that we were able to arrive at a closed-form expression in this case, because of the special nature of the input and impulse response. In general, it would be difficult or impossible to obtain such a closed-form result, but we can always use iteration of the difference equation to compute the output sample-by-sample.

8.3   SYSTEM FUNCTION OF AN IIR FILTER

We saw in Chapter 7 for the FIR case that the system function is the z-transform of the impulse response of the system, and that the system function and the frequency response are intimately related. Furthermore, the following result was shown:

Convolution in the n-domain corresponds to multiplication in the z-domain.

$$y[n] = h[n] * x[n] \qquad \Longleftrightarrow \qquad Y(z) = H(z)X(z)$$

The same is true for the IIR case. The system function for an FIR filter is always a polynomial; however, when the difference equation has feedback, it turns out that the system function $H(z)$ is the ratio of two polynomials. Ratios of polynomials are called *rational* functions. In this section we will determine the system function for the example of a first-order IIR system, and show how the system function, impulse response, and difference equation are related.

8.3.1   The General First-Order Case

The general form of the first-order difference equation with feedback is

$$y[n] = a_1 y[n-1] + b_0 x[n] + b_1 x[n-1] \qquad (8.3.1)$$

Since this equation must be satisfied for all values of n, we can take the z-transform of both sides of the equation to obtain

$$Y(z) = a_1 z^{-1} Y(z) + b_0 X(z) + b_1 z^{-1} X(z)$$

Subtracting the term $a_1 z^{-1} Y(z)$ from both sides of the equation leads to the following manipulations:

$$Y(z) - a_1 z^{-1} Y(z) = b_0 X(z) + b_1 z^{-1} X(z)$$

$$(1 - a_1 z^{-1}) Y(z) = (b_0 + b_1 z^{-1}) X(z)$$

Since the system is an LTI system, it should be true that $Y(z) = H(z)X(z)$, where $H(z)$ is the system function of the LTI system. Solving this equation for $H(z) = Y(z)/X(z)$, we obtain

$$H(z) = \frac{Y(z)}{X(z)} = \frac{b_0 + b_1 z^{-1}}{1 - a_1 z^{-1}} = \frac{B(z)}{A(z)} \tag{8.3.2}$$

Thus, we have shown that $H(z)$ for the first-order IIR system is a ratio of two polynomials. The numerator polynomial $B(z)$ is defined by the weighting coefficients $\{b_k\}$ that multiply the input signal $x[n]$ and its delayed versions; the denominator polynomial $A(z)$ is defined by the feedback coefficients $\{a_\ell\}$. That this correspondence is true in general should be clear from the analysis that leads to the formula for $H(z)$. Indeed, the following is true for IIR systems of *any* order:

The coefficients of the numerator polynomial of the system function of an IIR system are the coefficients of the feed-forward terms of the difference equation. For the denominator polynomial, the constant term is one, and the remaining coefficients are the negatives of the feedback coefficients.

In MATLAB, the `filter` function follows this same format. The statement

$$yy = filter(bb,aa,xx)$$

implements an IIR filter, where the vectors bb and aa hold the filter coefficients for the numerator and denominator polynomials, respectively.

Example 8.4 The following feedback filter:

$$y[n] = 0.5y[n-1] - 3x[n] + 2x[n-1].$$

would be implemented in MATLAB by

$$yy = filter([-3,2], [1,-0.5], xx)$$

where xx and yy are the input and output signal vectors, respectively. Notice that the aa vector has $-a_1$ for its second element, just like in the polynomial $A(z)$. We can imagine that the filter coefficient multiplying $y[n]$ is 1, so we always have 1 for the first element of aa. ◇

Exercise 8.6. Find the system function (i.e., z-transform) of the following feedback filter:

$$y[n] = 0.5y[n-1] - 3x[n] + 2x[n-1]$$

Exercise 8.7. Determine the system function of the system implemented by the following MATLAB statement:

```
yy = filter(5, [1,0.8], xx).
```

8.3.2 The System Function and Block-Diagram Structures

As we have seen, the system function displays the coefficients of the difference equation in a convenient way that makes it easy to move back and forth between the difference equation and the system function. In this section, we will show that this makes it possible to derive other difference equations, and thus other implementations, by simply manipulating the system function.

8.3.2.1   Direct Form I Structure   To illustrate the connection between the system function and the block diagram, let us return to the block diagram of Fig. 8.1, which is repeated in Fig. 8.5 for convenience. Block diagrams such as Fig. 8.5 are called *implementation structures*, or, more commonly, simply *structures*, because they give a pictorial representation of the difference equations that can be used to implement the system.

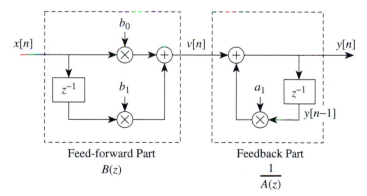

Figure 8.5   First-order IIR system showing one feedback coefficient a_1 and two feed-forward coefficients b_0 and b_1 in Direct Form I structure.

Recall that the product of two z-transform system functions corresponds to the cascade of two systems. The system function for the first-order feedback filter can be factored into an FIR piece and an IIR piece, as in

$$H(z) = \frac{b_0 + b_1 z^{-1}}{1 - a_1 z^{-1}} = \left(\frac{1}{1 - a_1 z^{-1}}\right)(b_0 + b_1 z^{-1}) = \left(\frac{1}{A(z)}\right) \cdot B(z)$$

The conclusion to be drawn from this algebraic manipulation is that a valid implementation for $H(z)$ is the pair of difference equations:

$$v[n] = b_0 x[n] + b_1 x[n-1] \tag{8.3.3}$$

$$y[n] = a_1 y[n-1] + v[n] \tag{8.3.4}$$

We see in Fig. 8.5 that the polynomial $B(z)$ is the system function of the feed-forward part of the block diagram, and that $1/A(z)$ is the system function of a feedback part that completes the system. The system implemented in this way is called the *Direct Form I* implementation because it is possible to go directly from the system function to this block diagram (or the difference equations that it represents) with no other manipulations than to write the numerator and denominator as polynomials in the variable z^{-1}.

8.3.2.2   Direct Form II Structure   We know that for an LTI cascade system, we can change the order of the systems without changing the overall system response. In other words,

$$H(z) = \left(\frac{1}{A(z)}\right) \cdot B(z) = B(z) \cdot \left(\frac{1}{A(z)}\right)$$

Using the correspondences that we have established leads to the block diagram shown in Fig. 8.6. Note that we have defined a new intermediate variable $w[n]$ as the output of the feedback part and input to the feed-forward part. Thus, the block diagram tells us that an equivalent implementation of the system is

$$w[n] = a_1 w[n-1] + x[n] \tag{8.3.5}$$

$$y[n] = b_0 w[n] + b_1 w[n-1] \tag{8.3.6}$$

Again, because there is such a direct and simple correspondence between Fig. 8.6 and $H(z)$, this implementation is called the *Direct Form II* implementation of the first-order IIR system with system function

$$H(z) = \frac{b_0 + b_1 z^{-1}}{1 - a_1 z^{-1}}$$

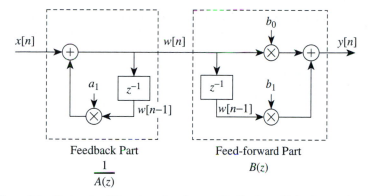

Figure 8.6   First-order IIR system showing one feedback coefficient a_1 and two feedforward coefficients b_0 and b_1 in a Direct Form II structure.

The block-diagram representation of Fig. 8.6 leads to a valuable insight. Notice that the input to each of the unit delay operators is the same signal $w[n]$. Thus, there is no need for two delay operations; they can be combined into a single delay, as in Fig. 8.7. Since delay operations are implemented with memory in a computer, the implementation of Fig. 8.7 would require less memory that the implementation of Fig. 8.6. Note, however, that both block diagrams represent the difference equations (8.3.5) and (8.3.6).

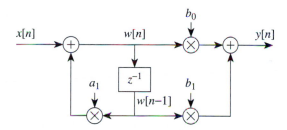

Figure 8.7   First-order IIR system in Direct Form II structure. This is identical to Fig. 8.6, except that the two delays have been merged into one.

Exercise 8.8. Find the z-transform system function of the following set of cascaded difference equations:

$$w[n] = -0.5w[n-1] + 7x[n]$$

$$y[n] = 2w[n] - 4w[n-1]$$

Draw the block diagrams of this system in both Direct Form I and Direct Form II.

8.3.2.3   The Transposed Form Structure   A somewhat surprising fact about block diagrams like Fig. 8.7 is that if the block diagram undergoes the following transformation:

1. All the arrows are reversed with multipliers being unchanged in value or location.

2. All branch points become summing points, and all summing points become branch points.

3. The input and the output are interchanged.

then the overall system has the same system function as the original system. We will not prove this, but it is true for the kinds of block diagrams that we have just introduced. However, we can use the z-transform to verify that this is true for our simple first-order system.

The feedback structure given in the signal-flow graph of Fig. 8.8 is the *transposed* form of the Direct Form II structure shown in Fig. 8.7. In order to derive the actual difference equations, we need to write the equations that are defined by the signal-flow graph. There is an orderly procedure for doing this if we follow two rules:

1. Assign variable names to the inputs of all delay elements—$v[n]$ is used in Fig. 8.8. The output of the delay is obviously $v[n-1]$.

2. Write equations at all of the summing nodes; there are two in this case.

$$y[n] = b_0 x[n] + v[n-1] \tag{8.3.7}$$

$$v[n] = b_1 x[n] + a_1 y[n] \tag{8.3.8}$$

The signal-flow graph specifies an actual computation, so (8.3.7) and (8.3.8) require three multiplies and two adds at each time step n. Equation (8.3.7) must be done first, because $y[n]$ is needed in (8.3.8).

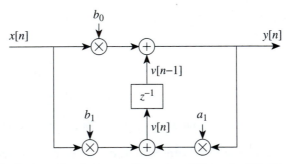

Figure 8.8   Computational structure for a general first-order IIR filter as a transposed Direct Form II structure.

Owing to the feedback, it is impossible to manipulate these equations into one of the other forms by eliminating variables. However, we can recombine these two equations in the z-transform domain to verify that we have the correct system function. First, we take the z-transform of each difference equation:

$$Y(z) = b_0 X(z) + z^{-1} V(z)$$

$$V(z) = b_1 X(z) + a_1 Y(z)$$

Now we eliminate $V(z)$ by substituting the second equation into the first.

$$Y(z) = b_0 X(z) + z^{-1}(b_1 X(z) + a_1 Y(z))$$

$$(1 - a_1 z^{-1})Y(z) = (b_0 + b_1 z^{-1})X(z)$$

so we get

$$H(z) = \frac{Y(z)}{X(z)} = \frac{b_0 + b_1 z^{-1}}{1 - a_1 z^{-1}}$$

Thus, because they have the same system function, (8.3.7) and (8.3.8) are equivalent to the Direct Form I (8.3.3) and (8.3.4), and to the Direct Form II (8.3.5) and (8.3.6).

Why are these different implementations of the same system function of interest to us? They all use the same number of multiplications and additions to compute exactly the same output from a given input. However, this is true only when the arithmetic is perfect. On a computer with finite precision (e.g., 16-bit words), each calculation will involve round-off errors, which means that each implementation will behave slightly differently. In practice, the implementation of high-quality digital filters in hardware demands correct engineering to control round-off errors and overflows.

8.3.3 Relation to the Impulse Response

In the analysis of Section 8.3.1 we implicitly assumed that the system function is the z-transform of the impulse response of an IIR system. While this is true, we have demonstrated only that it is true for the FIR case. In the IIR case, we need to be able to take the z-transform of an infinitely long sequence. As an example of such a sequence, consider $h[n] = a^n u[n]$. Applying the definition of the z-transform from equation (7.3.7), we would write

$$H(z) = \sum_{n=0}^{\infty} a^n z^{-n} = \sum_{n=0}^{\infty} (az^{-1})^n$$

which is the sum of all the terms in a geometric series where the ratio between successive terms is az^{-1}. Thus, we know that if $|az^{-1}| < 1$, then the sum is finite, and in fact is given by the closed-form expression

$$H(z) = \sum_{n=0}^{\infty} a^n z^{-n} = \frac{1}{1 - az^{-1}}$$

The condition for the infinite sum to equal the closed-form expression can be expressed as $|a| < z$. The values of z in the complex plane satisfying this condition are called the *region of convergence.* From the preceding analysis, we can state the following exceedingly important z-transform pair:

$$a^n u[n] \quad \Longleftrightarrow \quad \frac{1}{1 - az^{-1}} \qquad (8.3.9)$$

We will have many occasions to use this result in this chapter.

Example 8.5   As an example of the use of this result, recall that in Section 8.2.2 we showed by iteration that the impulse response of the system

$$y[n] = a_1 y[n-1] + b_0 x[n] + b_1 x[n-1] \qquad (8.3.10)$$

is

$$h[n] = b_0(a_1)^n u[n] + b_1(a_1)^{n-1} u[n-1] \qquad (8.3.11)$$

Thus, using the linearity property of the z-transform, the delay property of the z-transform, and the result of (8.3.9), the system function for this system is

$$H(z) = b_0 \left(\frac{1}{1 - a_1 z^{-1}} \right) + b_1 z^{-1} \left(\frac{1}{1 - a_1 z^{-1}} \right) = \frac{b_0 + b_1 z^{-1}}{1 - a_1 z^{-1}} \qquad (8.3.12)$$

which is what we obtained before by taking the z-transform of the difference equation and solving for $H(z) = Y(z)/X(z)$.   ◇

8.3.4 Summary of the Method

In this section, we have illustrated some important analysis techniques. We have seen that it is possible to go from the difference equation (8.3.10) directly to the system function (8.3.12). We have also seen that, in this simple example, it is possible by "taking the inverse z-transform" to go directly from the system function (8.3.12) to the impulse response (8.3.11) of the system without the tedious process of iterating the difference equation. We will see that it is possible to do this, in general, by a process of inverse z-transformation based on breaking the z-transform into a sum of terms like the right-hand side of (8.3.9). Before developing this technique, which will be applicable to higher-order systems, we will continue to focus on the first-order IIR system to illustrate some more important points about the z-transform and its relation to IIR systems.

8.4   POLES AND ZEROS

An interesting fact about the z-transform system function is that the numerator and denominator polynomials have zeros. These zero locations in the complex z-plane are very important for characterizing the system. Although we like to write the system function in terms of z^{-1} to facilitate the correspondence with the difference equation, it is probably better for finding roots to rewrite the polynomials as functions of z rather than z^{-1}. If we multiply the numerator and denominator by z, we obtain

$$H(z) = \frac{b_0 + b_1 z^{-1}}{1 - a_1 z^{-1}} = \frac{b_0 z + b_1}{z - a_1}$$

MATLAB GUI:
PeZ

In this form, it is easy to find the roots of the numerator and denominator polynomials. The numerator has one root at

$$b_0 z + b_1 = 0 \qquad \Longrightarrow \qquad z = -\frac{b_1}{b_0} \qquad \text{(Zero)}$$

and the denominator has its root at

$$z - a_1 = 0 \qquad \Longrightarrow \qquad z = a_1 \qquad \text{(Pole)}$$

If we consider $H(z)$ as a function of z over the entire complex z-plane, the root of the numerator is a *zero* of the function $H(z)$, i.e.,

$$H(z)\big|_{z=-(b_1/b_0)} = 0$$

Recall that the root of the denominator is a location in the z-plane where the function $H(z)$ "blows up"

$$H(z)\big|_{z=a_1} \to \infty$$

so this location ($z = a_1$) is called a "pole" of the system function $H(z)$.

Exercise 8.9. Find the poles and zeros of the following z-transform system function

$$H(z) = \frac{3 + 4z^{-1}}{1 + 0.5z^{-1}}$$

Exercise 8.10. For the z-transform of the following feedback filter

$$y[n] = 0.5y[n-1] - x[n] + 3x[n-1]$$

determine the locations of the pole and zeros.

8.4.1 Poles or Zeros at the Origin or Infinity

When the numerator and denominator polynomials have a different number of coefficients, we can have either zeros or poles at $z = 0$. We saw this in Chapter 7, where FIR systems, whose system functions have only a numerator polynomial, had a number of poles at $z = 0$ equal to the number of zeros of the polynomial. If we count all the poles and zeros at $z = \infty$, as well as $z = 0$, then we can assert that "the number of poles equals the number of zeros." Consider the following examples.

Example 8.6   The z-transform of the feedback filter

$$y[n] = 0.5y[n-1] + 2x[n]$$

is easily derived to be

$$H(z) = \frac{2}{1 - 0.5z^{-1}}$$

When we express $H(z)$ in positive powers of z

$$H(z) = \frac{2z}{z - 0.5}$$

we see that there is one pole at $z = 0.5$ and a zero at $z = 0$.   ◇

Example 8.7   For the z-transform of the feedback filter

$$y[n] = 0.5y[n-1] + 3x[n-1]$$

we can write $H(z)$ as

$$H(z) = \frac{3z^{-1}}{1 - 0.5z^{-1}} = \frac{3}{z - 0.5}$$

The system has one pole at $z = 0.5$, and if we take the limit of $H(z)$ as $z \to \infty$, we get $H(z) \to 0$. Thus, it also has one zero at $z = \infty$. $\diamond$

Both of the cases in Examples 8.6 and 8.7 have exactly one pole and one zero, if we count the zero at $z = \infty$.

Exercise 8.11. Determine the system function $H(z)$ of the following feedback filter:

$$y[n] = 0.5y[n - 1] + 3x[n - 2]$$

Show that $H(z)$ has a pole at $z = 0$, as well as $z = 0.5$. In addition, show that $H(z)$ has two zeros at $z = \infty$ by taking the limit as $z \to \infty$.

8.4.2 Pole Locations and Stability

The pole location of a first-order filter determines the shape of the impulse response. In Section 8.3.3 we showed that a system having system function

$$H(z) = b_0 \left(\frac{1}{1 - a_1 z^{-1}} \right) + b_1 z^{-1} \left(\frac{1}{1 - a_1 z^{-1}} \right)$$

$$= \frac{b_0 + b_1 z^{-1}}{1 - a_1 z^{-1}} = \frac{b_0(z + b_1/b_0)}{(z - a_1)}$$

has an impulse response

$$h[n] = b_0(a_1)^n u[n] + b_1(a_1)^{n-1} u[n - 1] = \begin{cases} 0 & \text{for } n < 0 \\ b_0 & \text{for } n = 0 \\ (b_0 + b_1 a_1^{-1})a_1^n & \text{for } n \geq 1 \end{cases}$$

That is, an IIR system with a single pole at $z = a_1$ has an impulse response that is proportional to a_1^n for $n \geq 1$. We see that if $|a_1| < 1$, the impulse response will die out as $n \to \infty$. On the other hand, if $|a_1| \geq 1$, the impulse response will not die out; in fact if $|a_1| > 1$, it will grow without bound. Since the pole of the system function is at $z = a_1$, we see that the location of the pole can tell us whether the impulse response will decay or grow. Clearly, it is desirable for the impulse response to die out, because an exponentially growing impulse response would produce unbounded outputs even if the input samples have finite size. Systems that produce bounded outputs when the input is bounded are called *stable systems*. If $|a_1| < 1$, the pole of the system function is *inside* the unit circle of the z-plane. It turns out that, for the IIR systems we have been discussing, the following is true in general:

> *A causal LTI IIR system with initial rest conditions is stable if all of the poles of its system function lie strictly inside the unit circle of the z-plane.*

Thus, stability of a system can be seen at a glance from a z-plane plot of the poles and zeros of the system function.

Example 8.8   The system whose system function is

$$H(z) = \frac{1 - 2z^{-1}}{1 - 0.8z^{-1}} = \frac{z - 2}{z - 0.8}$$

has a zero at $z = 2$ and a pole at $z = 0.8$. Therefore the system is stable. Note that the location of the zero, which is outside the unit circle, has nothing to do with stability of the system. Recall that the zeros correspond to an FIR system that is in cascade with an IIR system defined by the poles. Since FIR systems are always stable, it is not surprising that stability is determined solely by the poles of the system function. $\diamond$

Exercise 8.12.   An LTI IIR system has system function

$$H(z) = \frac{2 + 2z^{-1}}{1 - 1.25z^{-1}}$$

Plot the pole and zero in the z-plane, and state whether or not the system is stable.

8.5 FREQUENCY RESPONSE OF AN IIR FILTER

In Chapter 6, we introduced the concept of frequency response $\mathcal{H}(\hat{\omega})$ as the complex function that determines the amplitude and phase change experienced by a complex exponential input to an LTI system; i.e., if $x[n] = e^{j\hat{\omega}n}$, then

$$y[n] = \mathcal{H}(\hat{\omega})e^{j\hat{\omega}n} \tag{8.5.1}$$

In Chapter 7 we showed that the frequency response of an FIR system is related to the system function by

$$\mathcal{H}(\hat{\omega}) = H(e^{j\hat{\omega}}) = H(z)\big|_{z=e^{j\hat{\omega}}} \tag{8.5.2}$$

This relation between the system function and the frequency response also holds for IIR systems. However, we must add the provision that the system must be stable in order for the frequency response to exist and to be given by (8.5.2). This condition of stability is a general condition, but all FIR systems are stable, so up to now we have not had to be concerned with stability.

Recall that the system function for the general first-order IIR system has the form

$$H(z) = \frac{b_0 + b_1 z^{-1}}{1 - a_1 z^{-1}}$$

where the region of convergence of the system function is $|a_1 z^{-1}| < 1$ or $|a_1| < |z|$. If we wish to evaluate $H(z)$ for $z = e^{j\hat{\omega}}$, then the values of z on the unit circle should be in the region of convergence; i.e., we require $|z| = 1$ to be in the region of convergence of the z-transform. This means that $|a_1| < 1$, which was shown in Section 8.4.2 to be the condition for stability of the first-order system. In Section 8.8 we will give another interpretation of why stability is required for the frequency response to exist. Assuming stability in the first-order case, we get the following formula for the frequency response:

$$H(e^{j\hat{\omega}}) = H(z)\big|_{z=e^{j\hat{\omega}}} = \frac{b_0 + b_1 e^{-j\hat{\omega}}}{1 - a_1 e^{-j\hat{\omega}}} \qquad (8.5.3)$$

A simple evaluation will verify that (8.5.3) is a periodic function with a period equal to 2π. This must always be the case for the frequency response of a discrete-time system.

Remember that the frequency response $H(e^{j\hat{\omega}})$ is a complex-valued function of frequency $\hat{\omega}$. Therefore, we can reduce (8.5.3) to two separate real formulas for the magnitude and the phase as functions of frequency. For the magnitude response, it is expedient to compute the magnitude squared first, and then take a square root if necessary.

The magnitude squared can be formed by multiplying the complex $H(e^{j\hat{\omega}})$ in (8.5.3) by its conjugate (denoted by H^*). For our first-order example,

$$
\begin{aligned}
|H(e^{j\hat{\omega}})|^2 &= H(e^{j\hat{\omega}})H^*(e^{j\hat{\omega}}) \\[2mm]
&= \frac{b_0 + b_1 e^{-j\hat{\omega}}}{1 - a_1 e^{-j\hat{\omega}}} \cdot \frac{b_0^* + b_1^* e^{+j\hat{\omega}}}{1 - a_1^* e^{+j\hat{\omega}}} \\[2mm]
&= \frac{|b_0|^2 + |b_1|^2 + b_0 b_1^* e^{+j\hat{\omega}} + b_0^* b_1 e^{-j\hat{\omega}}}{1 + |a_1|^2 - a_1^* e^{+j\hat{\omega}} - a_1 e^{-j\hat{\omega}}} \\[2mm]
&= \frac{|b_0|^2 + |b_1|^2 + 2\Re e\{b_0^* b_1 e^{-j\hat{\omega}}\}}{1 + |a_1|^2 - 2\Re e\{a_1 e^{-j\hat{\omega}}\}}
\end{aligned}
$$

This derivation does not assume that the filter coefficients are real. If the coefficients were real, we would get the further simplification

$$|H(e^{j\hat{\omega}})|^2 = \frac{|b_0|^2 + |b_1|^2 + 2b_0 b_1 \cos(\hat{\omega})}{1 + |a_1|^2 - 2a_1 \cos(\hat{\omega})}$$

This formula is not particularly informative, because it is difficult to use it to visualize the shape of $|H(e^{j\hat{\omega}})|$. However, it could be used to write a program for evaluating and plotting the frequency response. The phase response is even messier. Arctangents would be used to extract the angles of the numerator and denominator, and then the two phases would be subtracted. When the filter coefficients are real, the phase is

$$\phi(\hat{\omega}) = \tan^{-1}\left(\frac{-b_1 \sin \hat{\omega}}{b_0 + b_1 \cos \hat{\omega}}\right) - \tan^{-1}\left(\frac{a_1 \sin \hat{\omega}}{1 - a_1 \cos \hat{\omega}}\right)$$

Again, the formula is so complicated that we cannot gain insight from it directly. In a later section, we will use the poles and zeros of the system function together with the relationship (8.5.2) to construct an approximate plot of the frequency response without recourse to formulas.

8.5.1 Frequency Response using MATLAB

Frequency responses can be computed and plotted easily by many signal processing software packages. In MATLAB, for example, the function freqz is provided for just that purpose. The frequency response is evaluated over an equally spaced grid in the $\hat{\omega}$ domain, and then its magnitude and phase can be plotted. In MATLAB, the functions abs and angle will extract the magnitude and the angle of each element in a complex vector.

Example 8.9   Consider the example

$$y[n] = 0.8y[n - 1] + 2x[n] + 2x[n - 1]$$

In order to define the filter coefficients in MATLAB, we put all the terms with $y[n]$ on one side of the equation, and the terms with $x[n]$ on the other.

$$y[n] - 0.8y[n - 1] = 2x[n] + 2x[n - 1]$$

Then we read off the filter coefficients and define the vectors aa and bb.

$$aa = [\ 1,\ -0.8\ ] \qquad\qquad bb = [\ 2,\ 2\ ]$$

Thus, the vectors aa and bb are in the same form as for the filter function. The following call to freqz will generate a 401-point vector HH containing the values of the frequency response at the vector of frequencies specified by the third argument, [-6:0.03:6].

$$HH = freqz(\ bb,\ aa,\ [-6:0.03:6]\ );$$

Plots of the resulting magnitude and phase are shown in Fig. 8.9. The frequency interval $-6 \le \hat{\omega} \le +6$ is shown so that the 2π-periodicity of $H(e^{j\hat{\omega}})$ will be evident.

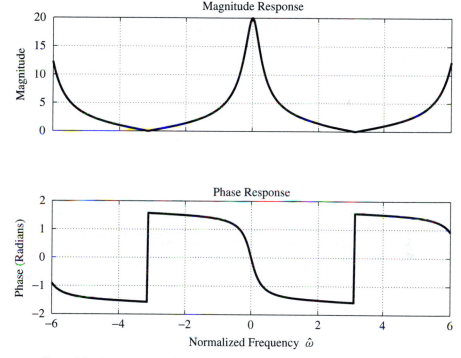

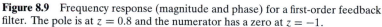

Figure 8.9   Frequency response (magnitude and phase) for a first-order feedback filter. The pole is at $z = 0.8$ and the numerator has a zero at $z = -1$.

◇

In this example, we can look for a connection between the poles and zeros and the shape of the frequency response. For this system, we have the system function

$$H(z) = \frac{2 + 2z^{-1}}{1 - 0.8z^{-1}}$$

which has a pole at $z = 0.8$ and a zero at $z = -1$. The point $z = -1$ is the same as $\hat{\omega} = \pi$ because $z = -1 = e^{j\pi} = e^{j\hat{\omega}}|_{\hat{\omega}=\pi}$. Thus, $H(e^{j\hat{\omega}})$ has the value zero at $\hat{\omega} = \pi$, since $H(z)$ is zero at $z = -1$. In a similar manner, the pole at $z = 0.8$ has an effect on the frequency response near $\hat{\omega} = 0$. Since $H(z)$ blows up at $z = 0.8$, the nearby points on the unit circle must have large values. The closest point on the unit circle is at $z = e^{j0} = 1$. In this case, we can evaluate the frequency response directly from the formula to get

$$H(e^{j\hat{\omega}})\Big|_{\hat{\omega}=0} = H(z)\Big|_{z=1} = \frac{2 + 2z^{-1}}{1 - 0.8z^{-1}}\Big|_{z=1} = \frac{2 + 2}{1 - 0.8} = \frac{4}{0.2} = 20$$

8.5.2 Three-Dimensional Plot of a System Function

The relationship between $H(e^{j\hat{\omega}})$ and the pole-zero locations of $H(z)$ can be illustrated by making a three-dimensional plot of $H(z)$ and then cutting out the frequency response. The frequency response $H(e^{j\hat{\omega}})$ is obtained by selecting the values of $H(z)$ along the unit circle, i.e., as $\hat{\omega}$ goes from $-\pi$ to $+\pi$, the equation $z = e^{j\hat{\omega}}$ defines the unit circle.

In this section, we use the system function

$$H(z) = \frac{1}{1 - 0.8z^{-1}}$$

to illustrate the relationship between the system function and the frequency response. Figures 8.10 and 8.11 are plots of the magnitude and phase of $H(z)$ over the region $[-1.4, 1.4] \times [-1.4, 1.4]$ of the z-plane. In the magnitude plot of Fig. 8.10, we observe that the pole (at $z = 0.8$) creates a large peak that makes all nearby values very large. At the precise location of the pole, $H(z) \to \infty$, but the grid in Fig. 8.10 does not contain the point $(z = 0.8)$, so the plot stays within a finite scale. The phase response in Fig. 8.11 also exhibits its most rapid transition at $\hat{\omega} = 0$ which is $z = 1$, the closest point on the unit circle to the pole at $z = 0.8$.

The frequency response $H(e^{j\hat{\omega}})$ is obtained by selecting the values of the z-transform along the unit circle in Figs. 8.10 and 8.11. Plots of $H(e^{j\hat{\omega}})$ versus $\hat{\omega}$ are given in Fig. 8.12. The shape of the frequency response can be explained in terms of the pole location by recognizing that in Fig. 8.10 the pole at $z = 0.8$ pushes $H(e^{j\hat{\omega}})$ up in the region near $\hat{\omega} = 0$ which is the same as $z = 1$. The unit circle values follow the ups and downs of $H(z)$ as $\hat{\omega}$ goes from $-\pi$ to $+\pi$.

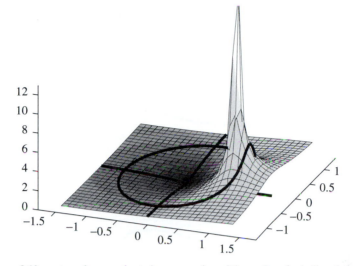

Figure 8.10 z-transform evaluated over a region of the z-plane including the unit circle. Values along the unit circle are shown as a dark line where the frequency response (magnitude) is evaluated. The view is from the fourth quadrant, so the point $z = 1$ is on the right. The first-order filter has a pole at $z = 0.8$ and a zero at $z = 0$.

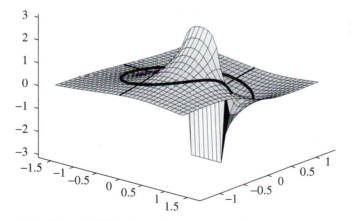

Figure 8.11   Phase of $H(z)$ evaluated over a region of the z-plane that includes the unit circle. View is from the fourth quadrant, so $z = 1$ lies to the right.

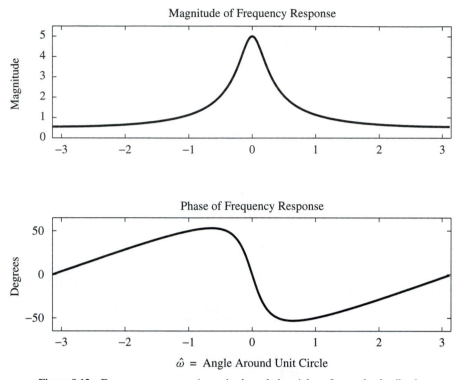

Figure 8.12 Frequency response (magnitude and phase) for a first-order feedback filter. The pole is at $z = 0.8$ and the numerator has a zero at $z = 0$. These are the values of $H(z)$ along the unit circle in the z-plane.

8.6 THREE DOMAINS

To illustrate the use of the analysis tools that we have developed, we consider the general second-order case. The three domains: n, z and $\hat{\omega}$ are depicted in Fig. 8.13. The defining equation for the IIR digital filter is the feedback difference equation, which, for the second-order case, is

$$y[n] = a_1 y[n-1] + a_2 y[n-2] + b_0 x[n] + b_1 x[n-1] + b_2 x[n-2]$$

This equation provides the algorithm for computing the output signal from the input signal by iteration using the filter coefficients $\{a_1, a_2, b_0, b_1, b_2\}$. It also defines the impulse response $h[n]$.

Following the procedures illustrated for the first-order case, we can also define the z-transform system function directly from the filter coefficients as

$$H(z) = \frac{b_0 + b_1 z^{-1} + b_2 z^{-2}}{1 - a_1 z^{-1} - a_2 z^{-2}}$$

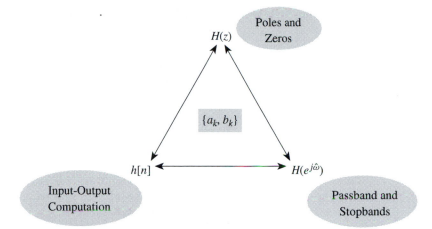

Figure 8.13   Relationship among the n-, z-, and $\hat{\omega}$-domains. The filter coefficients $\{a_k,\ b_k\}$ play a central role.

and we can also obtain the frequency response

$$H(e^{j\hat{\omega}}) = \frac{b_0 + b_1 e^{-j\hat{\omega}} + b_2 e^{-j2\hat{\omega}}}{1 - a_1 e^{-j\hat{\omega}} - a_2 e^{-j2\hat{\omega}}}$$

Since the system function is a ratio of polynomials, the poles and zeros of $H(z)$ make up a small set of parameters that completely define the filter.

Finally, the shapes of the passbands and stopbands of the frequency response are highly dependent on the pole and zero locations with respect to the unit circle, and the character of the impulse response can be related to the poles. To make this last point for the general case, we need to develop one more tool—a technique for getting $h[n]$ directly from $H(z)$. This process, which applies to any z-transform and its corresponding sequence, is called the *inverse z-transform*. The inverse z-transform is developed in Section 8.7.

8.7   THE INVERSE z-TRANSFORM AND SOME APPLICATIONS

We have seen how the three domains are connected for the first-order IIR system. Many of the concepts that we have introduced for the first-order system can be extended to higher-order systems in a straightforward manner. However, finding the impulse response from the system function is not an obvious extension of what we have done for the first-order case. We need to develop a process for inverting the z-transform that can be applied to systems with more than one pole. This process is called the *inverse z-transform*. In this section we will show how to find the inverse for a general rational z-transform. We will illustrate the process with some examples. The techniques that we develop will then be available for determining the impulse responses of second-order and higher-order systems.

8.7.1 Revisiting the Step Response of a First-Order System

In Section 8.2.4 we computed the step response of a first-order system by both iteration and convolution. Now we will show how the z-transform can be used for the same purpose. Consider a system whose system function is

$$H(z) = \frac{b_0 + b_1 z^{-1}}{1 - a_1 z^{-1}}$$

Recall that the z-transform of the output of this system is $Y(z) = H(z)X(z)$, so one approach to finding the output for a given input $x[n]$ is as follows:

1. Determine the z-transform $X(z)$ of the input signal $x[n]$.

2. Multiply $X(z)$ by $H(z)$ to get $Y(z)$.

3. Determine the inverse z-transform of $Y(z)$ to get the output $y[n]$.

Clearly, this procedure will work and will avoid both iteration and convolution if we can determine $X(z)$ and if we can perform the necessary inverse transformation. Our focus in this section will be on deriving a general procedure for step 3.

In the case of the step response, we see that the input $x[n] = u[n]$ is a special case of the more general sequence $a^n u[n]$; i.e., $a = 1$. Therefore, from (8.3.9) it follows that the z-transform of $x[n] = u[n]$ is

$$X(z) = \frac{1}{1 - z^{-1}}$$

so $Y(z)$ is

$$Y(z) = H(z)X(z) = \frac{b_0 + b_1 z^{-1}}{(1 - a_1 z^{-1})(1 - z^{-1})} = \frac{b_0 + b_1 z^{-1}}{1 - (1 + a_1)z^{-1} + a_1 z^{-2}} \qquad (8.7.1)$$

Now we need to go back to the n-domain by inverse transformation. A standard approach is to use a table of z-transform pairs and simply look up the answer in the table. Our previous discussions in Chapter 7 and earlier in this chapter have developed the basis for a simple version of such a table. A summary of the z-transform knowledge that we have developed so far is given in Table 8.1. Although more extensive tables can be constructed, the results that we have assembled in Table 8.1 are more than adequate for our purposes in this text.

SHORT TABLE OF z-TRANSFORMS		
$x[n]$	$\Longleftrightarrow$	$X(z)$
1. $ax_1[n] + bx_2[n]$	$\Longleftrightarrow$	$aX_1(z) + bX_2(z)$
2. $x[n - n_0]$	$\Longleftrightarrow$	$z^{-n_0} X(z)$
3. $y[n] = x[n] * h[n]$	$\Longleftrightarrow$	$Y(z) = H(z)X(z)$
4. $\delta[n]$	$\Longleftrightarrow$	1
5. $\delta[n - n_0]$	$\Longleftrightarrow$	z^{-n_0}
6. $a^n u[n]$	$\Longleftrightarrow$	$\dfrac{1}{1 - az^{-1}}$

Table 8.1: Summary of important z-transform properties and pairs.

Now let us return to the problem of finding $y[n]$ given $Y(z)$ in (8.7.1). The technique that we will use is based on the partial fraction expansion[2] of $Y(z)$. This technique is based on the observation that a rational function $Y(z)$ can be expressed as a sum of simpler rational functions, i.e.,

$$Y(z) = H(z)X(z) = \frac{b_0 + b_1 z^{-1}}{(1 - a_1 z^{-1})(1 - z^{-1})} = \frac{A}{1 - a_1 z^{-1}} + \frac{B}{1 - z^{-1}} \quad (8.7.2)$$

If the expression on the right is pulled together over a common denominator, it should be possible to find A and B so that the numerator of the resulting rational function will be equal to $b_0 + b_1 z^{-1}$. Equating the two numerators would give two equations in the two unknowns A and B. However, there is a much quicker way. A systematic procedure for finding the desired A and B is based on the observation that, for this example,

$$Y(z)(1 - a_1 z^{-1}) = \frac{b_0 + b_1 z^{-1}}{(1 - z^{-1})} = A + \frac{B(1 - a_1 z^{-1})}{1 - z^{-1}}$$

Then we can evaluate at $z = a_1$ to isolate A, i.e.,

$$Y(z)(1 - a_1 z^{-1})\Big|_{z=a_1} = \frac{b_0 + b_1 z^{-1}}{(1 - z^{-1})}\Big|_{z=a_1} = A + \frac{B(1 - a_1 z^{-1})}{1 - z^{-1}}\Big|_{z=a_1} = A$$

---

[2] The *partial fraction expansion* is an algebraic decomposition usually presented in calculus for evaluating certain types of integrals.

With this result, it is easy to see that

$$A = Y(z)(1 - a_1 z^{-1})\big|_{z=a_1} = \frac{b_0 + b_1 a_1^{-1}}{1 - a_1^{-1}}$$

Similarly, we can find B by

$$B = Y(z)(1 - z^{-1})\big|_{z=1} = \frac{b_0 + b_1}{1 - a_1}$$

Now using the superposition property of the z-transform (entry 1 in Table 8.1), and the exponential z-transform pair (entry 6 in Table 8.1), we can write down the desired answer as

$$y[n] = \left(\frac{b_0 + b_1 a_1^{-1}}{1 - a_1^{-1}}\right) a_1^n u[n] + \left(\frac{b_0 + b_1}{1 - a_1}\right) u[n]$$

which, after some manipulation becomes

$$y[n] = \left(\frac{(b_0 + b_1) - (b_0 a_1 + b_1)a_1^n}{1 - a_1}\right) u[n] \tag{8.7.3}$$

If we substitute the value $b_1 = 0$ into (8.7.3), we get

$$y[n] = b_0 \left(\frac{1 - a_1^{n+1}}{1 - a_1}\right) u[n]$$

which is the same result obtained in Section 8.2.4 both by iteration of the difference equation (8.2.13) and by convolution (8.2.15).

   With this example, we have established the framework for using the basic properties of z-transforms together with a few basic z-transform pairs to perform inverse z-transformation for any rational z-transform. We summarize this procedure in the following subsection.

8.7.2 A General Procedure for Inverse z-Transformation

Let $X(z)$ be any rational z-transform of degree N in the denominator and M in the numerator. Assuming that $M < N$, we can find the sequence $x[n]$ that corresponds to $X(z)$ by the following procedure:

PROCEDURE FOR INVERSE z-TRANSFORMATION ($M < N$)

1. Factor the denominator polynomial of $H(z)$ and express the pole factors in the form $(1 - p_k z^{-1})$ for $k = 1, 2, \ldots, N$.

2. Make a partial fraction expansion of $H(z)$ into a sum of terms of the form

$$H(z) = \sum_{k=1}^{N} \frac{A_k}{1 - p_k z^{-1}} \quad \text{where} \quad A_k = H(z)(1 - p_k z^{-1})\big|_{z=p_k}$$

3. Write down the answer as

$$h[n] = \sum_{k=1}^{N} A_k (p_k)^n u[n]$$

This procedure will always work if the poles, p_k, are distinct. Repeated poles complicate the process, but can be handled systematically. We will restrict our attention to the case of non-repeated poles. Furthermore, this procedure can be applied to the inversion of any rational z-transform, not just a system function. We will illustrate the use of this procedure with two examples.

Example 8.10   Let a z-transform $X(z)$ be

$$X(z) = \frac{1 - 2.1z^{-1}}{1 - 0.3z^{-1} - 0.4z^{-2}} = \frac{1 - 2.1z^{-1}}{(1 + 0.5z^{-1})(1 - 0.8z^{-1})}$$

We wish to write $X(z)$ in the form

$$X(z) = \frac{A}{1 + 0.5z^{-1}} + \frac{B}{1 - 0.8z^{-1}}$$

Continuing the procedure for partial fraction expansion we obtain

$$A = X(z)(1 + 0.5z^{-1})\big|_{z=-0.5} = \frac{1 - 2.1z^{-1}}{1 - 0.8z^{-1}}\bigg|_{z=-0.5} = \frac{1 + 4.2}{1 + 1.6} = 2$$

and

$$B = X(z)(1 - 0.8z^{-1})\big|_{z=0.8} = \frac{1 - 2.1z^{-1}}{1 + 0.5z^{-1}}\bigg|_{z=0.8} = \frac{1 - 2.1/0.8}{1 + 0.5/0.8} = -1$$

Therefore,

$$X(z) = \frac{2}{1 + 0.5z^{-1}} - \frac{1}{1 - 0.8z^{-1}} \tag{8.7.4}$$

and

$$x[n] = 2(-0.5)^n u[n] - (0.8)^n u[n]$$

Note that the poles at $z = p_1 = -0.5$ and $z = p_2 = 0.8$ give rise to terms in $x[n]$ of the form p_k^n. ◇

In Example 8.10, the degree of the numerator is $M = 1$ and the degree of the denominator is $N = 2$. This is important because the partial fraction expansion works only for rational functions such that $M < N$. The next example shows why this is so, and illustrates a method of dealing with this complication.

Example 8.11   Let $Y(z)$ be

$$Y(z) = \frac{2 - 2.4z^{-1} - 0.4z^{-2}}{1 - 0.3z^{-1} - 0.4z^{-2}} = \frac{2 - 2.4z^{-1} - 0.4z^{-2}}{(1 + 0.5z^{-1})(1 - 0.8z^{-1})}$$

Now we must add a constant term to the partial fraction expansion, otherwise, we cannot generate the term $-0.4z^{-2}$ in the numerator when we combine the partial fractions over a common denominator. That is, we must assume the following form for $Y(z)$:

$$Y(z) = \frac{A}{1 + 0.5z^{-1}} + \frac{B}{1 - 0.8z^{-1}} + C$$

How can we determine the constant C? One way is to perform long division of the denominator polynomial into the numerator polynomial until we get a remainder whose degree is lower than that of the denominator. In this case, the polynomial long division looks as follows:

$$
\begin{array}{r}
1 \\
-0.4z^{-2} - 0.3z^{-1} + 1 \;\overline{\smash{)}\; -0.4z^{-2} - 2.4z^{-1} + 2} \\
\underline{-0.4z^{-2} - 0.3z^{-1} + 1} \\
-2.1z^{-1} + 1
\end{array}
$$

Thus, if we place the remainder $(1 - 2.1z^{-1})$ over the denominator (in factored form), we can write $Y(z)$ as a rational part (fraction) plus the constant 1, i.e.,

$$Y(z) = \frac{1 - 2.1z^{-1}}{(1 + 0.5z^{-1})(1 - 0.8z^{-1})} + 1$$

The next step would be to apply the partial fraction expansion technique to the rational part of $Y(z)$. Since the rational part turns out to be identical to $X(z)$ in (8.7.4) from Example 8.10, the results would be the same as in that example, so we can write $Y(z)$ as

$$Y(z) = \frac{2}{1 + 0.5z^{-1}} - \frac{1}{1 - 0.8z^{-1}} + 1$$

Therefore, from Table 8.1,

$$y[n] = 2(-0.5)^n u[n] - (0.8)^n u[n] + \delta[n]$$

Notice again that the time–domain sequence has terms of the form p_k^n. The constant term in the system function generates an impulse, which is nonzero only at $n = 0$ (entry 4 in Table 8.1). ◇

8.8   STEADY-STATE RESPONSE AND STABILITY

A stable system is one that does not "blow up." This intuitive statement can be formalized by saying that the output of a stable system can always be bounded $(|y[n]| < M_y)$ whenever the input is bounded $(|x[n]| < M_x)$.[3]

We can use the inverse z-transform method developed in Section 8.7 to demonstrate an important point about stability, the frequency response, and the sinusoidal steady-state response. To illustrate this point, consider the LTI system defined by

$$y[n] = a_1 y[n - 1] + b_0 x[n]$$

From our discussion so far, we can state without further analysis that the system function of this system is

$$H(z) = \frac{b_0}{1 - a_1 z^{-1}}$$

and that the impulse response is

$$h[n] = b_0 a_1^n u[n]$$

We can state also that the frequency response is

$$H(e^{j\hat{\omega}}) = H(z)\big|_{z=e^{j\hat{\omega}}} = \frac{b_0}{1 - a_1 e^{-j\hat{\omega}}}$$

but this is true only if the system is stable $(|a_1| < 1)$. The objective of this section is to define the concept of stability and demonstrate its impact on the response to a sinusoid applied at $n = 0$.

---

[3] This definition for stability is called bounded-input, bounded-output stability. The constants M_x and M_y can be different.

Recall from Section 8.5 and equations (8.5.1) and (8.5.3) that the output of this system for a complex exponential input is

$$y[n] = H(e^{j\hat{\omega}_0})e^{j\hat{\omega}_0 n} = \left(\frac{b_0}{1 - a_1 e^{-j\hat{\omega}_0}}\right) e^{j\hat{\omega}_0 n} \qquad -\infty < n < \infty$$

What if the complex exponential input sequence is suddenly applied instead of existing for all n? The z-transform tools that we have developed make it easy to solve this problem. Indeed, the z-transform is ideally suited for situations where the sequences are either finite-length sequences or suddenly applied exponentials. For the suddenly applied complex exponential sequence with frequency $\hat{\omega}_0$

$$x[n] = e^{j\hat{\omega}_0 n} u[n]$$

the z-transform is found from entry 6 of Table 8.1 to be

$$X(z) = \frac{1}{1 - e^{j\hat{\omega}_0} z^{-1}}$$

and the z-transform of the output of the LTI system is

$$Y(z) = H(z)X(z) = \left(\frac{b_0}{1 - a_1 z^{-1}}\right)\left(\frac{1}{1 - e^{j\hat{\omega}_0} z^{-1}}\right) = \frac{b_0}{(1 - a_1 z^{-1})(1 - e^{j\hat{\omega}_0} z^{-1})}$$

Using the technique of partial fraction expansion, we can easily show that

$$Y(z) = \frac{\left(\frac{b_0 a_1}{a_1 - e^{j\hat{\omega}_0}}\right)}{1 - a_1 z^{-1}} + \frac{\left(\frac{b_0}{1 - a_1 e^{-j\hat{\omega}_0}}\right)}{1 - e^{j\hat{\omega}_0} z^{-1}}$$

Therefore, the output due to the suddenly applied complex exponential sequence is

$$y[n] = \left(\frac{b_0 a_1}{a_1 - e^{j\hat{\omega}_0}}\right)(a_1)^n u[n] + \left(\frac{b_0}{1 - a_1 e^{-j\hat{\omega}_0}}\right) e^{j\hat{\omega}_0 n} u[n] \qquad (8.8.1)$$

Equation (8.8.1) shows that the output consists of two terms. One term is proportional to an exponential sequence a_1^n that is solely determined by the pole at $z = a_1$. If $|a_1| < 1$, this term will die out with increasing n, in which case it would be called the *transient component*. The second term is proportional to the input complex exponential signal, and the constant of proportionality term is $H(e^{j\hat{\omega}_0})$, the frequency response of the system evaluated at the frequency of the suddenly applied complex sinusoid. This complex exponential component is the *sinusoidal steady-state component* of the output.

Now we see that the location of the pole of $H(z)$ is crucial if we want the output to reach the sinusoidal steady state. Clearly, if $|a_1| < 1$, then the system is stable and the pole is inside the unit circle. For this condition, the exponential a_1^n dies out and we can state that the limiting value for large n

$$y[n] \rightarrow \left(\frac{b_0}{1 - a_1 e^{-j\hat{\omega}_0}} \right) e^{j\hat{\omega}_0 n} = H(e^{j\hat{\omega}_0}) e^{j\hat{\omega}_0 n}$$

Otherwise, if $|a_1| \geq 1$, the term proportional to a_1^n will grow with increasing n and soon dominate the output. The following example gives a specific numerical illustration.

Example 8.12   If $b_0 = 5$, $a_1 = -0.8$, and $\hat{\omega}_0 = 2\pi/10$, the transient component is

$$y_t[n] = \left(\frac{-4}{-0.8 - e^{j0.2\pi}} \right) (-0.8)^n u[n] = 2.3351 e^{-j0.3502} (-0.8)^n u[n]$$

$$= 2.1933(-0.8)^n u[n] - j0.8012(-0.8)^n u[n]$$

Similarly, the steady-state component is

$$y_{ss}[n] = \left(\frac{5}{1 + 0.8 e^{-j0.2\pi}} \right) e^{j0.2\pi n} u[n] = 2.9188 e^{j0.2781} e^{j0.2\pi n} u[n]$$

$$= 2.9188 \cos\left(\tfrac{2\pi}{10} n + 0.2781 \right) u[n] + j\, 2.9188 \sin\left(\tfrac{2\pi}{10} n + 0.2781 \right) u[n]$$

Figure 8.14 shows the real part of the total output (top panel), and also the transient component (middle panel), and the steady-state component (bottom panel). The signals all start at $n = 0$

when the complex exponential is applied. Note how the transient component oscillates, but dies away, which explains why the steady-state component eventually equals the total output. In Fig. 8.14, $y_{ss}[n]$ and $y[n]$ look identical for $n > 15$.   ◇

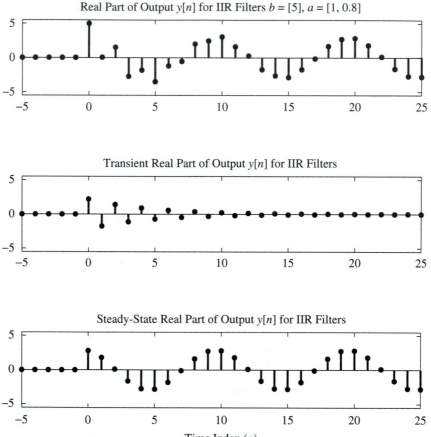

Figure 8.14 Illustration of transient and steady-state responses of an IIR system. The bottom panel is the complex exponential component (steady-state response); the middle panel is the decaying transient component; and the top panel is the total output.

On the other hand, if the pole were at $z = 1.1$, the system would be unstable and the output would "blow up" as shown in Fig. 8.15. In this case, the output contains a term $(1.1)^n$ that eventually dominates and grows without bound.

The result of Example 8.12 can be generalized by observing that the only difference between this example and a system with a higher-order system function is that the total output would include one exponential factor for each pole of the system function as well as the term $H(e^{j\hat{\omega}_0})e^{j\hat{\omega}_0 n}u[n]$. That is, it can be shown that for

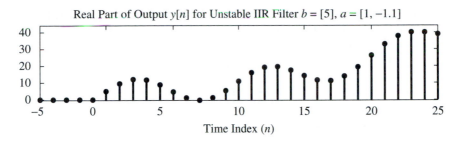

Figure 8.15   Illustration of an unstable IIR system. Pole is at $z = 1.1$.

a suddenly applied exponential input sequence $x[n] = e^{j\hat{\omega}_0 n}u[n]$, the output of a Nth-order IIR system will always be of the form

$$y[n] = \sum_{k=1}^{N} A_k(p_k)^n u[n] + H(e^{j\hat{\omega}_0})e^{j\hat{\omega}_0 n}u[n]$$

where the p_ks are the poles of the system function. Therefore, the sinusoidal steady state will exist and dominate in the total response if the poles of the system function all lie strictly inside the unit circle. This makes the concept of frequency response useful in a practical setting where all signals must have a beginning point at some finite time.

8.9   SECOND-ORDER FILTERS

We now turn our attention to filters with two feedback coefficients, a_1 and a_2. The general difference equation (8.1.1) becomes the second-order difference equation

$$y[n] = a_1 y[n - 1] + a_2 y[n - 2] + b_0 x[n] + b_1 x[n - 1] + b_2 x[n - 2] \quad (8.9.1)$$

MATLAB GUI:
PeZ

As before, we can characterize the second-order filter (8.9.1) in each of the three domains: time domain, frequency domain, and z-domain. We start with the z-transform domain because by now we have demonstrated that the poles and zeros of the system function give a great deal of insight into most aspects of both the time and frequency responses.

8.9.1 z-Transform of Second-Order Filters

Using the approach followed in Section 8.3.1 for the first-order case, we can take the z-transform of the second-order difference equation (8.9.1) by replacing each delay with z^{-1} (second entry in Table 8.1), and also replacing the input and output signals with their z-transforms:

$$Y(z) = a_1 z^{-1} Y(z) + a_2 z^{-2} Y(z) + b_0 X(z) + b_1 z^{-1} X(z) + b_2 z^{-2} X(z)$$

In the z-transform domain, the input-output relationship is $Y(z) = H(z)X(z)$, so we can solve for $H(z)$ by finding $Y(z)/X(z)$. For the second-order filter we get

$$Y(z) - a_1 z^{-1} Y(z) - a_2 z^{-2} Y(z) = b_0 X(z) + b_1 z^{-1} X(z) + b_2 z^{-2} X(z)$$

$$(1 - a_1 z^{-1} - a_2 z^{-2})Y(z) = (b_0 + b_1 z^{-1} + b_2 z^{-2})X(z)$$

which can be solved for $H(z)$ as

$$H(z) = \frac{Y(z)}{X(z)} = \frac{b_0 + b_1 z^{-1} + b_2 z^{-2}}{1 - a_1 z^{-1} - a_2 z^{-2}} \qquad (8.9.2)$$

Thus, the system function $H(z)$ for an IIR filter is a ratio of two second-degree polynomials, where the numerator polynomial depends on the feed-forward coefficients $\{b_k\}$ and the denominator depends on the feedback coefficients $\{a_\ell\}$. It should be possible to work problems such as Exercise 8.13 by simply reading the filter coefficients from the difference equation and then substituting them directly into the z-transform expression for $H(z)$.

Exercise 8.13. Find system function $H(z)$ of the following IIR filter:

$$y[n] = 0.5y[n-1] + 0.3y[n-2] - x[n] + 3x[n-1] - 2x[n-2]$$

Similarly, given the system function $H(z)$, it is a simple matter to write down the difference equation.

Exercise 8.14. For the system function

$$H(z) = \frac{1 + 2z^{-1} + z^{-2}}{1 - 0.8z^{-1} + 0.64z^{-2}}$$

write down the difference equation that relates the input $x[n]$ to the output $y[n]$.

Example 8.13   The connection between $H(z)$ and the difference equation can be generalized to higher-order filters. If we are given a fourth-order system

$$H(z) = \frac{1 - 3z^{-2}}{1 - 0.8z^{-1} + 0.6z^{-3} + 0.3z^{-4}}$$

the corresponding difference equation is

$$y[n] = 0.8y[n-1] - 0.6y[n-3] - 0.3y[n-4] + x[n] - 3x[n-2]$$

As before, note the sign change in the feedback coefficients, $\{a_k\}$.   ◇

8.9.2 Structures for Second-Order IIR Systems

The difference equation (8.9.1) can be interpreted as an algorithm for computing the output sequence from the input. Other computational orderings are possible, and the z-transform has the power to derive alternative structures through polynomial manipulations. Two alternative computational orderings that will implement the system defined by $H(z)$ in (8.9.2) are given in Fig. 8.16.

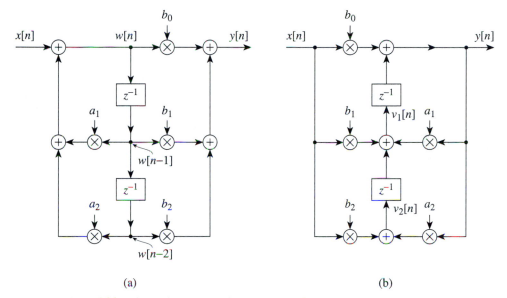

(a)   (b)

Figure 8.16   Alternative computational structures for the second-order recursive filter: (a) Direct Form II (DF-II), (b) Transposed Direct Form II (TDF-II).

In order to verify that each block diagram in Fig. 8.16 has the correct system function, we need to write the equations of the structure at the adders, and then eliminate the internal variable(s). For the case of the Direct Form II in Fig. 8.16(a), the equations at the output of the summing nodes are

$$y[n] = b_0 w[n] + b_1 w[n-1] + b_2 w[n-2]$$
$$w[n] = x[n] + a_1 w[n-1] + a_2 w[n-2]$$

(8.9.3)

It is impossible to eliminate $w[n]$ in these two equations, unless we work in the z-transform domain. The corresponding z-transform equations are

$$Y(z) = b_0 W(z) + b_1 z^{-1} W(z) + b_2 z^{-2} W(z)$$
$$W(z) = X(z) + a_1 z^{-1} W(z) + a_2 z^{-2} W(z)$$

which can be rearranged into the form

$$Y(z) = (b_0 + b_1 z^{-1} + b_2 z^{-2}) W(z)$$

$$X(z) = (1 - a_1 z^{-1} - a_2 z^{-2}) W(z)$$

Since the system function $H(z)$ is the ratio of $Y(z)$ to $X(z)$, we get

$$H(z) = \frac{Y(z)}{X(z)} = \frac{b_0 + b_1 z^{-1} + b_2 z^{-2}}{1 - a_1 z^{-1} - a_2 z^{-2}}$$

Thus we have proved that the Direct Form II (DF-II) structure in Fig. 8.16(a) which implements the pair of difference equations (8.9.3) is identical to the system defined by the single difference equation (8.9.1).

The Transposed Direct Form-II (Fig. 8.16(b)) can be worked out similarly. The difference equations represented by the block diagram are

$$y[n] = b_0 x[n] + v_1[n - 1]$$

$$v_1[n] = b_1 x[n] + a_1 y[n] + v_2[n - 1] \qquad (8.9.4)$$

$$v_2[n] = b_2 x[n] + a_2 y[n]$$

Taking the z-transform of each of the three equations gives

$$Y(z) = b_0 X(z) + z^{-1} V_1(z)$$

$$V_1(z) = b_1 X(z) + a_1 Y(z) + z^{-1} V_2(z)$$

$$V_2(z) = b_2 X(z) + a_2 Y(z)$$

Using these equations, we can eliminate $V_1(z)$ and $V_2(z)$ as follows:

$$Y(z) = b_0 X(z) + z^{-1}(b_1 X(z) + a_1 Y(z) + z^{-1} V_2(z))$$

$$Y(z) = b_0 X(z) + z^{-1}(b_1 X(z) + a_1 Y(z) + z^{-1}(b_2 X(z) + a_2 Y(z)))$$

Moving all the $X(z)$ terms to the right-hand side and the $Y(z)$ terms to the left-hand side gives

$$(1 - a_1 z^{-1} - a_2 z^{-2}) Y(z) = (b_0 + b_1 z^{-1} + b_2 z^{-2}) X(z)$$

so we get by division

$$H(z) = \frac{Y(z)}{X(z)} = \frac{b_0 + b_1 z^{-1} + b_2 z^{-2}}{1 - a_1 z^{-1} - a_2 z^{-2}}$$

Thus we have shown that the Transposed Direct Form II (TDF-II) is equivalent to the system function for the basic Direct Form-I difference equation in (8.9.1). Both examples illustrate the power of the z-transform approach in manipulating polynomials that correspond to different structures.

In theory, the system with system function given by (8.9.2) can be implemented by iterating any of the equations (8.9.1), (8.9.3), or (8.9.4). For example, the MATLAB function `filter` uses the TDF-II structure. However, as mentioned before, the reason for having different block diagram structures is that the order of calculation defined by equations (8.9.1), (8.9.3), and (8.9.4) is different. In a hardware implementation, the different structures will behave differently, especially when using fixed-point arithmetic where rounding error is fed back into the structure. With double-precision floating-point arithmetic as in MATLAB, there is little difference.

Exercise 8.15. Draw the block diagram of the Direct Form I difference equation defined by (8.9.1), and compare it to the other block diagrams in Fig. 8.16.

8.9.3 Poles and Zeros

Finding the poles and zeros of $H(z)$ is less confusing if we rewrite the polynomials as functions of z rather than z^{-1}. Thus the general second-order rational z-transform would become

$$H(z) = \frac{b_0 + b_1 z^{-1} + b_2 z^{-2}}{1 - a_1 z^{-1} - a_2 z^{-2}} = \frac{b_0 z^2 + b_1 z + b_2}{z^2 - a_1 z - a_2}$$

after multiplying the numerator and denominator by z^2. Recall from algebra the following important property of polynomials:

PROPERTY OF REAL POLYNOMIALS

A polynomial of degree N has N roots. If all the coefficients of the polynomial are real, the roots either must be real or must occur in complex conjugate pairs.

Therefore, in the second-order case, the numerator and denominator polynomials each have two roots, and there are two possibilities: Either the roots are complex conjugates of each other, or they are both real. We will now concentrate on the roots of the denominator, but exactly the same results hold for the numerator. From the quadratic formula, we get two poles at

$$\frac{a_1 \pm \sqrt{a_1^2 + 4a_2}}{2}$$

When $a_1^2 + 4a_2 \geq 0$, both poles are real; when $a_1^2 + 4a_2 = 0$, they are real and equal. However, when $a_1^2 + 4a_2 < 0$, the square root produces an imaginary result and we have complex-conjugate poles with values

$$p_1 = \tfrac{1}{2}a_1 + j\tfrac{1}{2}\sqrt{-a_1^2 - 4a_2} \qquad \text{and} \qquad p_2 = \tfrac{1}{2}a_1 - j\tfrac{1}{2}\sqrt{-a_1^2 - 4a_2}$$

In polar form, the complex poles can be expressed as $p_1 = re^{j\theta}$ and $p_2 = re^{-j\theta}$, where the radius r is

$$r = \sqrt{(\tfrac{1}{2}a_1)^2 + \tfrac{1}{4}(-a_1^2 - 4a_2)} = \sqrt{\tfrac{1}{4}a_1^2 - \tfrac{1}{4}a_1^2 - a_2} = \sqrt{-a_2}$$

and the angle θ satisfies

$$r\cos\theta = \tfrac{1}{2}a_1 \qquad \Longrightarrow \qquad \theta = \cos^{-1}\left(\frac{a_1}{2\sqrt{-a_2}}\right)$$

Example 8.14   The following $H(z)$ has two poles and two zeros.

$$H(z) = \frac{2 + 2z^{-1}}{1 - z^{-1} + z^{-2}} = 2\frac{z^2 + z}{z^2 - z + 1}$$

The poles $\{p_1, p_2\}$ and zeros $\{z_1, z_2\}$ are

$$p_1 = \tfrac{1}{2} + j\tfrac{1}{2}\sqrt{3} = e^{j\pi/3}$$
$$p_2 = \tfrac{1}{2} - j\tfrac{1}{2}\sqrt{3} = e^{-j\pi/3}$$
$$z_1 = 0$$
$$z_2 = -1$$

The system function can be written in factored form as either of the two forms

$$H(z) = \frac{2z(z + 1)}{(z - e^{j\pi/3})(z - e^{-j\pi/3})} = \frac{2(1 + z^{-1})}{(1 - e^{j\pi/3}z^{-1})(1 - e^{-j\pi/3}z^{-1})}$$

Since the numerator has no z^{-2} term, we have one zero at the origin. As is our custom, we plot these locations in the z-plane and mark the pole locations with **x** and the zeros with **o**. See Fig. 8.17.   ◇

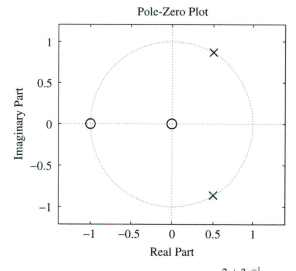

Figure 8.17   Pole-zero plot for system with $H(z) = \dfrac{2 + 2z^{-1}}{1 - z^{-1} + z^{-2}}$. The unit circle is shown for reference.

8.9.4 Impulse Response of a Second-Order IIR System

We have derived the general z-transform system function for the second-order filter

RECURSIVE
FILTERS

$$H(z) = \frac{B(z)}{A(z)} = \frac{b_0 + b_1 z^{-1} + b_2 z^{-2}}{1 - a_1 z^{-1} - a_2 z^{-2}} \tag{8.9.5}$$

and we have seen that the denominator polynomial $A(z)$ has two roots that define the poles of the second-order filter. Using the partial fraction expansion technique developed in Section 8.7, we can express the system function (8.9.5) as

$$H(z) = (-b_2/a_2) + \frac{A_1}{1 - p_1 z^{-1}} + \frac{A_2}{1 - p_2 z^{-1}}$$

where A_1 and A_2 can be evaluated by $A_k = H(z)(1 - p_k z^{-1})|_{z=p_k}$. Therefore, the impulse response will have the form

$$h[n] = (-b_2/a_2)\delta[n] + A_1(p_1)^n u[n] + A_2(p_2)^n u[n]$$

Furthermore, the poles may both be real or they may be a pair of complex conjugates. We will examine these two cases separately.

8.9.4.1   Real Poles   If p_1 and p_2 are real, the impulse response is composed of two real exponentials of the form p_k^n. This case is illustrated by the following example:

Example 8.15   Assume that

$$H(z) = \frac{1}{1 - \frac{5}{6}z^{-1} + \frac{1}{6}z^{-2}} = \frac{1}{(1 - \frac{1}{2}z^{-1})(1 - \frac{1}{3}z^{-1})} \tag{8.9.6}$$

from which we see that the poles are at $z = \frac{1}{2}$ and $z = \frac{1}{3}$ and that there are two zeros at $z = 0$. The poles and zeros of $H(z)$ are plotted in Fig. 8.18. We can extract the filter coefficients from $H(z)$ and write the following difference equation

$$y[n] = \frac{5}{6}y[n-1] - \frac{1}{6}y[n-2] + x[n] \tag{8.9.7}$$

which must be satisfied for any input and its corresponding output. Specifically, the impulse response would satisfy the difference equation

$$h[n] = \frac{5}{6}h[n-1] - \frac{1}{6}h[n-2] + \delta[n] \tag{8.9.8}$$

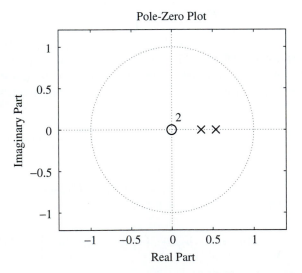

Figure 8.18   Pole-zero plot for system of Example 8.15. The poles are at $z = \frac{1}{2}$ and $z = \frac{1}{3}$; there are two zeros at $z = 0$.

which can be iterated to compute $h[n]$ if we know the values of $h[-1]$ and $h[-2]$, i.e., the values of the impulse response sequence just prior to $n = 0$ where the impulse first becomes nonzero. These values are supplied by the initial rest condition, which means that $h[-1] = 0$ and $h[-2] = 0$. The following table shows the computation of a few values of the impulse response:

n	$n < 0$	0	1	2	3	4	$\cdots$
$x[n]$	0	1	0	0	0	0	$\cdots$
$h[n-2]$	0	0	0	1	$\frac{5}{6}$	$\frac{19}{36}$	$\cdots$
$h[n-1]$	0	0	1	$\frac{5}{6}$	$\frac{19}{36}$	$\frac{65}{216}$	$\cdots$
$h[n]$	0	1	$\frac{5}{6}$	$\frac{19}{36}$	$\frac{65}{216}$	$\frac{211}{1296}$	$\cdots$

In contrast to the simpler first-order case, it is very difficult to guess the general nth term for the impulse response sequence. Fortunately, we can rely on the inverse z-transform technique to give us the general formula. Applying the partial fraction expansion to (8.9.6), we get

$$H(z) = \frac{3}{1 - \frac{1}{2}z^{-1}} - \frac{2}{1 - \frac{1}{3}z^{-1}}$$

which implies that

$$h[n] = 3(\tfrac{1}{2})^n \, u[n] - 2(\tfrac{1}{3})^n \, u[n] = \begin{cases} 3(\tfrac{1}{2})^n - 2(\tfrac{1}{3})^n & \text{for } n \geq 0 \\ 0 & \text{for } n < 0 \end{cases}$$

Since both poles are inside the unit circle, the impulse response dies out for n large, i.e., the system is stable. $\diamondsuit$

Exercise 8.16. Find the impulse response of the following second-order system:

$$y[n] = \tfrac{1}{4} y[n-2] + 5x[n] - 4x[n-1]$$

Plot the resulting signal $h[n]$ versus n.

8.9.5 Complex Poles

Now let us assume that the coefficients a_1 and a_2 in the second-order difference equation are such that the poles of $H(z)$ are complex. If we express the poles in polar form

$$p_1 = r\,e^{j\theta} \quad \text{and} \quad p_2 = r\,e^{-j\theta} = p_1^*$$

it is convenient to rewrite the denominator polynomial in terms of the parameters r and θ. Basic algebra allows us to start from the factored form and derive the polynomial coefficients:

$$
\begin{aligned}
A(z) &= \left(1 - p_1 z^{-1}\right)\left(1 - p_2 z^{-1}\right) \\
&= \left(1 - r e^{j\theta} z^{-1}\right)\left(1 - r e^{-j\theta} z^{-1}\right) \\
&= 1 - \left(r e^{j\theta} + r e^{-j\theta}\right) z^{-1} + r^2 z^{-2} \\
&= 1 - (2r \cos \theta) z^{-1} + r^2 z^{-2}
\end{aligned}
\tag{8.9.9}
$$

The system function is therefore

$$
H(z) = \frac{b_0 + b_1 z^{-1} + b_2 z^{-2}}{(1 - r e^{j\theta} z^{-1})(1 - r e^{-j\theta} z^{-1})} = \frac{b_0 + b_1 z^{-1} + b_2 z^{-2}}{1 - 2r \cos \theta z^{-1} + r^2 z^{-2}}
\tag{8.9.10}
$$

We can also identify the two feedback filter coefficients as

$$
a_1 = 2r \cos \theta \quad \text{and} \quad a_2 = -r^2
\tag{8.9.11}
$$

so the corresponding difference equation is

$$
y[n] = (2r \cos \theta) y[n-1] - r^2 y[n-2] + b_0 x[n] + b_1 x[n-1] + b_2 x[n-2]
\tag{8.9.12}
$$

This parameterization is significant because it allows us to see directly how the poles define the feedback terms of the difference equation (8.9.12). For example, if we want to change the angle of the pole, then we vary the coefficient a_1. Finally, we must remember that (8.9.11) is valid only for the special case of complex-conjugate poles; when the poles (p_1, p_2) are both real, the filter coefficients are

$$
a_1 = p_1 + p_2 \quad \text{and} \quad a_2 = p_1 p_2
$$

Example 8.16   Consider the following system

$$
y[n] = y[n-1] - y[n-2] + 2x[n] + 2x[n-1]
\tag{8.9.13}
$$

whose system function is

$$
H(z) = \frac{2 + 2z^{-1}}{1 - z^{-1} + z^{-2}} = \frac{2(1 + z^{-1})}{(1 - e^{j\pi/3} z^{-1})(1 - e^{-j\pi/3} z^{-1})}
\tag{8.9.14}
$$

A pole-zero plot for $H(z)$ was already given in Fig. 8.17. Using the partial fraction expansion technique, we can write $H(z)$ in the form

$$H(z) = \frac{\left(\frac{2+2e^{-j\pi/3}}{1-e^{-j2\pi/3}}\right)}{1 - e^{j\pi/3}z^{-1}} + \frac{\left(\frac{2+2e^{j\pi/3}}{1-e^{j2\pi/3}}\right)}{1 - e^{-j\pi/3}z^{-1}}$$

$$= \frac{2e^{-j\pi/3}}{1 - e^{j\pi/3}z^{-1}} + \frac{2e^{j\pi/3}}{1 - e^{-j\pi/3}z^{-1}}$$

so $h[n]$ is

$$h[n] = 2e^{-j\pi/3}e^{j(\pi/3)n}u[n] + 2e^{j\pi/3}e^{-j(\pi/3)n}u[n]$$

$$= 4\cos\left(\tfrac{2\pi}{6}(n-1)\right)u[n]$$

The two complex exponentials with frequencies $\pm\pi/3$ combine to form the cosine. The impulse response is plotted in Fig. 8.19.

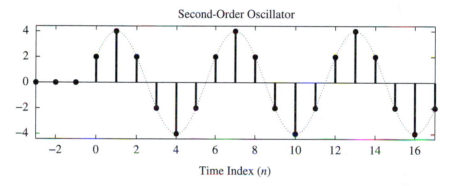

Figure 8.19   Impulse response for system with $A(z) = 1 - z^{-1} + z^{-2}$.

$\diamondsuit$

An important observation about the system in Example 8.16 is that it produces a pure sinusoid when stimulated by an impulse. Such a system is an example of a *sine wave oscillator*. After being stimulated by the single input sample from the impulse, the system continues indefinitely to produce a sinusoid of frequency $\hat{\omega}_0 = 2\pi(\tfrac{1}{6})$. This frequency is equal to the angle of the poles. A first-order filter (or a filter with all real poles) can only decay (or grow) as $(p)^n$ or oscillate up and down as $(-1)^n$, but a second-order system can oscillate with different periods. This is important when modeling physical signals such as speech, music, or other sounds.

Note that in order to produce the continuing sinusoidal output, the system must have its poles on the unit circle[4] of the z-plane, i.e., $r = 1$. Also note that the angle of

---

[4] Strictly speaking, a system with poles on the unit circle is unstable, so for some inputs it may blow up, but not for the impulse input.

the poles is exactly equal to the radian frequency of the sinusoidal output. Thus, we can control the frequency of the sinusoidal oscillator by adjusting the a_1 coefficient of the difference equation (8.9.12) while leaving a_2 fixed at $a_2 = -1$.

Example 8.17   As an example of an oscillator with a different frequency, we can use (8.9.12) to define a difference equation with prescribed pole locations. If we take $r = 1$ and $\theta = \pi/2$, as shown in Fig. 8.20, we get $a_1 = 2r\cos\theta = 0$ and $a_2 = -r^2 = -1$.

$$y[n] = -y[n-2] + x[n] \tag{8.9.15}$$

The system function of this system is

$$H(z) = \frac{1}{1 + z^{-2}}$$

$$= \frac{1}{(1 - e^{j\pi/2}z^{-1})(1 - e^{-j\pi/2}z^{-1})} = \frac{\tfrac{1}{2}}{1 - e^{j\pi/2}z^{-1}} + \frac{\tfrac{1}{2}}{1 - e^{-j\pi/2}z^{-1}}$$

The inverse z-transform gives a general formula for $h[n]$:

$$h[n] = \tfrac{1}{2}e^{j(\pi/2)n}u[n] + \tfrac{1}{2}e^{-j(\pi/2)n}u[n] = \begin{cases} \cos\left(2\pi\left(\tfrac{1}{4}\right)n\right) & \text{for } n \geq 0 \\ 0 & \text{for } n < 0 \end{cases} \tag{8.9.16}$$

Once again, the frequency of the cosine term in the impulse response is equal to the angle of the pole, $\pi/2 = 2\pi(\tfrac{1}{4})$.   ◇

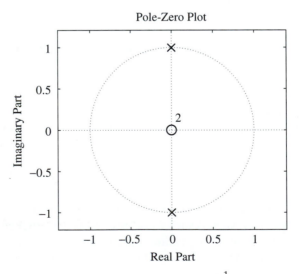

Pole-Zero Plot

Figure 8.20   Pole-zero plot for system with $H(z) = \dfrac{1}{1 + z^{-2}}$. The unit circle is shown for reference.

If the complex conjugate poles of the second-order system lie on the unit circle, the output oscillates sinusoidally and does not decay to zero. If the poles lie outside the unit circle, the output grows exponentially, whereas if they are inside the unit circle, the output decays exponentially to zero.

Example 8.18   As an example of a stable system, if we take $r = \frac{1}{2}$ and $\theta = \pi/3$, as shown in Fig. 8.21, we get $a_1 = 2r \cos \theta = 2(\frac{1}{2})(\frac{1}{2}) = \frac{1}{2}$ and $a_2 = -r^2 = -(\frac{1}{2})^2 = -\frac{1}{4}$, and the difference equation (8.9.12) becomes

$$y[n] = \tfrac{1}{2}y[n-1] - \tfrac{1}{4}y[n-2] + x[n] \tag{8.9.17}$$

The system function of this system is

$$H(z) = \frac{1}{1 - \frac{1}{2}z^{-1} + \frac{1}{4}z^{-2}} = \frac{\frac{e^{-j\pi/6}}{\sqrt{3}}}{1 - \frac{1}{2}e^{j\pi/3}z^{-1}} + \frac{\frac{e^{j\pi/6}}{\sqrt{3}}}{1 - \frac{1}{2}e^{-j\pi/3}z^{-1}}$$

and the general formula for $h[n]$ is

$$h[n] = \frac{2}{\sqrt{3}}(\tfrac{1}{2})^n \cos\left(2\pi(\tfrac{1}{6})n - \tfrac{\pi}{6}\right)u[n] \tag{8.9.18}$$

In this case, the general formula for $h[n]$ has a decay of $(\frac{1}{2})^n$ multiplying a periodic cosine with period 6. The frequency of the cosine term in the impulse response (8.9.18) is again the angle of the pole, $\pi/3 = 2\pi/6$; while the decaying term is controlled by the radius of the pole, i.e., $r^n = (\frac{1}{2})^n$.   ◇

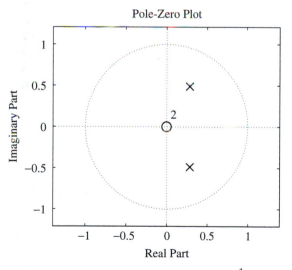

Figure 8.21   Pole-zero plot for system with $H(z) = \dfrac{1}{1 - \frac{1}{2}z^{-1} + \frac{1}{4}z^{-2}}$. The unit circle is shown for reference.

8.10   FREQUENCY RESPONSE OF SECOND-ORDER IIR FILTER

Since the frequency response of a stable system is related to the z-transform by

$$H(e^{j\hat{\omega}}) = H(z)\big|_{z=e^{j\hat{\omega}}}$$

we get the following formula for the frequency response of a second-order system:

$$H(e^{j\hat{\omega}}) = \frac{b_0 + b_1 e^{-j\hat{\omega}} + b_2 e^{-j2\hat{\omega}}}{1 - a_1 e^{-j\hat{\omega}} - a_2 e^{-j2\hat{\omega}}} \tag{8.10.1}$$

Since (8.10.1) contains terms like $e^{-j\hat{\omega}}$ and $e^{-j2\hat{\omega}}$, $H(e^{j\hat{\omega}})$ is guaranteed to be a periodic function with a period of 2π.

The magnitude squared of the frequency response can be formed by multiplying the complex $H(e^{j\hat{\omega}})$ by its conjugate (denoted by H^*). Rather than work out a general formula, we take a specific numerical example to show the kind of formula that results.

Example 8.19   Consider the case where the system function is

$$H(z) = \frac{1 - z^{-2}}{1 - 0.9z^{-1} + 0.81z^{-2}}$$

The magnitude squared is derived by multiplying out all the terms in the numerator and denominator of $H(e^{j\hat{\omega}})H^*(e^{j\hat{\omega}})$, and then collecting terms where the inverse Euler formula applies.

$$|H(e^{j\hat{\omega}})|^2 = H(e^{j\hat{\omega}})H^*(e^{j\hat{\omega}})$$

$$= \frac{1 - e^{-j2\hat{\omega}}}{1 - 0.9e^{-j\hat{\omega}} + 0.81e^{-j2\hat{\omega}}} \cdot \frac{1 - e^{j2\hat{\omega}}}{1 - 0.9e^{j\hat{\omega}} + 0.81e^{j2\hat{\omega}}}$$

$$= \frac{2 + 2\cos(2\hat{\omega})}{2.4661 - 3.258\cos\hat{\omega} + 1.62\cos(2\hat{\omega})}$$

This formula is useful because it is expressed completely in terms of cosine functions. The procedure is general, so a similar formula could be derived for any IIR filter. Since the cosine is an even function, we can state that any magnitude-squared function $|H(e^{j\hat{\omega}})|^2$ will always be even, i.e.,

$$|H(e^{-j\hat{\omega}})|^2 = |H(e^{j\hat{\omega}})|^2$$

The phase response is a bit messier. If arctangents are used to extract the angle of the numerator and denominator, then the two phases must be subtracted. The filter coefficients in this example are real, so the phase is

$$\phi(\hat{\omega}) = \tan^{-1}\left(\frac{\sin(2\hat{\omega})}{1 - \cos(2\hat{\omega})}\right) - \tan^{-1}\left(\frac{0.9\sin\hat{\omega} - 0.81\sin(2\hat{\omega})}{1 - 0.9\cos\hat{\omega} + 0.81\cos(2\hat{\omega})}\right)$$

which is an odd function of $\hat{\omega}$.   ◇

The formulas obtained in this example are too complicated to provide much insight directly. In a later section we will see how to use the poles and zeros of the system function to construct an approximate plot of the frequency response without recourse to such formulas.

8.10.1 Frequency Response via MATLAB

Tedious calculation and plotting of $H(e^{j\hat{\omega}})$ by hand is usually unnecessary if a computer program such as MATLAB is available. The MATLAB function freqz is provided for just that purpose. The frequency response can be evaluated over a grid in the $\hat{\omega}$ domain, and then its magnitude and phase can be plotted. In MATLAB, the functions abs and angle will extract the magnitude and the angle of each element in a complex vector.

Example 8.20   Consider the system introduced in Example 8.19:

$$y[n] = 0.9y[n-1] - 0.81y[n-2] + x[n] - x[n-2]$$

In order to define the filter coefficients in MATLAB, we put all the terms with $y[n]$ on one side of the equation, and the terms with $x[n]$ on the other.

$$y[n] - 0.9y[n-1] + 0.81y[n-2] = x[n] - x[n-2]$$

Then we read off the filter coefficients and define the vectors aa and bb.

$$aa = [\ 1,\ -0.9,\ 0.81\ ] \qquad bb = [\ 1,\ 0,\ -1\ ]$$

The following call to freqz will generate a vector HH containing the values of the frequency response at the vector of frequencies specified by the third argument, [-6:(pi/100):6].

$$HH = freqz(\ bb,\ aa,\ [-6:(pi/100):6]\ );$$

A plot of the resulting magnitude and phase is shown in Fig. 8.22. The frequency interval $-6 \le \hat{\omega} \le +6$ is shown so that the periodicity of $H(e^{j\hat{\omega}})$ will be evident; usually just the interval $-\pi \le \hat{\omega} \le +\pi$ or $-0.5 < \hat{f} \le 0.5$ is shown.

For this example, we can look for a connection between the poles and zeros and the shape of the frequency response. For this $H(z)$ we have

$$H(z) = \frac{1 - z^{-2}}{1 - 0.9z^{-1} + 0.81z^{-2}}$$

which has its poles at $z = 0.9e^{\pm j\pi/3}$ and its zeros at $z = 1$ and $z = -1$. Since $z = -1$ is the same as $z = e^{j\pi}$, we conclude that $H(e^{j\hat{\omega}})$ is zero at $\hat{\omega} = \pi$, because $H(z) = 0$ at $z = -1$; likewise, the zero of $H(z)$ at $z = +1$ is a zero of $H(e^{j\hat{\omega}})$ at $\hat{\omega} = 0$. The poles have angles of $\pm\pi/3$ rad, so the poles have an effect on the frequency response near $\hat{\omega} = \pm\pi/3$. Since $H(z)$

blows up at $z = 0.9e^{\pm j\pi/3}$, the nearby points on the unit circle (at $z = e^{\pm j\pi/3}$) must have large values. In this case, we can evaluate the frequency response directly from the formula to get

$$
H(e^{j\hat{\omega}})\Big|_{\hat{\omega}=\pi/3} = H(z)\Big|_{z=e^{j\pi/3}}
$$

$$
= \frac{1 - z^{-2}}{1 - 0.9z^{-1} + 0.81z^{-2}}\Bigg|_{z=e^{j\pi/3}}
$$

$$
= \frac{1 - (-\tfrac{1}{2} - j\tfrac{1}{2}\sqrt{3})}{1 - 0.9(\tfrac{1}{2} - j\tfrac{1}{2}\sqrt{3}) + 0.81(-\tfrac{1}{2} - j\tfrac{1}{2}\sqrt{3})}
$$

$$
= \frac{|1.5 + j0.5(\sqrt{3})|}{|0.145 + j0.045(\sqrt{3})|} = 10.522
$$

This value of the frequency response magnitude is a good approximation to the true maximum value, which actually occurs at $\hat{\omega} = 0.334\pi$.   ◇

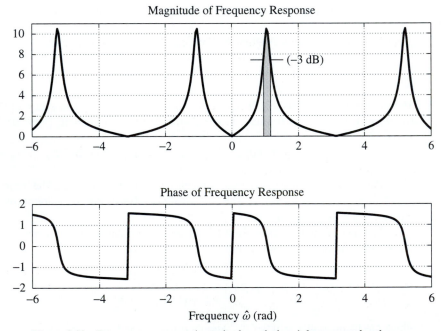

Figure 8.22   Frequency response (magnitude and phase) for a second-order feedback filter. The poles are at $z = 0.9e^{\pm j\pi/3}$ and the numerator has zeros at $z = 1$ and $z = -1$. The gray region shows the 3-dB bandwidth around the peak at $\hat{\omega} = \pi/3$.

8.10.2 3-dB Bandwidth

The width of the peak of the frequency response in Fig. 8.22 is called the *bandwidth*. It must be measured at some standard point on the plot of $|H(e^{j\hat{\omega}})|$. The most common practice is to use the 3-dB width, which is calculated as follows:

> Determine the peak value for $|H(e^{j\hat{\omega}})|$ and then find the nearest frequency on each side of the peak where the value of the frequency response is $(1/\sqrt{2})H_{\text{peak}}$. The 3-dB width is the difference $\Delta\hat{\omega}$ between these two frequencies.

In Fig. 8.22, the true peak value is 10.526 at $\hat{\omega} = 0.334\pi$, so we look for points where $|H(e^{j\hat{\omega}})| = (1/\sqrt{2})H_{\text{peak}} = (0.707)(10.526) = 7.443$. These occur at $\hat{\omega} = 0.302\pi$ and $\hat{\omega} = 0.369\pi$, so the bandwidth is $\Delta\hat{\omega} = 0.067\pi = 2\pi(0.0335) = 0.2105$ rad.

The 3-dB bandwidth calculation can be carried out efficiently with a computer program, but it is also helpful to have an approximate formula that can give quick "back-of-the-envelope" calculations. An excellent approximation for the second-order case with narrow peaks is given by the formula

$$\Delta\hat{\omega} \approx 2\frac{|1 - r|}{\sqrt{r}} \tag{8.10.2}$$

which shows that the distance of the pole from the unit circle $|1 - r|$ controls the bandwidth.[5] In Fig. 8.22, the bandwidth (8.10.2) evaluates to

$$\Delta\hat{\omega} = 2\frac{(1 - 0.9)}{0.95} = \frac{0.2}{0.95} \approx 0.2108 \text{ rad}$$

Thus we see that the approximation is quite good in this case, where the pole is rather close to the unit circle (radius = 0.9).

8.10.3 Three-Dimensional Plot of System Functions

Since the frequency response $H(e^{j\hat{\omega}})$ is the system function evaluated on the unit circle, we can illustrate the connection between the z and $\hat{\omega}$ domains with a three-dimensional plot such as the one shown in Fig. 8.23.

---

[5] This approximate formula for bandwidth is good only when the poles are isolated from one another. The approximation breaks down, for example, when a second-order system has two poles with small angles.

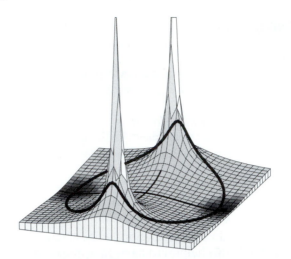

Figure 8.23 z-transform evaluated over a region of the z-plane including the unit circle. The view is from the fourth quadrant, so the $z = 1$ point is at the right. Values along the unit circle are the frequency response (magnitude) for a second-order feedback filter. The poles are at $z = 0.85e^{\pm j\pi/2}$ and the numerator has zeros at $z = \pm 1$.

Z-to-FREQ
FLYING
THRU
Z-PLANE

Figure 8.23 shows a plot of the system function $H(z)$ at points inside, outside, and on the unit circle. The peaks located at the poles, $0.85e^{\pm j\pi/2}$, determine the frequency response behavior near $\hat{\omega} = \pm\pi/2$. If the poles were moved closer to the unit circle, the frequency response would have a higher and narrower peak. The zeros at $z = \pm 1$ create valleys that lie on the unit circle at $\hat{\omega} = 0, \pi$.

We can estimate any value of $|H(e^{j\hat{\omega}})|$ directly from the poles and zeros. This can be done systematically by writing $H(z)$ in the following form:

$$H(z) = G \frac{(z - z_1)(z - z_2)}{(z - p_1)(z - p_2)}$$

where z_1 and z_2 are the zeros and p_1 and p_2 are the poles of the second-order filter. The parameter G is a gain term that may have to be factored out. Then the magnitude of the frequency response is

$$|H(e^{j\hat{\omega}})| = G \frac{|e^{j\hat{\omega}} - z_1| \, |e^{j\hat{\omega}} - z_2|}{|e^{j\hat{\omega}} - p_1| \, |e^{j\hat{\omega}} - p_2|} \tag{8.10.3}$$

Equation (8.10.3) has a simple geometric interpretation. Each term $|e^{j\hat{\omega}} - z_i|$ or $|e^{j\hat{\omega}} - p_i|$ is the vector length from the zero z_i or the pole p_i to the unit circle position $e^{j\hat{\omega}}$, shown in Fig. 8.24. The frequency response is the product of the lengths of the vectors to the zeros divided by the product of the lengths of the vectors to the poles. As we go around the unit circle, these vector lengths change. When we are on top of a zero, one of the numerator lengths is zero, so $|H(e^{j\hat\omega})| = 0$ at that frequency. When we are close to a pole, one of the denominator lengths is very small, so $|H(e^{j\hat\omega})|$ will be large at that frequency.

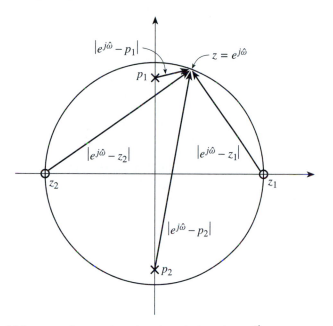

Figure 8.24 z-transform evaluated on the unit circle ($z = e^{j\hat\omega}$) by using a product of vector lengths from the poles and zeros.

We can apply this geometric reasoning to estimate the magnitude of $H(e^{j\hat\omega})$ at $\hat\omega = \pi/2$ in Fig. 8.23. We begin by estimating the lengths of the vectors from the zeros and poles to the point $z = e^{j\pi/2}$, which is the same as $z = j$. The lengths of the vectors from the zeros are then divided by the lengths of the vectors from the poles, so we get

$$|H(e^{j\pi/2})| = |H(j)| = \frac{|j - 1|\,|j + 1|}{|j - 0.85j|\,|j - (-0.85j)|} = \frac{2}{0.15 \times 1.85} = 7.207$$

The gain G was assumed to be 1.

8.11   EXAMPLE OF AN IIR LOWPASS FILTER

First-order and second-order IIR filters are useful and provide simple examples, but, in many cases, we use higher-order IIR filters because they can realize frequency responses with flatter passbands and stopbands and sharper transition regions. The `butter`, `cheby1`, `cheby2`, and `ellip` functions in MATLAB's *Signal Processing Toolbox* can be used to design filters with prescribed frequency-selective characteristics. As an example, consider the system with system function

$$H(z) = \frac{0.0798(1 + z^{-1} + z^{-2} + z^{-3})}{1 - 1.556z^{-1} + 1.272z^{-2} - 0.398z^{-3}} \tag{8.11.1}$$

$$= \frac{0.0798(1 + z^{-1})(1 - e^{j\pi/2}z^{-1})(1 - e^{-j\pi/2}z^{-1})}{(1 - 0.556z^{-1})(1 - 0.846e^{j0.3\pi}z^{-1})(1 - 0.846e^{-j0.3\pi}z^{-1})} \tag{8.11.2}$$

$$= -\frac{1}{5} + \frac{0.62}{1 - .556z^{-1}} + \frac{0.17e^{j0.96\pi}}{1 - .846e^{j0.3\pi}z^{-1}} + \frac{0.17e^{-j0.96\pi}}{1 - .846e^{-j0.3\pi}z^{-1}} \tag{8.11.3}$$

This system is an example of a lowpass *elliptic filter* whose numerator and denominator coefficients were obtained using the MATLAB function ellip. Each of the three different forms above are useful: (8.11.1) for identifying the filter coefficients, (8.11.2) for sketching the pole-zero plot and the frequency response, and (8.11.3) for finding the impulse response.[6] Figure 8.25 shows the poles and zeros of this filter. Note that all the zeros are on the unit circle and that the poles are strictly inside the unit circle, as they must be for a stable system.

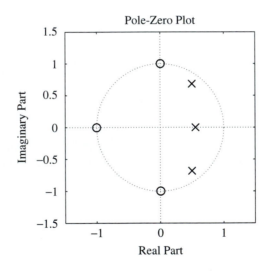

Figure 8.25   Pole-zero plot for a third-order IIR filter (8.11.2).

Exercise 8.17. From (8.11.1) determine the difference equation (Direct Form I) for implementing this system.

---

[6] Factoring polynomials and obtaining the partial fraction expansion was done in MATLAB using the functions roots and residuez, respectively.

The system function was evaluated on the unit circle using the MATLAB function `freqz`. A plot of this result is shown in Fig. 8.26. Note that the frequency response is large in the vicinity of the poles and small around the zeros. In particular, the passband of the frequency response is $|\hat{\omega}| \leq 2\pi(0.15)$, which corresponds to the poles with angles at $\pm 0.3\pi$. Also, the frequency response is exactly zero at $\hat{\omega} = \pm 0.5\pi$ and $\hat{\omega} = \pi$ since the zeros of $H(z)$ are at these angles and lie on the unit circle.

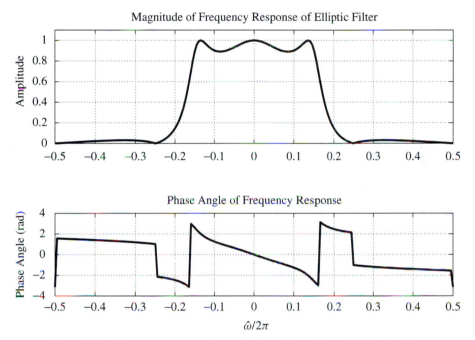

Figure 8.26   Frequency response (magnitude and phase) for a third-order IIR filter.

Exercise 8.18. From (8.11.1) or (8.11.2), determine the value of the frequency response at $\hat{\omega} = 0$.

Finally, Fig. 8.27 shows the impulse response of the system. Note that it oscillates and dies out with increasing n because of the two complex conjugate poles at angles $\pm 0.3\pi$ and radius 0.846. The decaying envelope is $(0.846)^n$.

Exercise 8.19. Use the partial fraction form (8.11.3) to determine an equation for the impulse response of the filter. *Hint:* Apply the inverse z-transform.

The elliptic filter example described in this section is a simple example of a practical IIR lowpass filter. Higher-order filters can exhibit much better frequency-selective filter characteristics.

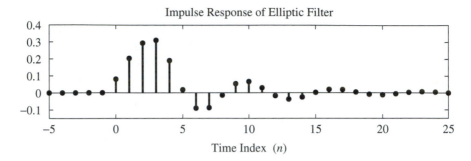

Figure 8.27   Impulse response for a third-order IIR filter.

8.12 SUMMARY AND LINKS

The class of IIR filters was introduced in this chapter, along with the z-transform method for filters with poles. The z-transform changes problems about impulse responses, frequency responses, and system structures into algebraic manipulations of polynomials and rational functions. Poles of the system function $H(z)$ turn out to be the most important elements for IIR filters because properties such as the shape of the frequency response or the form of the impulse response can be inferred quickly from the pole locations.

We also continued to stress the important concept of "domains of representation." The n-domain or time domain, the $\hat{\omega}$-domain or frequency domain, and the z-domain give us three domains for thinking about the characteristics of a system. We completed the ties between domains by introducing the inverse z-transform for constructing a signal from its z-transform. As a result, even difficult problems such as convolution can be simplified by working in the most convenient domain (z) and then transforming back to the original domain (n).

pez.m

Lab C.10 is devoted to IIR filters. This lab uses a MATLAB user interface tool called PeZ that supports an interactive exploration of the three domains. The PeZ tool has the potential for widespread use because it presents the user with multiple views of an LTI system: pole-zero domain, frequency response and impulse response. Similar capabilities are now being incorporated into many commercial software packages, e.g., sptool in MATLAB.

Three-Domain Demos

The CD-ROM also contains the following demonstrations of the relationship between the z-plane and the frequency domain and time domain:

1. A set of "three-domain" movies that show how the frequency response and impulse response of an IIR filter change as a pole location is varied. Several different filters are demonstrated.

2. A movie that animates the relationship between the z-plane and the unit circle where the frequency response lies.

HW Problems and Solutions

The reader is again reminded of the large number of solved homework problems on the CD-ROM that are available for review and practice.

PROBLEMS

8.1   Find the impulse response of the second-order system

$$y[n] = \sqrt{2}\,y[n-1] - y[n-2] + x[n]$$

Express your answer as separate formulas for the cases where $n < 0$ and $n \geq 0$, thus covering the entire range of n.

8.2   Determine a general formula for the Fibonacci sequence by finding the impulse response of the following second-order system:

$$y[n] = y[n-1] + y[n-2] + x[n]$$

8.3   For the following feedback filters, determine the locations of the poles and zeros and plot the positions in the z-plane.

$$y[n] = \tfrac{1}{2}y[n-1] + \tfrac{1}{3}y[n-2] - x[n] + 3x[n-1] - 2x[n-2]$$
$$y[n] = \tfrac{1}{2}y[n-1] - \tfrac{1}{3}y[n-2] - x[n] + 3x[n-1] + 2x[n-2]$$

In the second case, only the signs on $y[n-2]$ and $x[n-2]$ were changed. Compare the two results.

8.4   In this problem, the degrees of the numerator and denominator polynomials are different, so there should be zeros (or poles) at $z = 0$ or $z = \infty$. Determine the poles and zeros of the following filters:

$$y[n] = \tfrac{1}{2}y[n-1] - \tfrac{1}{3}y[n-2] - x[n]$$
$$y[n] = \tfrac{1}{2}y[n-1] - \tfrac{1}{3}y[n-2] - x[n-2]$$
$$y[n] = \tfrac{1}{2}y[n-1] - \tfrac{1}{3}y[n-2] - x[n-4]$$

Plot the positions of the poles and zeros in the z-plane. In all cases, make sure that the number of poles and zeros is the same, by allowing zeros (or poles) at $z = \infty$ or $z = 0$.

8.5   Given an IIR filter defined by the difference equation

$$y[n] = -\tfrac{1}{2}y[n-1] + x[n]$$

(a) Determine the system function $H(z)$. What are its poles and zeros?

(b) When the input to the system is three successive impulses

$$x[n] = \begin{cases} +1 & \text{for } n = 0, 1, 2 \\ 0 & \text{for } n < 0 \text{ and } n \geq 3 \end{cases}$$

determine the functional form for the output signal $y[n]$. Assume that the output signal $y[n]$ is zero for $n < 0$. (*Hint:* Use linearity to find the output as the sum of three terms, each related to the impulse response of the system. Recall that the impulse response of a first-order IIR filter has the form $b_0 a^n$ for $n \geq 0$.)

8.6   A linear time-invariant filter is described by the difference equation

$$y[n] = -0.8y[n-1] + 0.8x[n] + x[n-1]$$

(a) Determine the system function $H(z)$ for this system. Express $H(z)$ as a ratio of polynomials in z^{-1} (negative powers of z) and also as a ratio of polynomials in positive powers of z.

(b) Plot the poles and zeros of $H(z)$ in the z-plane.

(c) From $H(z)$, obtain an expression for $H(e^{j\hat{\omega}})$, the frequency response of this system.

(d) Show that $|H(e^{j\hat{\omega}})|^2 = 1$ for all $\hat{\omega}$.

8.7   Given an IIR filter defined by the difference equation

$$y[n] = -y[n-5] + x[n] \qquad (8.13.1)$$

(a) Determine the system function $H(z)$.

(b) How many poles does the system have? Compute and plot the pole locations.

(c) When the input to the system is the two-point pulse signal:

$$x[n] = \begin{cases} +1 & \text{when } n = 0, 1 \\ 0 & \text{when } n \neq 0, 1 \end{cases}$$

determine the output signal $y[n]$, so that you can make a plot of its general form. Assume that the output signal is zero for $n < 0$.

(d) The output signal is periodic for $n > 0$. Determine the period.

8.8   Given an IIR filter defined by the difference equation

$$y[n] = -0.9\,y[n-6] + x[n] \qquad (8.13.2)$$

(a) Find the z-transform system function for the system.

(b) Find the poles of the system and plot their location in the z-plane.

8.9   Given an IIR filter defined by the difference equation

$$y[n] = -\tfrac{1}{2}y[n-1] + x[n]$$

(a) Determine the system function $H(z)$. What are its poles and zeros?

(b) When the input to the system is

$$x[n] = \delta[n] + \delta[n-1] + \delta[n-2]$$

determine the output signal $y[n]$. Assume that $y[n]$ is zero for $n < 0$.

8.10   Determine the inverse z-transform of the following:

(a) $H_a(z) = \dfrac{1 - z^{-1}}{1 + 0.77z^{-1}}$

(b) $H_b(z) = \dfrac{1 + 0.8z^{-1}}{1 - 0.9z^{-1}}$

(c) $H_c(z) = \dfrac{z^{-2}}{1 - 0.9z^{-1}}$

(d) $H_d(z) = 1 - z^{-1} + 2z^{-3} - 3z^{-4}$

8.11 Determine the inverse z-transform of the following:

(a) $X_a(z) = \dfrac{1 - z^{-1}}{1 - \frac{1}{6}z^{-1} - \frac{1}{6}z^{-2}}$

(b) $X_b(z) = \dfrac{1 + z^{-2}}{1 + 0.9z^{-1} + 0.81z^{-2}}$

(c) $X_c(z) = \dfrac{1 + z^{-1}}{1 - 0.1z^{-1} - 0.72z^{-2}}$

8.12 Given an IIR filter defined by the difference equation

$$y[n] = \tfrac{1}{2}y[n - 1] + x[n]$$

(a) When the input to the system is a unit-step sequence, $u[n]$, determine the functional form for the output signal $y[n]$. Use the inverse z-transform method. Assume that the output signal $y[n]$ is zero for $n < 0$.

(b) Find the output when $x[n]$ is a complex exponential that starts at $n = 0$:

$$x[n] = e^{j(\pi/4)n}u[n]$$

(c) From (b), identify the steady-state component of the response, and compare its magnitude and phase to the frequency response at $\hat{\omega} = \pi/4$.

8.13 For each of the pole-zero plots on the next page, determine which of the following systems (specified by either an $H(z)$ or a difference equation) matches the pole-zero plot.

S_1 : $y[n] = 0.77y[n - 1] + x[n] + x[n - 1]$

S_2 : $y[n] = 0.77y[n - 1] + 0.77x[n] - x[n - 1]$

S_3 : $H(z) = \dfrac{1 - z^{-1}}{1 + 0.77z^{-1}}$

S_4 : $H(z) = 1 - z^{-1} + z^{-2} - z^{-3} + z^{-4} - z^{-5}$

S_5 : $y[n] = \displaystyle\sum_{k=0}^{7} x[n - k]$

S_6 : $H(z) = 3 - 3z^{-1}$

S_7 : $y[n] = x[n] + x[n - 1] + x[n - 2] + x[n - 3] + x[n - 4] + x[n - 5]$

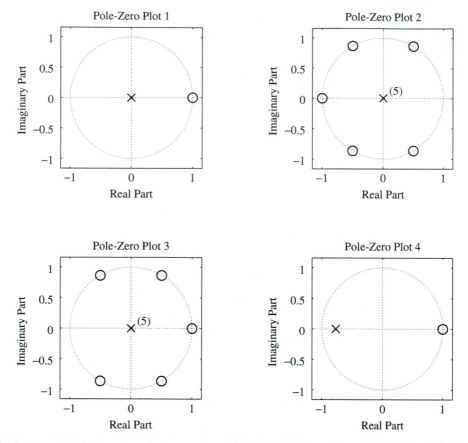

Pole-Zero Plot 1

Pole-Zero Plot 2

Pole-Zero Plot 3

Pole-Zero Plot 4

8.14 For each of these frequency-response plots (A–F), determine which of the following systems (specified by either an $H(z)$ or a difference equation) matches the frequency response. *Note:* The frequency axis for each plot extends over the range $-\pi \leq \hat{\omega} \leq \pi$.

$S_1:$ $y[n] = 0.77y[n-1] + x[n] + x[n-1]$

$S_2:$ $y[n] = 0.77y[n-1] + 0.77x[n] - x[n-1]$

$S_3:$ $H(z) = \dfrac{1 - z^{-1}}{1 + 0.77z^{-1}}$

$S_4:$ $H(z) = 1 - z^{-1} + z^{-2} - z^{-3} + z^{-4} - z^{-5}$

$S_5:$ $y[n] = \displaystyle\sum_{k=0}^{7} x[n-k]$

$S_6:$ $H(z) = 3 - 3z^{-1}$

$S_7:$ $y[n] = x[n] + x[n-1] + x[n-2] + x[n-3] + x[n-4] + x[n-5]$

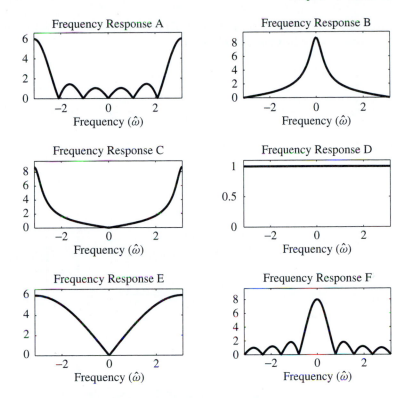

8.15 A linear time-invariant filter is described by the difference equation

$$y[n] = 0.8y[n-1] - 0.8x[n] + x[n-1]$$

(a) Determine the system function $H(z)$ for this system. Express $H(z)$ as a ratio of polynomials in z^{-1} and as a ratio of polynomials in z.

(b) Plot the poles and zeros of $H(z)$ in the z-plane.

(c) From $H(z)$, obtain an expression for $H(e^{j\hat{\omega}})$, the frequency response of this system.

(d) Show that $|H(e^{j\hat{\omega}})|^2 = 1$ for all $\hat{\omega}$.

(e) If the input to the system is

$$x[n] = 4 + \cos[(\pi/4)n] - 3\cos[(2\pi/3)n]$$

what can you say, without further calculation, about the form of the output $y[n]$?

8.16 Match the frequency responses (A–E) with the correct pole-zero plots (PZ 1–6). Poles are denoted by **x** and zeros by **o**.

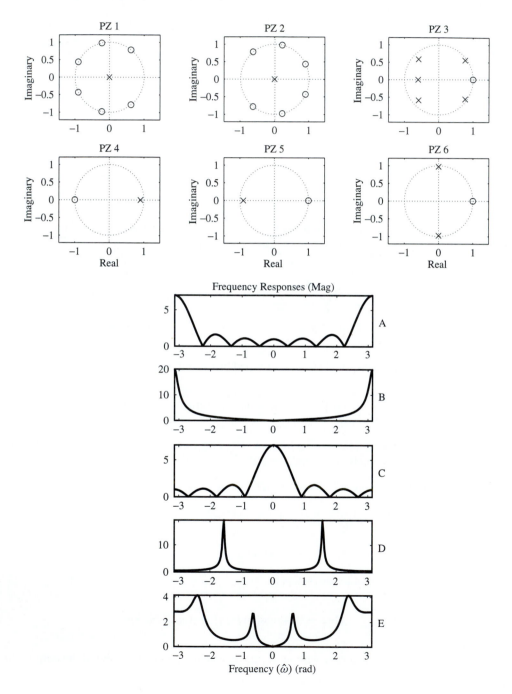

8.17 Match the impulses responses (J–N) with the correct pole-zero plots (PZ 1–6). Poles are denoted by **x** and zeros by **o**.

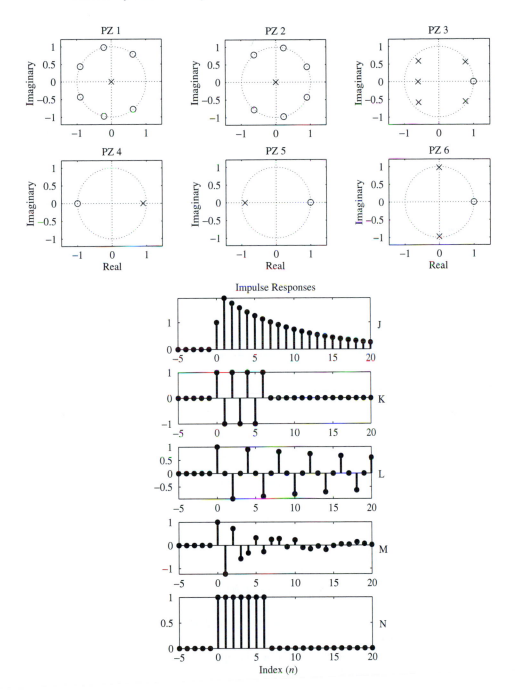

8.18 The system function of the linear time-invariant system shown here is given by the formula

$$H(z) = \frac{(1 - z^{-1})(1 - e^{j\pi/2}z^{-1})(1 - e^{-j\pi/2}z^{-1})}{(1 - 0.9e^{j2\pi/3}z^{-1})(1 - 0.9e^{-j2\pi/3}z^{-1})}$$

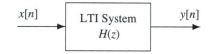

(a) Write the difference equation that gives the relation between the input $x[n]$ and the output $y[n]$. (Hint: Multiply out the factors of $H(z)$.)

(b) Plot the poles and zeros of $H(z)$ in the complex z-plane. (*Hint:* Express $H(z)$ as a ratio of factors in z instead of z^{-1}.)

(c) If the input is of the form $x[n] = Ae^{j\phi}e^{j\hat{\omega}n}$, for what values of $-\pi \le \hat{\omega} \le \pi$ will $y[n] = 0$?

8.19 The input to the C-to-D converter in the system below is

$$x(t) = 4 + \cos(500\pi t) - 3\cos[(2000\pi/3)t]$$

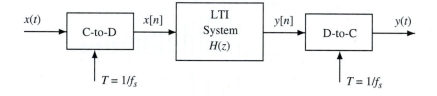

The system function for the LTI system is

$$H(z) = \frac{(1 - z^{-1})(1 - e^{j\pi/2}z^{-1})(1 - e^{-j\pi/2}z^{-1})}{(1 - 0.9e^{j2\pi/3}z^{-1})(1 - 0.9e^{-j2\pi/3}z^{-1})}$$

If $f_s = 1000$ samples/sec, determine an expression for $y(t)$, the output of the D-to-C converter.

8.20 Answer the following questions about the system whose z-transform system function is

$$H(z) = \frac{1 + 0.8z^{-1}}{1 - 0.9z^{-1}}$$

(a) Determine the poles and zeros of $H(z)$.

(b) Determine the difference equation relating the input and output of this filter.

(c) Derive a simple expression (purely real) for the magnitude squared of the frequency response $|H(e^{j\hat{\omega}})|^2$.

(d) Is this filter a lowpass filter or a highpass filter? Explain your answer in terms of the poles and zeros of $H(z)$.

8.21 The diagram in Fig. 8.28 depicts a *cascade connection* of two linear time-invariant systems, i.e., the output of the first system is the input to the second system, and the overall output is the output of the second system.

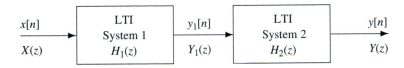

Figure 8.28   Cascade connection of two LTI systems.

(a) Use z-transforms to show that the system function of the overall system (from $x[n]$ to $y[n]$) is $H(z) = H_2(z)H_1(z)$, where $Y(z) = H(z)X(z)$.

(b) Suppose that System 1 is an FIR filter described by the difference equation $y_1[n] = x[n] + \frac{5}{6}x[n-1]$, and System 2 is described by the system function $H_2(z) = 1 - 2z^{-1} + z^{-2}$. Determine the system function of the overall cascade system.

(c) For the systems in (b), obtain a single difference equation that relates $y[n]$ to $x[n]$ in Fig. 8.28.

(d) For the systems in (b), plot the poles and zeros of $H(z)$ in the complex z-plane.

(e) Derive a condition on $H(z)$ that guarantees that the output signal will always be equal to the input signal.

(f) If System 1 is the difference equation $y_1[n] = x[n] + \frac{5}{6}x[n-1]$, find a system function $H_2(z)$ so that output of the cascaded system will always be equal to its input. In other words, find $H_2(z)$, which will undo the filtering action of $H_1(z)$. This is called *deconvolution* and $H_2(z)$ is the *inverse* of $H_1(z)$.

(g) Suppose that $H_1(z)$ represents a general FIR filter. What conditions on $H_1(z)$ must hold if $H_2(z)$ is to be a stable and causal inverse filter for $H_1(z)$?

8.22 Define a discrete-time signal using the formula:

$$y[n] = (0.99)^n \cos(2\pi(0.123)n + \phi)$$

(a) Make a sketch of $y[n]$ versus n, as a "stem" plot. Take the range of n to be $0 \le n \le 20$.

(b) Design an IIR filter that will synthesize $y[n]$. Give your answer in the form of a difference equation with numerical values for the coefficients. Assume that the synthesis will be accomplished by using an impulse input to "start" the difference equation (which is at rest, i.e., has zero initial conditions).

9

Spectrum Analysis

This chapter introduces the basic ideas of spectrum analysis. *Analysis* means that we calculate a spectrum representation from a set of signal values $x[n]$. In a practical situation, the signal $x[n]$ is a digital recording of a signal such as speech or music. Therefore, we can use digital computation to derive the spectrum of the signal. The best-known analysis method is the fast Fourier transform (FFT), which is a very efficient computational tool for spectrum analysis. A major goal of this chapter is to present the essential ideas of spectrum analysis so that the reader can interpret the output of an FFT computation, and understand how it relates to the "true" spectral content of a signal.

A second goal is to present the rudimentary ideas of what is commonly called time-frequency analysis. In one type of time-frequency analysis, we analyze a very long signal by doing many short FFTs, which are then assembled into a single gray-scale image, called a *spectrogram*. The resulting image shows the spectrum analyst how a localized frequency spectrum is evolving with time. True understanding of the spectrogram requires study at an advanced level, but we will present two viewpoints that allow the reader to appreciate the limitations as well as the power of time-frequency spectrum analysis, and perhaps provide motivation for further study.

The frequency spectrum tells how a signal is composed of complex exponential signals. When we first discussed the frequency spectrum in Chapter 3, we concentrated on *synthesis*. A variety of complicated signals were synthesized by summing the contributions due to a few complex exponentials with different magnitudes and phases. Armed with the concept of spectrum representation, we were then able to define the frequency response $H(e^{j\hat{\omega}})$ for an LTI system, which tells us that an input signal of the form $e^{j\hat{\omega}_0 n}$ produces the output $H(e^{j\hat{\omega}_0})e^{j\hat{\omega}_0 n}$. The principle of superposition then leads to the idea that the frequency response treats each spectral component individually, so a filter's output can be calculated by summing the contributions due to each spectral component in the input signal.

When the make-up of a signal is unknown, we need a theory that tells us whether it is appropriate to represent the signal as a sum of complex exponential components, and, if so, how to compute the magnitude and phase of these spectral components. For example, if we sample a speech waveform, can we decompose it into a weighted sum of sinusoids? If so, we have a much simpler way to represent the waveform, because we have uncovered its hidden spectral content. Fortunately, it is true that *almost any signal can be represented* as a superposition of complex exponential signals. This was shown by Fourier[1] in 1807, and his result is the basis for much of the modern theory of signals and systems. We will not attempt any substantial presentation of Fourier analysis because the mathematical complexity would divert us from our goals. Instead, we take the pragmatic approach of demonstrating that discrete-time signals can be *analyzed* to determine their frequency spectrum, and then show how this analysis can be calculated by FIR filtering or by FFTs.

9.1 INTRODUCTION AND REVIEW

The basic idea of spectrum analysis is to create a system that will compute the values of spectral components from the numerical values of the signal $x[n]$. Describing the rough form of this computational structure is our first task.

9.1.1 Review of the Frequency Spectrum

The frequency spectrum of a signal is the information required to represent the signal as an additive combination of complex exponential signals. Each complex exponential has a magnitude (A_k), phase (ϕ_k), and frequency ($\hat{\omega}_k$). The general form for representing a real signal $x[n]$ as a sum of complex exponentials is

$$x[n] = X_0 + \sum_{k=1}^{N} \left(X_k e^{j\hat{\omega}_k n} + X_k^* e^{-j\hat{\omega}_k n} \right) \qquad (9.1.1)$$

where $X_k = A_k e^{j\phi_k}$ and $0 < \hat{\omega}_k \leq \pi$.[2] Figure 9.1 is a graphical representation of the complex amplitudes X_k versus frequency for the case $N = 3$. An equivalent representation is

$$x[n] = X_0 + \sum_{k=1}^{N} 2\Re e\left\{ X_k e^{j\hat{\omega}_k n} \right\} = X_0 + \sum_{k=1}^{N} 2A_k \cos(\hat{\omega}_k n + \phi_k)$$

which is obtained from (9.1.1) by combining the positive and negative frequency terms.

---

[1] Grattan-Guiness, *Joseph Fourier, 1768–1830,* The MIT Press, Cambridge, MA, 1972.

[2] Note that we have used a definition of the complex phasors X_k that differs from the definition in Chapter 3 by a factor of 2. This is necessary for consistency with the standard definition of the discrete Fourier transform.

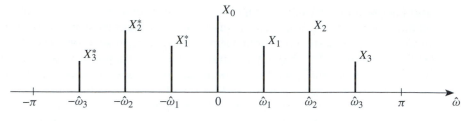

Figure 9.1   The spectrum of a real signal over the frequency range $-\pi < \hat{\omega} \le \pi$.

Equation (9.1.1) represents a signal as a DC component of amplitude X_0, plus complex exponential components with frequencies $\hat{\omega}_k$ and complex amplitudes X_k. Because $x[n]$ is assumed to be real, X_0 is real and the complex amplitude of the component at frequency $\hat{\omega} = -\hat{\omega}_k$ must be the complex conjugate of the complex amplitude of the component at frequency $\hat{\omega} = \hat{\omega}_k$. This symmetry is shown in Fig. 9.1, which is a graphical presentation of the spectrum with three nonzero frequencies ($N = 3$), and a DC component at $\hat{\omega} = 0$.

Note that for a discrete-time signal, all frequencies are normally assumed to lie between $-\pi$ and $+\pi$. However, an equivalent plot could be constructed using only positive frequencies if we translate the negative frequencies $-\pi \le \hat{\omega}_k < 0$ to their positive alias frequencies $\pi \le 2\pi - \hat{\omega}_k < 2\pi$. This translation follows directly from the formula

$$e^{-j\hat{\omega}_k n} = e^{j2\pi n}e^{-j\hat{\omega}_k n} = e^{j(2\pi-\hat{\omega}_k)n}$$

because $e^{j2\pi n} = 1$ when n is an integer. In other words, the positive alias frequencies $2\pi - \hat{\omega}_k$ are indistinguishable from the negative frequencies $-\hat{\omega}_k$. The spectrum plot equivalent to Fig. 9.1, but using only positive frequencies $0 \le -\hat{\omega} < 2\pi$, is shown in Fig. 9.2. We will find this form of the spectrum useful when we discuss the fast Fourier transform algorithm.

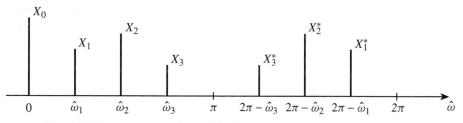

Figure 9.2   The spectrum of a real signal using only positive frequencies. The frequency range is $0 \le \hat{\omega} < 2\pi$.

9.1.2 A Spectrum Analyzer

A spectrum analysis system must extract the frequency, magnitude, and phase of each spectral component in the signal. The frequency selection step is the most difficult part, so we propose a two-part structure for the spectrum analyzer, as shown in Fig. 9.3. In Fig. 9.3, the frequencies are calculated first, and then separate channels are used for each spectral component. The kth channel would extract the magnitude and phase of the spectrum value at frequency $\hat{\omega}_k$. One possible implementation for the kth channel would be to use a bandpass filter centered at $\hat{\omega} = \hat{\omega}_k$.

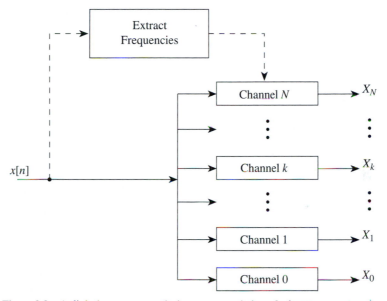

Figure 9.3   A digital spectrum analysis system consisting of a frequency extraction algorithm followed by separate channels for each frequency component. The notation X_k is defined in (9.1.1), where X_k is the complex amplitude at $\hat{\omega} = \hat{\omega}_k$.

The frequency extraction step is essential. However, algorithms for calculating the optimal frequency extraction in Fig. 9.3 are relatively complicated and beyond the scope of our introductory textbook. Even if we do not calculate the best frequencies in the signal, we must still choose the frequencies for the analysis channels. Our approach will be to pick the frequencies ahead of time, independent of the signal. Since there must be a finite number of channels ($N + 1$), one obvious choice is to cover the total 2π frequency range with equally spaced frequencies:

$$\hat{\omega}_k = \frac{2\pi}{N}k \qquad \text{for } k = 0, 1, 2, 3, \ldots, N \qquad (9.1.2)$$

This choice, in effect, assumes no *a priori* knowledge of the signal content. Another commonly used coverage is a logarithmic spacing of the frequencies, i.e., $\hat{\omega}_{k+1}/\hat{\omega}_k = $ constant, especially when the analysis must be done over many octaves. Some physical systems, such as human hearing, can be modeled as a spectrum analyzer with this constant ratio between frequencies.

Even though the exact frequency content of the signal may be unknown, we can use the equally spaced frequencies to approximate the signal. One simple approach is to start with a large number of fixed frequencies, i.e., a very large value for N in (9.1.2), and then check the output of the spectrum analysis channels to find which frequency components have large magnitudes. If we retain these large components and discard the rest, the spectrum is only an approximate representation, but, with fixed frequencies, the computation of the amplitudes and phases can be reduced to a simple filtering structure, or to the FFT.

There is a limit to how much we can say about the spectrum analyzer without using some mathematical formulas. We will approach the problem by using all of the knowledge that we have developed in previous chapters.

9.2 SPECTRUM ANALYSIS BY FILTERING

The main theme of this chapter is computing the spectrum values X_k from the signal values $x[n]$. In this section, we will adopt a filtering viewpoint for the spectrum analysis calculation, because it allows us to build upon our knowledge of the frequency response. Each channel of the spectrum analyzer (Fig. 9.3) must do two jobs: First, isolate the frequency of interest; second, compute the complex amplitude for that frequency. The job of isolating the frequency of interest is ideally suited to a band-pass filter (BPF) with a narrow passband. Once $x[n]$ is passed through the BPF, we can measure the output for its amplitude and phase. Rather than design individual BPFs for each channel, we introduce a *frequency-shifting* property of the spectrum, so that all channels can share the same filter.

9.2.1 Frequency Shifting

As we observed in our discussion of amplitude modulation in Chapter 3, the frequency of a complex exponential signal can be shifted if it is multiplied by another complex exponential signal. For example, let the signal $x[n]$ given by (9.1.1) be multiplied by the complex exponential signal $e^{-j\hat{\omega}_s n}$. The complex exponential signal $e^{-j\hat{\omega}_s n}$ is *modulated* by the signal $x[n]$, and the process of forming $x[n]e^{-j\hat{\omega}_s n}$ is called *amplitude modulation*. The new signal is:

$$
\begin{aligned}
x_{\hat{\omega}_s}[n] &= x[n]e^{-j\hat{\omega}_s n} \\
&= \left[X_0 + \sum_{k=1}^{N} \left(X_k e^{j\hat{\omega}_k n} + X_k^* e^{-j\hat{\omega}_k n} \right) \right] e^{-j\hat{\omega}_s n} \\
&= X_0 e^{-j\hat{\omega}_s n} + \sum_{k=1}^{N} \left(X_k e^{j(\hat{\omega}_k - \hat{\omega}_s)n} + X_k^* e^{-j(\hat{\omega}_k + \hat{\omega}_s)n} \right) \quad (9.2.1)
\end{aligned}
$$

It is clear from this equation that the frequencies in $x_{\hat{\omega}_s}[n]$ are those of $x[n]$ shifted left (in the negative direction) by $\hat{\omega}_s$. This is shown in Fig. 9.4 for the special case where $\hat{\omega}_s = \hat{\omega}_1$. The entire spectrum of the signal $x[n]$ from Fig. 9.1 has been translated to the left by $\hat{\omega}_s$; i.e., the original spectral component at frequency $\pm\hat{\omega}_k$ is shifted to location $\pm\hat{\omega}_k - \hat{\omega}_s$. When the frequency shift $\hat{\omega}_s$ is set equal to $\hat{\omega}_1$, the original component at $\hat{\omega}_1$ moves to zero frequency, and the original DC component moves to $-\hat{\omega}_1$. In Section 9.2.2, we will use this fact in a scheme to measure X_k, given knowledge of $\hat{\omega}_k$.

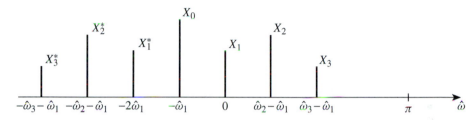

Figure 9.4   Illustration of a frequency-shifted spectrum for the case where the frequency shift $\hat{\omega}_s$ is equal to $\hat{\omega}_1$.

Exercise 9.1. Take the shifted spectrum found in Fig. 9.4 and redo its plot using only positive frequencies as in Fig. 9.2. Assume that $\hat{\omega}_1 = 0.1\pi$, $\hat{\omega}_2 = 0.4\pi$, and $\hat{\omega}_3 = 0.8\pi$. Label all the frequency locations. Repeat for $\hat{\omega}_1 = 0.35\pi$, $\hat{\omega}_2 = 0.7\pi$, and $\hat{\omega}_3 = 0.8\pi$.

9.2.2 Measuring the Average Value

The signal defined by (9.1.1) contains one constant component (X_0); the rest are oscillating components at frequencies other than zero. The average value of $x[n]$ is equal to X_0, the DC component at $\hat{\omega} = 0$. This average value can be extracted by passing $x[n]$ through a lowpass filter (LPF) that rejects all the other frequency components. The frequency response of the LPF must be 1 at $\hat{\omega} = 0$ and 0 at all the other frequencies, $\hat{\omega}_k$.

Exercise 9.2. Suppose that $x[n]$ has a spectrum with components at $\hat{\omega}_1 = 0.25\pi$, $\hat{\omega}_2 = 0.5\pi$, and $\hat{\omega}_3 = 0.75\pi$, and a DC value of $X_0 = 3$. Use the 8-point FIR filter with coefficients $\{b_k\} = \{\frac{1}{8}, \frac{1}{8}, \frac{1}{8}, \frac{1}{8}, \frac{1}{8}, \frac{1}{8}, \frac{1}{8}, \frac{1}{8}\}$ to process this signal. Show that the output is $y[n] = 3$ for all n. In addition, make a plot of $|H(e^{j\hat{\omega}})|$ versus $\hat{\omega}$ to verify that $H(e^{j\hat{\omega}})$ is a lowpass filter.

9.2.3 Channel Filters

The fact that frequency shifting can be used to translate a specific frequency component to zero frequency is convenient, because measurement at $\hat{\omega} = 0$ amounts to finding the average value of the signal with a lowpass filter. When the lowpass filter extracts the DC component of the shifted spectrum, it is actually measuring the complex amplitude of one of the shifted spectral components in $x[n]$. A system for doing this is depicted in Fig. 9.5, where the lowpass filter must not only measure the average value, but also null out all other competing sinusoidal components.

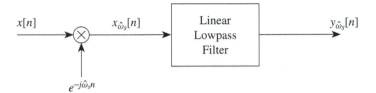

Figure 9.5   System for measuring the amplitude of a complex exponential component at frequency $\hat{\omega} = \hat{\omega}_s$, based on frequency shifting.

The operation of the system of Fig. 9.5 can be illustrated for the simple case of two cosines:

$$x[n] = X_0 + 2A_1 \cos(\hat{\omega}_1 n + \phi_1) + 2A_2 \cos(\hat{\omega}_2 n + \phi_2)$$

$$= X_0 + X_1 e^{j\hat{\omega}_1 n} + X_1^* e^{-j\hat{\omega}_1 n} + X_2 e^{j\hat{\omega}_2 n} + X_2^* e^{-j\hat{\omega}_2 n}$$

If we pick the frequency shift to be $\hat{\omega}_s = \hat{\omega}_1$, the input to the lowpass filter is:

$$x_{\hat{\omega}_1}[n] = x[n]e^{-j\hat{\omega}_1 n} = X_0 e^{-j\hat{\omega}_1 n} + X_1 + X_1^* e^{-j2\hat{\omega}_1 n} + X_2 e^{j(\hat{\omega}_2 - \hat{\omega}_1)n} + X_2^* e^{-j(\hat{\omega}_2 + \hat{\omega}_1)n}$$

In this case, the result of the amplitude modulation is to create a new DC component equal to X_1 plus interfering components at frequencies $-\hat{\omega}_1$, $-2\hat{\omega}_1$, $\hat{\omega}_2 - \hat{\omega}_1$, and $-(\hat{\omega}_2 + \hat{\omega}_1)$. The filter's frequency response $H(e^{j\hat{\omega}})$ gives a simple formula for the output of the linear filter

$$y_{\hat{\omega}_1}[n] = H(e^{-j\hat{\omega}_1})X_0 e^{-j\hat{\omega}_1 n} + H(e^{j0})X_1 + H(e^{-j2\hat{\omega}_1})X_1^* e^{-j2\hat{\omega}_1 n}$$

$$+ H(e^{j(\hat{\omega}_2 - \hat{\omega}_1)})X_2 e^{j(\hat{\omega}_2 - \hat{\omega}_1)n} + H(e^{-j(\hat{\omega}_2 + \hat{\omega}_1)})X_2^* e^{-j(\hat{\omega}_2 + \hat{\omega}_1)n} \quad (9.2.2)$$

From (9.2.2) we can see that if $H(e^{j0}) \neq 0$, and $H(e^{j\hat{\omega}}) = 0$, for $\hat{\omega} = -\hat{\omega}_1$, $-2\hat{\omega}_1$, $\hat{\omega}_2 - \hat{\omega}_1$, and $-(\hat{\omega}_2 + \hat{\omega}_1)$, then the output $y_{\hat{\omega}_1}[n]$ will be a constant proportional to the complex constant we want to measure, X_1. In other words, the system of Fig. 9.5 can extract the spectral component at frequency $\hat{\omega}_1$, if the filter's frequency response is zero at all of the other shifted frequencies. The trick is that we must design the lowpass filter without knowing where these shifted frequencies will lie.

Figure 9.6 shows one way to visualize the LPF multiplying the shifted spectrum. In Fig. 9.6 the passband of a narrow lowpass filter is drawn as an ideal filter where the passband gain is constant for $-\hat{\omega}_c < \hat{\omega} < \hat{\omega}_c$. The ideal stopband is zero over the region $|\hat{\omega}| > \hat{\omega}_c$. This ideal LPF will filter out the component at $\hat{\omega} = 0$ and remove all interfering components above $\hat{\omega}_c$, if we pick $\hat{\omega}_c$ to be less than the minimum frequency separation, e.g., $\hat{\omega}_2 - \hat{\omega}_1$ in Fig. 9.6.

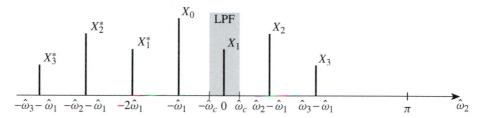

Figure 9.6   The shaded region shows the passband of an ideal LPF for extracting the frequency-shifted spectral component at $\hat{\omega} = 0$. This component was originally located at $\hat{\omega} = \hat{\omega}_1$.

What if the signal $x[n]$ has more spectral components than the simple example discussed above? First of all, the modulator in Fig. 9.5 would have to be systematically adjusted to each frequency known (or suspected) to be present in the signal. Second, the lowpass filter's frequency response might have to be changed for each frequency shift, because the interfering shifted frequencies move to different locations after modulation, and the filter must null them out. If both of these problems could be solved, the entire frequency spectrum could be measured. Unfortunately, there appear to be at least two major problems with the general application of this idea:

1. We must know in advance which frequencies are present in the signal.

2. The frequency response of the lowpass filter in Fig. 9.5 must be zero at all frequencies except $\hat{\omega} = 0$.

One solution to the first problem would be to try all possible frequencies between 0 and π. Theoretically, it is possible to do the measurement at an infinite number of frequencies, and this leads to the mathematical definition of the Fourier transform, which is a valuable tool in the mathematical analysis of signals and systems. However, in practice, it is often sufficient to compute the spectrum at only a finite set of frequencies, such as an equally spaced set.

The second problem is related to the first. If we try to measure the spectrum at all frequencies, we need a filter that has zero gain at all frequencies except $\hat{\omega} = 0$. Such a filter can be defined mathematically, but it cannot be implemented in hardware or software.

Thus, spectrum analysis seems to be a difficult problem with only a theoretical solution. Fortunately, this is not the case. In the Section 9.3 we will explain how a practical filter such as the running-sum filter can provide the *exact* spectrum measurement for the interesting class of periodic signals.

9.3 SPECTRUM ANALYSIS OF PERIODIC SIGNALS

Spectrum analysis usually involves approximations, but there is one case where we can do a perfect job. In this section, we will derive equations for a spectrum analyzer that works perfectly when the input is a periodic signal. The same approach will then be used for other cases, such as nonperiodic signals, but the analysis will involve approximation errors.

9.3.1 Periodic Signals

A periodic signal is defined by the shifting property

$$x[n - N] = x[n] \qquad \text{for all } n$$

where the parameter N is the *period*. In other words, when $x[n]$ is delayed by N samples, it is exactly the same signal. A complex exponential can be periodic with period equal to N, but its frequency must then be an integer multiple of $2\pi/N$. This is proved by the following analysis:

$$e^{j\hat{\omega}_0(n-N)} = e^{j\hat{\omega}_0 n}$$

$$e^{j\hat{\omega}_0 n} e^{-j\hat{\omega}_0 N} = e^{j\hat{\omega}_0 n}$$

The last equation requires that

$$e^{-j\hat{\omega}_0 N} = 1 = e^{-j2\pi k}$$

where k is an integer. Equating exponents, we conclude that $\hat{\omega}_0 N = 2\pi k$, and we can write any periodic complex exponential as

$$e^{j\hat{\omega}_0 n} = e^{j(2\pi k/N)n}$$

Once the period is specified, we can create different complex exponential signals by changing the value of k. However, there are actually only N different frequencies, because

$$e^{j(2\pi(k+N)/N)n} = e^{j(2\pi k/N)n} e^{j2\pi kn} = e^{j(2\pi k/N)n}$$

so we may as well restrict k to be in the range $k = 0, 1, 2, \ldots, N - 1$.

9.3.2 Spectrum of a Periodic Signal

Now we are ready to write down the spectrum representation for a general periodic discrete-time signal. If $x[n]$ has a period equal to N, then its spectrum can contain only complex exponentials with the same period. Hence the general representation of a periodic $x[n]$ by a sum of complex exponentials is

$$x[n] = \sum_{k=0}^{N-1} X_k e^{j(2\pi k/N)n} \tag{9.3.1}$$

The range on the sum is 0 to $N-1$, because there are only N different frequencies for the complex exponentials. The frequencies are all multiples of a common fundamental frequency $2\pi/N$, which is further evidence of the fact that $x[n]$ is periodic. The coefficient X_k is the complex amplitude for the kth frequency component in the spectrum of $x[n]$.

At this point, we are going to shift notation in anticipation of a standard formula for an operation known as the discrete Fourier transform (DFT). In this new notation, we rewrite (9.3.1) by replacing X_k with $X[k]/N$

$$x[n] = \frac{1}{N} \sum_{k=0}^{N-1} X[k]\, e^{j(2\pi k/N)n} \tag{9.3.2}$$

Note that we have included a scaling constant $1/N$, and we have eliminated the subscript in favor of the [ ] notation for indexing the coefficients $X[k]$. Later on, we will see that (9.3.2) conforms to the standard definition of the inverse discrete Fourier transform.

The signal defined in (9.3.2) is periodic with period N. We prove this by substituting $n - N$ for n in (9.3.2), which gives

$$x[n - N] = \frac{1}{N} \sum_{k=0}^{N-1} X[k] e^{j(2\pi k/N)(n-N)}$$

$$= \frac{1}{N} \sum_{k=0}^{N-1} X[k] e^{j(2\pi k/N)n} e^{-j2\pi k} = x[n]$$

since $e^{-j2\pi k} = 1$ when k is an integer.

The signal defined in (9.3.2) has equally spaced frequencies $(2\pi/N)k$ over the positive frequency range, i.e.,

$$0 < (2\pi/N)k \le \pi \quad \text{for} \quad 0 < k \le N/2$$

$$\pi < (2\pi/N)k < 2\pi \quad \text{for} \quad N/2 < k < N-1$$

Although the index $k = N - 1$ corresponds to the positive frequency $\hat{\omega} = 2\pi(N-1)/N = 2\pi(1-1/N)$, it is also the positive alias frequency of the negative frequency $\hat{\omega} = -(2\pi/N)$. Likewise, $k = N - 2$ is the positive alias frequency of $\hat{\omega} = -(4\pi/N)$, etc. See Figs. 9.1 and 9.2 for an example of this aliasing relationship. When we have a real signal $x[n]$, there is a symmetry in the spectrum, so we can conclude that the DFT coefficients satisfy the following constraints: $X[N-1] = X^*[1]$, $X[N-2] = X^*[2]$, or, in general, $X[N-k] = X^*[k]$ for $k = 0, 1, \ldots, N-1$.

Example 9.1   The representation of (9.3.2) is a general form that applies to all periodic discrete-time signals with period N; these signals differ only in the choice of the coefficients $X[k]$. For a numerical example, consider the signal

$$x[n] = 8 + 10\sin((2\pi/10)n) = 8 - j5e^{j(2\pi/10)n} + j5e^{-j(2\pi/10)n} \qquad (9.3.3)$$

which is plotted in Fig. 9.7. This signal is periodic with a period $N = 10$. The same signal expressed in terms of only positive frequencies is

$$x[n] = 8 - j5e^{j(2\pi/10)n} + j5e^{j(2\pi/10)9n}$$

The frequency spectrum of this signal is shown in Fig. 9.9(a).   ◇

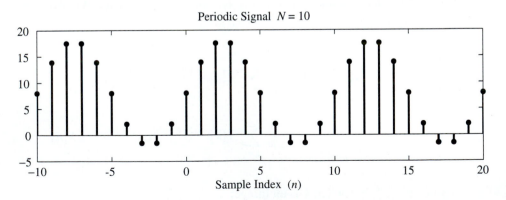

Figure 9.7   Waveform of a periodic signal with period 10.

9.3.3 Filtering with a Running Sum

In Section 6.7 and Section 7.7, we showed examples of FIR filters that calculated either a running average or a running sum. Both filters are quite simple, and either one can do the filtering job needed in Fig. 9.5. We will use the L-point running-sum linear filter which is defined by the difference equation[3]

$$y[n] = \sum_{\ell=0}^{L-1} x[n - \ell] \tag{9.3.4}$$

The filter coefficients in (9.3.4) are all equal to one, so the frequency response of the running-sum filter is

$$H(e^{j\hat{\omega}}) = \sum_{m=0}^{L-1} e^{-j\hat{\omega}m} = \frac{\sin(\hat{\omega}L/2)}{\sin(\hat{\omega}/2)} e^{-j\hat{\omega}(L-1)/2} \tag{9.3.5}$$

A plot of $|H(e^{j\hat{\omega}})|$ for $L = 10$ is shown in Fig. 9.9(b), together with the shifted spectrum (dotted lines). Equation (9.3.5) and the frequency-response plot in Fig. 9.9(b) give two important properties of $H(e^{j\hat{\omega}})$:

- The DC gain of the running-sum filter is L, because $H(e^{j0}) = L$.
- The nulls of $H(e^{j\hat{\omega}})$ are equally spaced at multiples of $2\pi/L$, because $H(e^{j2\pi k/L}) = 0$ for $k = 1, 2, \ldots, L - 1$.

The fact that the zeros of $H(e^{j\hat{\omega}})$ are equally spaced at $\hat{\omega} = 2\pi k/L$ suggests that the running-sum filter will work very well for spectrum analysis of signals whose constituent frequencies are also equally spaced. As we have just seen, this is the case for a periodic signal, so the nulling property shown in Fig. 9.9(b) works if we use a running-sum filter whose length L is equal to the period N of $x[n]$.

9.3.4 Spectrum Analysis Using Running-Sum Filtering

If the N-point running-sum filter is used as the lowpass filter in Fig. 9.5, we obtain the system depicted in Fig. 9.8. The signal out of the multiplier

$$x_k[n] = x[n]e^{-j(2\pi k/N)n}$$

is frequency-shifted by $(2\pi k/N)$.

---

[3] The running-average filter would have a factor of $1/L$ to average the L samples involved in computing the output. We do not need this factor because we are going to be interested in the case $L = N$, and the factor of $1/N$ is already part of the DFT synthesis formula (9.3.2).

Figure 9.8 System for measuring the magnitude and phase of a complex exponential component at frequency $2\pi k/N$. The output will be a constant signal whenever the input is periodic with period N.

Figure 9.9(b) shows how the running-sum filter works for spectrum analysis when the filter length matches the period of the input signal, i.e., $L = N = 10$ for the example in (9.3.3). The dotted lines in Fig. 9.9(b) show the shifted spectrum for $x_1[n]$ which is obtained by multiplying $x[n]$ by $e^{-j(2\pi/10)n}$. The spectral component of $x[n]$ at frequency $(2\pi/10)$ is shifted to zero frequency where the gain of the filter is 10, and the other two spectral components are shifted to frequencies where the filter has nulls. The output of the filter always contains just one spectral line at $\hat{\omega} = 0$, so it will be a complex constant. For the case where $k = 1$, the output signal will be the constant value $-j50$, which is 10 times the value of the complex amplitude of the spectral component at frequency $2\pi/10$. If we now substitute the value $X[1] = -j50$ into (9.3.2) with $N = 10$, the appropriate spectral component $-j5e^{j(2\pi/10)n}$ will be synthesized. The other spectral components can be extracted by changing the value of k in the modulating exponential. Most will be zero, except for $k = 9$ and $k = 0$. The spectral component $X[N - 1] = X[9]$ is also known by virtue of conjugate symmetry, i.e., $X[N - 1] = X^*[1]$. The coefficient $X[0]$ will be extracted directly by the running sum without frequency shifting, because the lowpass filter satisfies $H(e^{\pm j2\pi/10}) = 0$.

As suggested by the preceding example, the system in Fig. 9.8 gives exact results for periodic signals having period N. In fact, the output of the kth channel is the constant $X[k]$. In order to prove that $y_k[n]$ will be equal to $X[k]$, we start with the representation for the input signal as a sum of complex exponentials of frequencies $(2\pi k/N)$.[4]

$$x[n] = \frac{1}{N} \sum_{\ell=0}^{N-1} X[\ell]e^{j(2\pi/N)\ell n} \qquad (9.3.6)$$

After the multiplier, we have

$$x_k[n] = \frac{1}{N} \sum_{\ell=0}^{N-1} X[\ell]e^{j(2\pi/N)\ell n} e^{-j(2\pi/N)kn}$$

---

[4] We use the dummy summation index ℓ in writing (9.3.6) so as not to confuse it with k, which in this case stands for the general frequency index in Fig. 9.8.

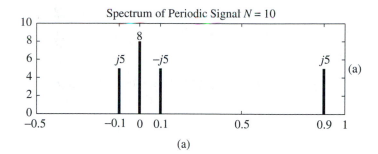

(a)

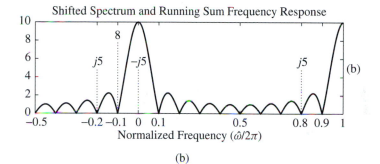

(b)

Figure 9.9 (a) Frequency spectrum of the periodic signal of Fig. 9.7. (b) Frequency spectrum left-shifted by $2\pi(1/10)$ (dotted lines) together with the frequency response of a 10-point running-sum filter (solid line).

Then $y_k[n]$ can be obtained by multiplying each spectral component in $x_k[n]$ by the appropriate value of the frequency response of the running-sum filter. The resulting formula is

$$y_k[n] = \frac{1}{N}\sum_{\ell=0}^{N-1}H(e^{j2\pi(\ell-k)/N})X[\ell]e^{j(2\pi/N)(\ell-k)n}$$

$$= \frac{1}{N}H(e^{j0})X[k] + \frac{1}{N}\sum_{\substack{\ell=0\\ \ell\neq k}}^{N-1}H(e^{j2\pi(\ell-k)/N})X[\ell]e^{j(2\pi/N)(\ell-k)n}$$

When $k = \ell$, the exponent of one of the terms becomes zero, so we get a constant term that has been pulled out of the summation. Since the frequency response of the N-point running-sum filter has the properties $H(e^{j0}) = N$ and $H(e^{j2\pi(\ell-k)/N}) = 0$ whenever $(\ell - k) \neq 0$, it follows that for a periodic input signal as represented by (9.3.6), the output will be a constant:

$$y_k[n] = X[k] \quad \text{for all } n \tag{9.3.7}$$

If we combine the formula for the running-sum output (9.3.4) with the complex exponential multiplier, we get

$$y_k[n] = \sum_{\ell=0}^{N-1} x_k[n - \ell] = \sum_{m=n-N+1}^{n} x_k[m]$$

$$= \sum_{m=n-N+1}^{n} x[m]e^{-j(2\pi/N)km} \qquad (9.3.8)$$

Since we know that the output is constant, we can evaluate $y_k[n]$ in Fig. 9.8 at any single time index to get the value $X[k]$. If we choose this time to be $n = N - 1$, then we can equate (9.3.7) and (9.3.8) to obtain the standard computational formula

$$X[k] = \sum_{m=0}^{N-1} x[m]e^{-j(2\pi/N)km} \qquad k = 0, 1, 2, \ldots, N - 1 \qquad (9.3.9)$$

which states that each of the the spectral components, $X[k]$, can be obtained by summing over one period. Any period would work, but we choose the base period, $0 \le m \le N - 1$.

Figure 9.10 shows a complete block diagram of all N channels used in the spectrum analyzer based on linear filtering. Each channel is composed of a modulator-filter combination that produces one spectral coefficient. The structure is usually called a *filter bank*. Even though each filter could produce a continuous output stream, we need only one value when the input is periodic with period N, because the output of each channel filter is then constant.

9.3.5 The DFT: Discrete Fourier Transform

The foregoing discussion is a roundabout way to derive the analysis and synthesis equations of a famous transform, called the *discrete Fourier transform*, or DFT:

$$x[n] = \frac{1}{N} \sum_{k=0}^{N-1} X[k]e^{j(2\pi/N)kn} \qquad n = 0, 1, 2, \ldots, N - 1 \qquad (9.3.10)$$

$$X[k] = \sum_{n=0}^{N-1} x[n]e^{-j(2\pi/N)kn} \qquad k = 0, 1, 2, \ldots, N - 1 \qquad (9.3.11)$$

Equation (9.3.11) is (forward) DFT and (9.3.10) is the inverse DFT, or IDFT. Equation (9.3.11) is also the *analysis* equation for calculating the spectrum from the signal,

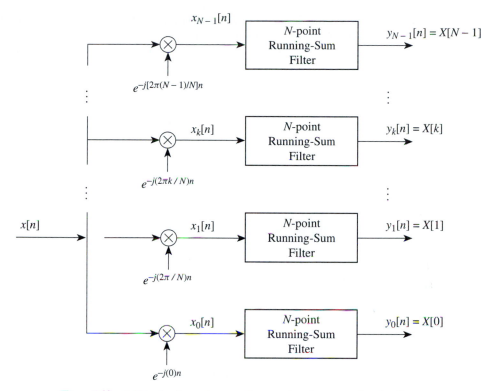

Figure 9.10   A filter-bank spectrum analyzer system for computing the discrete Fourier transform $X[k]$ for $k = 0, 1, \ldots, N - 1$.

and (9.3.10) is the *synthesis* equation used to re-create the signal from its spectrum.[5] This pair of analysis and synthesis equations holds for *any* discrete-time signal that is periodic with period N. The reason we obtain these exact formulas for the periodic case is that the input period and the length of the impulse response of the running-sum filter are the same. Thus, the filter always sums over one complete period of the input $x_k[n]$ regardless of the time at which the output is evaluated.

For emphasis, we restate a very important idea:

The filter-bank implementation of Fig. 9.10 is equivalent to evaluating the DFT analysis formula (9.3.11) for each different value of k.

---

[5] Note that we have used n as both the dummy summation index in (9.3.11) and the independent variable in (9.3.10), and vice versa for k, which is the dummy variable in (9.3.10). This usually causes no problem, but in cases like (9.3.6) it is always possible to label the dummy index differently.

The upshot of this equivalence is that we will use the DFT formulas for computation, but the filter bank for interpretation and insight. This is true especially in light of the fast Fourier transform (FFT), a remarkably efficient algorithm for calculating the DFT.

9.3.6 DFT Examples

The DFT can be applied to *transform* a time-domain signal into its frequency-domain representation. In this section, we present three examples of common signals.

Example 9.2   First we consider the case of a complex exponential with a special frequency. Let the signal $x_1[n]$ be defined as:

$$x_1[n] = e^{j(2\pi k_0/N)n} \qquad \text{for } n = 0, 1, 2, \ldots, N-1$$

Its frequency is an integer multiple of $2\pi/N$. The N-point DFT of $x[n]$ is

$$X_1[k] = \sum_{n=0}^{N-1} x_1[n]e^{-j(2\pi/N)kn}$$

$$= \sum_{n=0}^{N-1} e^{j(2\pi k_0/N)n}e^{-j(2\pi/N)kn}$$

$$= \sum_{n=0}^{N-1} e^{-j(2\pi/N)(k-k_0)n}$$

$$= 1 + e^{-j(2\pi/N)(k-k_0)} + e^{-j(2\pi/N)(k-k_0)2} + \ldots + e^{-j(2\pi/N)(k-k_0)(N-1)}$$

$$= \frac{1 - e^{-j(2\pi/N)(k-k_0)N}}{1 - e^{-j(2\pi/N)(k-k_0)}}$$

The numerator in this last formula evaluates to $1 - 1 = 0$ for all k. However, when $k = k_0$, the denominator is also zero. For $k = k_0$, we can evaluate the DFT sum directly to get the value N. Therefore a compact mathematical expression for the DFT is

$$X_1[k] = N\delta[k - k_0]$$

i.e., a scaled impulse at $k = k_0$. $\diamond$

Example 9.3   A second example can be generated easily from Example 9.2. Compute the DFT of $x_2[n] = \cos(2\pi k_0 n/N)$. Since the cosine can be expressed as the sum of two complex exponentials

$$x_2[n] = \cos(2\pi k_0 n/N) = \tfrac{1}{2}e^{j(2\pi k_0/N)n} + \tfrac{1}{2}e^{-j(2\pi k_0/N)n}$$

we can take the DFT of each exponential and add the results. This approach takes advantage of the fact that the DFT, like its cousin the z-transform, is a linear operation. The answer is:

$$X_2[k] = \tfrac{1}{2}N\delta[k - k_0] + \tfrac{1}{2}N\delta[k - (-k_0)]$$

or equivalently

$$X_2[k] = \tfrac{1}{2}N\delta[k - k_0] + \tfrac{1}{2}N\delta[k + k_0]$$

$\diamond$

Example 9.4   A third example is a complex exponential whose frequency might not be a multiple of $2\pi/N$:

$$x_3[n] = e^{j(\hat{\omega}_0 n + \phi)} \qquad \text{for } n = 0, 1, 2, \ldots, N-1$$

The N-point DFT of $x[n]$ is

$$X_3[k] = \sum_{n=0}^{N-1} e^{j(\hat{\omega}_0 n + \phi)} e^{-j(2\pi/N)kn}$$

$$= e^{j\phi} \sum_{n=0}^{N-1} e^{-j(2\pi k/N - \hat{\omega}_0)n}$$

$$= e^{j\phi} \left(e^{-j(0)} + e^{-j(2\pi k/N - \hat{\omega}_0)} + \ldots + e^{-j(2\pi k/N - \hat{\omega}_0)(N-1)} \right)$$

$$= e^{j\phi} \frac{1 - e^{-j(2\pi k/N - \hat{\omega}_0)N}}{1 - e^{-j(2\pi k/N - \hat{\omega}_0)}}$$

Notice that, in computing this result, we have ignored the fact that $x_3[n]$ is *not* periodic with period N.

If we define the exponent to be $\theta = (2\pi k/N - \hat{\omega}_0)$, then this last formula can be simplified somewhat by factoring out $e^{-j\theta N/2}$ from the numerator and $e^{-j\theta/2}$ from the denominator. The final result, after a couple of algebraic steps, is:

$$X_3[k] = e^{j\phi} \; e^{-j(2\pi k/N - \hat{\omega}_0)(N-1)/2} \frac{\sin((2\pi k/N - \hat{\omega}_0)N/2)}{\sin((2\pi k/N - \hat{\omega}_0)/2)}$$

This result says that $|X_3[k]|$ is composed of samples of a Dirichlet function,[6] as shown in Fig. 9.11 for the case when $N = 16$ and $\hat{\omega}_0 = 9\pi/32 = 2\pi/N(2.25)$. The envelope of the magnitude of the Dirichlet function has been plotted so that it is obvious where the samples

---

[6] See Section 6.7 for a definition of the Dirichlet function.

are being taken. Notice that the peak of the envelope is at 2.25. In contrast to the first example, none of the DFT values are exactly zero.   ◇

The signal $x_1[n]$ in Example 9.2 is a special case of $x_3[n]$ in Example 9.4 when $\hat{\omega}_0 = 2\pi k_0/N$. Thus the formula for $X_1[k]$ also can be obtained by evaluating $X_3[k]$ at $\hat{\omega}_0 = 2\pi k_0/N$. Referring to Fig. 9.11, we can see that the DFT of $X_1[k]$ is very simple, because the sampling of the Dirichlet function envelope occurs exactly at the peak and at the zero crossings of the Dirichlet function. The peak value is N, and it is the only non-zero value in the DFT.

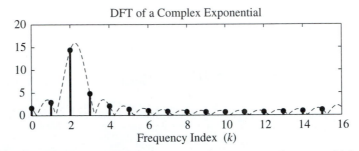

Figure 9.11   DFT of a complex exponential whose frequency is *not a multiple of* $2\pi/N$.

There is one other important difference between $x_1[n]$ and $x_3[n]$ in the previous examples. Notice that $x_1[n]$ is a periodic signal with period equal to N, while $x_3[n]$ either is not periodic, or does not repeat with a period equal to N, unless its frequency is one of the magic values, $2\pi k_0/N$. The synthesis formula of the DFT will always give a periodic signal with period equal to N. Therefore, using the synthesis formula to re-create $x_3[n]$ from its spectrum $X_3[k]$ will give the correct values of $x_3[n]$ over the range $n = 0, 1, 2, \ldots, N-1$, but the values outside that range will be a periodic extension, forcing the reconstructed $x_3[n]$ to have a period equal to N. This is illustrated in Fig. 9.12.

Each of these examples illustrates the algebraic complexity of working with the DFT summation formula. Fortunately, such analysis is rarely needed since most DFTs are computed numerically. In the next section, we present the basic ideas of the FFT algorithm, which is a very efficient computational method for the DFT.

9.3.7 The Fast Fourier Transform (FFT)

Both the DFT (9.3.11) and the IDFT summations (9.3.10) can simply be regarded as a computational method for taking N numbers in the time domain and creating N (complex) numbers in the frequency domain. Equation (9.3.11) is really N separate summations, one for each value of k. To evaluate one of the $X[k]$ coefficients, we need N multiplies and $N - 1$ additions. If we count up the arithmetic operations required to evaluate all of the $X[k]$ coefficients, the total is N^2 complex multiplications and $N^2 - N$ complex additions.

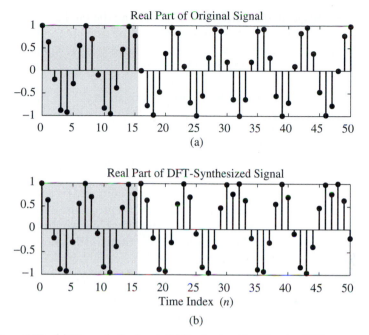

Figure 9.12   (a) Nonperiodic signal. (b) Synthesis with a 16-point DFT creates a periodic result with period equal to 16, even when the input is not periodic. The shaded region shows that over the range $0 \leq n \leq 15$ the two are identical, but the periodic extension of the DFT synthesis cannot create the rest of the original signal (a), which never repeats.

One of the most important discoveries[7] in the field of digital signal processing was the *fast Fourier transform,* or FFT, a set of algorithms that can evaluate (9.3.11) or (9.3.10) with a number of operations proportional to $N \log_2 N$ rather than N^2. When N is a power of two, the FFT computes the set of coefficients $X[k]$ with approximately $(N/2) \log_2 N$ complex operations. The $N \log_2 N$ behavior becomes increasingly significant for large N. For example, if $N = 1024$, the FFT will compute the set of coefficients $X[k]$ with $(N/2) \log_2 N = 5120$ complex multiplications rather than $N^2 = 1,048,576$, as required by direct evaluation of (9.3.11). The algorithm works best when the DFT length N is a power of two, and it also works efficiently if N has many small-integer factors. On the other hand, when N is a prime number, the FFT offers no savings over a direct evaluation of the DFT summation. FFT algorithms of many different variations are widely available in most computer languages and for almost any machine. In MATLAB, the command is simply `fft`, and most other spectral analysis functions in MATLAB call `fft` to do the bulk of their work.

---

[7] J. W. Cooley and J. W. Tukey, "An Algorithm for the Machine Computation of Complex Fourier Series," *Mathematics of Computation,* vol. 19, pp. 297–301, April 1965. The basic idea of the FFT has been traced back as far as Gauss at the beginning of the 19th Century.

Finally, consider once more Fig. 9.10 and the DFT formula in (9.3.11). We have emphasized that these two are equivalent. The existence of the FFT algorithm means that we prefer formula (9.3.11) when computing the spectrum values. However, we will find that Fig. 9.10 is still extremely useful when interpreting the results of spectrum analysis.

More details on the FFT and its derivation can be found in Section 9.8 at the end of this chapter.

9.4   SPECTRUM ANALYSIS OF SAMPLED PERIODIC SIGNALS

The FFT algorithm is a method to efficiently compute the spectrum of a discrete-time (periodic) signal, or the DFT. On the other hand, most of the signals that we record have their origin as continuous-time signals, so we would hope to understand their frequency content in the continuous-time domain. If we want to use the DFT, then we must first sample the continuous-time signal $x_c(t)$ with a C-to-D converter, as depicted in Fig. 9.13. Then the issue becomes how to relate the "true" frequency content of $x_c(t)$ to that calculated by the combination of sampling followed by the DFT of $x[n]$.

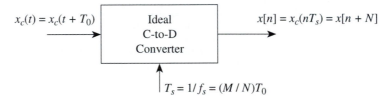

Figure 9.13   Sampling a periodic continuous-time signal with period $T_0 = 1/f_0$ to obtain a periodic discrete-time signal with period $N = M(f_s/f_0)$.

Since the DFT (as we understand it so far) applies only to periodic signals, we consider first the analysis of discrete-time periodic signals obtained by sampling periodic continuous-time signals. The ratio of the sampling period, T_s, to the period of the continuous-time signal cannot be arbitrary if we are to obtain a periodic $x[n]$, because the period of $x[n]$ must be an integer. To derive a specific condition in Fig. 9.13, consider a continuous-time periodic signal such that $x_c(t + T_0) = x_c(t)$ for all t. If $x_c(t)$ is sampled at the rate $f_s = 1/T_s$, and we desire a periodic discrete-time signal, then the following equations must hold:

$$x[n] = x_c(nT_s) \quad = \quad x_c(nT_s + T_0)$$
$$x[n] = x[n + N] \quad = \quad x_c(nT_s + NT_s)$$

One obvious choice for T_0 is $T_0 = NT_s$, where N is an integer. In this case, one period of $x[n]$ corresponds to one period of $x_c(t)$. But this is not the only possibility; it may be that one period of $x[n]$ corresponds to two periods of $x_c(t)$, or three periods, and so on. We can express this general condition by writing

$$MT_0 = NT_s$$

where M is the integer number of periods of $x_c(t)$ that correspond to one period of $x[n]$. Rearranging, we get

$$\frac{T_0}{T_s} = \frac{N}{M}$$

so the ratio of the fundamental period to the sampling period must be a rational number. If T_0/T_s is irrational, the sequence $x[n]$ will never be periodic even if the continuous-time input is periodic.

Once the period of the discrete-time signal is established to be N, we can calculate its spectrum by computing a DFT whose length is N. As an example, consider the periodic signal

$$
\begin{aligned}
x_c(t) = \ & 0.0472 \cos(2\pi(200)t + 1.5077) + 0.1362 \cos(2\pi(400)t + 1.8769) \\
& + 0.4884 \cos(2\pi(500)t - 0.1852) + 0.2942 \cos(2\pi(1600)t - 1.4488) \\
& + 0.1223 \cos(2\pi(1700)t) \hspace{6cm} (9.4.1)
\end{aligned}
$$

The waveform of this signal is plotted in the top graph of Fig. 9.14, from which it is clear that the fundamental period of $x_c(t)$ is $T_0 = 10$ msec. Likewise, the formula (9.4.1) makes it clear that all the frequencies of $x_c(t)$ are integer multiples of $2\pi(100)$, so its fundamental frequency is $f_0 = 100$ Hz. The magnitude spectrum of this signal is plotted in the middle part of Fig. 9.14, and the phase spectrum is shown in the lower part. Both are plotted as a function of continuous-time cyclic frequency $f = \omega/2\pi$. We can verify that the frequencies are all multiples of 100 Hz, even though there is no spectral component at 100 Hz. Also note that since the waveform is real, the phase angles at the negative frequencies are the negative of the phase angles of the corresponding positive frequency components, i.e., the complex amplitudes for the negative frequencies are the complex conjugates of the complex amplitudes for the corresponding positive frequencies.

Note that the sampling theorem imposes an additional constraint on the choice of sampling period. To avoid aliasing, we must choose $f_s = 1/T_s$ to be greater than twice the highest frequency present in $x(t)$. In this case we require $f_s > 2(1700)$, but we are free to choose any sampling frequency greater than 3400 samples/sec that will give a convenient value for N. For example, if the signal $x_c(t)$ is sampled at a sampling rate of $f_s = 4000$ samples/sec ($T_s = 0.25$ msec), the corresponding discrete-time signal is that shown in Fig. 9.15 (top). Notice that the period of the sequence $x[n]$ is $N = T_0/T_s = f_s/f_0 = 4000/100 = 40$ samples. If one period of $x[n]$ from the upper panel of Fig. 9.15 is used in a 40-point DFT (9.3.11), we obtain the 40 spectrum coefficients $X[k]$ shown in Fig. 9.15 (middle and bottom). These 40 spectrum coefficients represent the sequence $x[n]$ in the synthesis formula of the IDFT (9.3.10).

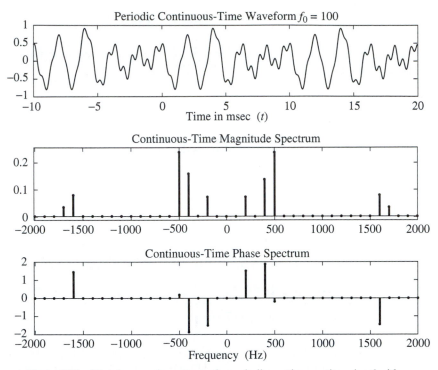

Figure 9.14   Waveform and spectrum of a periodic continuous-time signal with fundamental frequency 100 Hz.

It is essential that we be able to tie Figs. 9.14 and 9.15 together. The spectrum of the discrete-time sequence is indexed from $k = 0$ to $k = 39$, but each index k also corresponds to a continuous-time frequency in Hz. The nonzero values of $X[k]$ correspond to frequency components that are actually in the continuous-time signal. Recall that the relationship between continuous-time frequencies ω and discrete-time frequencies $\hat{\omega}$ is

$$\hat{\omega} = \omega T_s$$

For example, a frequency of $\omega = 2\pi(400)$ for the continuous-time signal corresponds to $\hat{\omega} = 2\pi(400)/4000 = 2\pi(4/40)$ for the discrete-time signal, and therefore to index $k = 4$ and in the representation of (9.3.10). Also, observe that the negative frequencies in Fig. 9.14 are manifest at the positive alias frequencies in Fig. 9.15. As a specific example, -400 Hz for the continuous-time signal corresponds to $\hat{\omega} = 2\pi + 2\pi(-400)/4000 = 2\pi(40 - 4)/40$ for the discrete-time signal, and therefore to index $k = 36$ in the representation of (9.3.10).

Exercise 9.3.   Derive a formula that converts an analog frequency ω into the correct frequency index k. Assume that T_s is known, and that $|\omega| < \pi/T_s$, so there is no aliasing. Also, assume that the DFT length N is known.

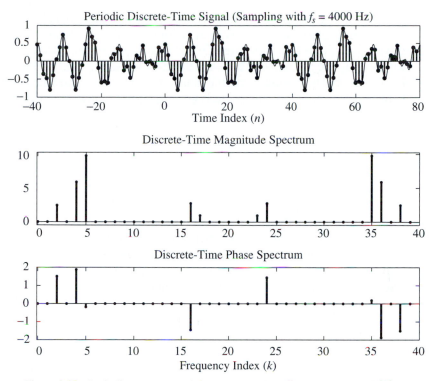

Figure 9.15 Periodic sequence and the corresponding discrete spectrum $X[k]$ obtained by sampling the waveform of Fig. 9.14 with sampling rate $f_s = 4000$ samples/sec.

9.5 SPECTRUM ANALYSIS OF NONPERIODIC SIGNALS

We have seen that if a discrete-time signal is periodic, it is possible to calculate its spectrum exactly using the DFT. Although this is an elegant result, in most cases of practical interest we do not have periodic signals. Usually we have recordings that give a very long sequence obtained by sampling for many minutes or even hours. For example, when sampling an audio signal at a sampling rate of 44.1 kHz, one hour of stereophonic music would be represented by $44100 \times 2 \times 60 \times 60 = 317,520,000$ samples. We might treat the long sequence as infinitely long, but that does no good for FFT computations. We may hope that the long sequence is periodic, but it is unlikely that a long recording would be exactly periodic, or even contain short segments that are exactly periodic.

Instead, we will apply our DFT analysis to finite-length signals. We could treat any recording, no matter how long, as having finite length, but this would lead to an enormously long FFT. For approximately one hour of audio at 44.1 kHz, the closest power-of-two FFT needed would be $2^{28} = 268,435,456$ per channel. A better approach is to break the long signal into small segments and analyze each one with an FFT. This is reasonable because the very long recording probably contains short passages where the spectral content does not vary, so there may be a natural segment length. Indeed, we have already seen that music can be thought of in just this way. In this section, we will study how the DFT can be applied to finite-length signals, and we will develop the relationship between finite-length signals and periodic signals for the DFT. Our goal is to understand the limits of the DFT method when applied to finite-length signals, and how this analysis gives an approximate spectrum representation.

9.5.1 Spectrum Analysis of Finite-Length Signals

How can we adapt the DFT to finite-length signals? Since the formula (9.3.11) is simply a sum over the range 0 to $N-1$, why not just reuse the formula? One way to justify using the DFT technique for finite-length sequences is to think about constructing a periodic signal from the finite-length signal. Then the techniques of Section 9.3.4 provide an exact spectrum analysis for the artificial periodic signal. We can construct a periodic sequence by repeating the finite-length portion. Mathematically, we assume that $x[n] = 0$ for $n < 0$ and $n \geq L$; i.e., $x[n]$ has L nonzero samples. Then we use the repetition formula

$$\tilde{x}[n] = \sum_{r=-\infty}^{\infty} x[n + rN]$$

which is guaranteed to produce a signal $\tilde{x}[n]$ whose period is N. Note that we do not actually have to create the signal $\tilde{x}[n]$ since we need only one period for analysis. However, it is important to acknowledge the implicit existence of this periodic signal in the DFT computation. Clearly, if $N \geq L$, then $\tilde{x}[n] = x[n]$ for $0 \leq n \leq N-1$; i.e., one period of $\tilde{x}[n]$ is identical to $x[n]$. Since $\tilde{x}[n]$ is periodic, now we can apply the N-point DFT (9.3.11) to one period of $\tilde{x}[n]$

$$X[k] = \sum_{n=0}^{N-1} \tilde{x}[n]e^{-j(2\pi/N)kn} \qquad k = 0, 1, 2, \ldots, N-1$$

But that period of $\tilde{x}[n]$ is really $x[n]$, so we get exactly the same set of complex numbers as if we had just used the DFT directly on $x[n]$ over its nonzero portion $n = 0, 1, 2, \ldots, L-1$.

$$X[k] = \sum_{n=0}^{N-1} x[n]e^{-j(2\pi/N)kn} = \sum_{n=0}^{L-1} x[n]e^{-j(2\pi/N)kn} \qquad (9.5.1)$$

Changing the upper limit in (9.5.1) from $N-1$ to $L-1$ is permissible because $x[n] = 0$ for $L \leq n \leq N-1$ if $L < N$.

FFT spectrum analysis computation: The spectrum of a length-L signal can be computed with an N-point DFT, if we *zero-pad* the signal; i.e., prior to computing the N-point FFT, we must increase the signal length from L to N by appending $N-L$ zero samples to the nonzero samples of $x[n]$. The MATLAB function `fft` has this behavior for its default. If we create a vector `xx` with L elements, we can specify the FFT length, e.g., $N = 512$, when we write the statement

$$\texttt{XX = fft(xx, 512);}$$

If $L < 512$ the vector `xx` will be zero-padded automatically.

Whenever the DFT values $X[k]$ are used to reconstruct a discrete-time signal using the inverse DFT (9.3.10), what will be the result? The IDFT formula (9.3.10) must give back the periodic signal

$$\frac{1}{N} \sum_{k=0}^{N-1} X[k] e^{j(2\pi/N)kn} = \tilde{x}[n] = \sum_{r=-\infty}^{\infty} x[n + rN] \tag{9.5.2}$$

which always reconstructs a periodic signal $\tilde{x}[n]$. Since we have chosen $N \geq L$, we can, of course, obtain $x[n]$ from $\tilde{x}[n]$ because we know that

$$x[n] = \begin{cases} \tilde{x}[n] & 0, 1, 2, \ldots, N-1 \\ 0 & \text{otherwise} \end{cases}$$

In other words, we should evaluate the sum in (9.5.2) only for $0 \leq n \leq N - 1$ and ignore the values outside this interval.

The question that naturally arises out of the foregoing discussion is, "What is the best value of N to use for a given finite-length sequence?" It is not possible to give a simple answer to this question. Clearly, choosing any N so that $N \geq L$ guarantees that the original finite-length sequence can be recovered from its spectrum by using (9.5.2), but is there any reason to use a DFT length greater than $N = L$? As an illustration, consider the finite-length sequence

$$x[n] = \begin{cases} 0.5[1 - \cos(2\pi n/L)] & 0 \leq n \leq L-1 \\ 0 & \text{otherwise} \end{cases} \tag{9.5.3}$$

Figure 9.16 shows the sequence of (9.5.3) for $L = 20$ and repeated periodically with period $N = 40$. With $N = 40$ we have spread out the nonzero parts of $x[n]$ (zero-padded) by placing 20 zero samples in between. Since the period is $N = 40$, it takes 40 spectral components to represent the signal, as shown in the bottom panel of Fig. 9.16. Note that only a few of these are large, and that about half of them are actually equal to zero. If we now increase the value of N to 50, we spread the copies of $x[n]$ farther apart, as shown in Fig. 9.17 for $L = 20$ and $N = 50$. With a period of $N = 50$, it now takes 50 spectral components to represent the signal, but only those for $0 \leq k \leq 4$ and $45 < k < 50$ seem to be large.

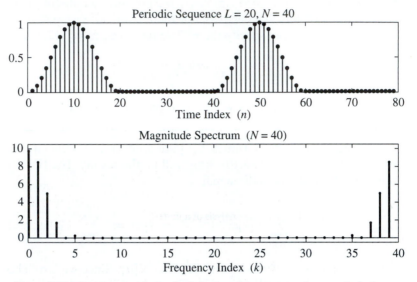

Figure 9.16   Spectrum analysis of a finite-length sequence using a period of $N = 40$. (Only the magnitude is shown.)

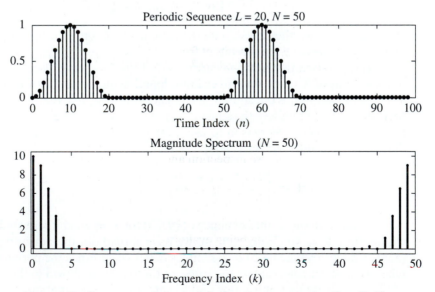

Figure 9.17   Spectrum analysis of a finite-length sequence using $N = 50$. The magnitude spectrum $|X[k]|$ is the 50-point DFT of the 20-point signal, padded with 30 zeros.

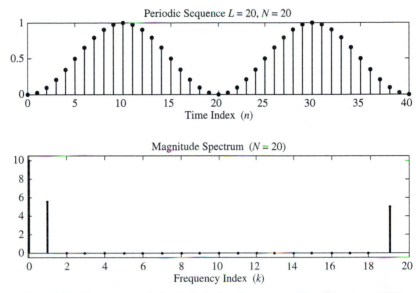

Figure 9.18   Spectrum analysis of a finite-length sequence ($L = 20$) using a DFT length of $N = 20$.

However, if we decrease the value of N in (9.5.1) and move the copies of $x[n]$ closer together, we can eliminate all the zeros in between. This is shown in Fig. 9.18 for $L = 20$ and $N = 20$. The bottom panel of Fig. 9.18 shows the corresponding magnitude spectrum of the periodic signal, which has just three nonzero components at $k = 0, 1, 19$. Notice that in this case, the periodic repetition of $x[n]$ creates a signal that is equal to $0.5[1 - \cos(2\pi n/20)]$ for all n, so we should not be surprised that (9.5.1) gives an exact representation with three components: a DC component and complex exponential components at frequencies $\hat{\omega} = +2\pi/20$, and $-2\pi/20$, which is the same as $\hat{\omega} = 2\pi - 2\pi/20 = 2\pi(19)/20$.

9.5.2 Frequency Sampling

As we change N, the spectral representations appear somewhat different, but are actually quite similar. Close inspection and comparison of Figs. 9.16 and 9.18 reveals that the 20 spectral components shown in Fig. 9.18 are actually equal to every other component of the spectrum for $N = 40$ in Fig. 9.16. Furthermore, Figs. 9.16, 9.17, and 9.18 all seem to have a similar shape. In each case, the spectrum was computed by using the FFT with zero padding, so the computation is based on the same L data values. Indeed, the formula being evaluated is always

$$X[k] = \sum_{n=0}^{L-1} x[n]e^{-j(2\pi k/N)n} \qquad k = 0, 1, \ldots, N-1$$

with $L = 20$ and different values for N. When the value of N is changed, we are using a different set of frequencies $\hat{\omega}_k = (2\pi/N)k$. This observation leads us to state the following general property:

Frequency Sampling Property of the DFT: The N-point DFT is an evaluation of the formula

$$X(e^{j\hat{\omega}}) = \sum_{n=0}^{L-1} x[n]e^{-j\hat{\omega}n}$$

at the frequencies $\hat{\omega}_k = (2\pi/N)k$, for $k = 0, 1, 2, \ldots, N - 1$, i.e., $X[k] = X(e^{j(2\pi/N)k})$.

We can vary the frequency spacing by changing the value of N. Figure 9.19 shows the result for $N = 120$, where we see enough frequency samples to identify convergence to the function $X(e^{j\hat{\omega}})$.

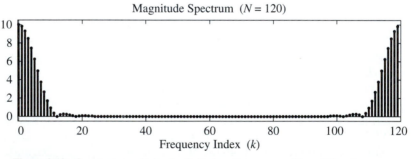

Figure 9.19   Spectrum analysis of a finite-length sequence ($L = 20$) using a 120-point DFT, i.e., $N = 120$.

If we increase N even more, we get points on the function $X(e^{j\hat{\omega}})$ that are closer and closer together. Figure 9.20 shows the result of using $N = 1024$. In this case, we have drawn a continuous plot,[8] because the vertical lines to the points would be crowded too close together to see individual lines. For all practical purposes, this evaluation of $|X(e^{j\hat{\omega}})|$ with frequency spacing $2\pi/1024$ is adequate for plotting this function as a continuous curve. Therefore, it is reasonable to label the frequency axis in Fig. 9.20 in terms of $\hat{\omega}$, in the range $0 \le \hat{\omega} < 2\pi$, instead of k, in the range $0 \le k \le N - 1$. As we let $N \to \infty$, the spacing between frequency components becomes infinitesimally small, and we are in effect taking a limit. The frequency samples are $X[k] = X(e^{j2\pi k/N})$, so $X(e^{j2\pi k/N}) \to X(e^{j\hat{\omega}})$ as $N \to \infty$, and we obtain the continuous function $X(e^{j\hat{\omega}})$ of the continuous variable $\hat{\omega}$.

---

[8] In MATLAB, the `plot` command connects the points by straight lines.

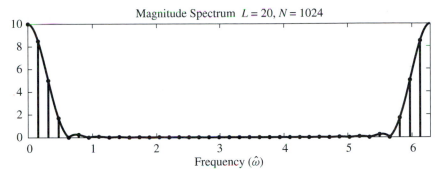

Figure 9.20   A large value for the DFT length N gives a dense sampling of the spectrum. The result converges to $X(e^{j\hat{\omega}})$, a function of $\hat{\omega}$. Frequency samples at $\hat{\omega} = 2\pi k/40$ are also shown to illustrate the relationship between the 40-point DFT and the continuous function $X(e^{j\hat{\omega}})$.

Superimposed on $X(e^{j\hat{\omega}})$ in Fig. 9.20 is the magnitude spectrum for $N = 40$ as previously plotted in the bottom panel of Fig. 9.16. If we denote the values of $X[k]$ for $N = 40$ as $X_{40}[k]$, then $|X_{40}[k]| = |X(e^{j\hat{\omega}})|$ evaluated at $\hat{\omega} = 2\pi k/40$. Likewise, the values of $|X_{20}[k]|$ are the samples at $\hat{\omega} = 2\pi k/20$. This explains why the spectral component for $k = 2$ in Fig. 9.16 is identical to the spectral component for $k = 1$ in Fig. 9.18. Both are sampling the same formula; the spectral values in Fig. 9.18 are $X_{20}[k] = X(e^{j2\pi k/20})$ for $k = 0, 1, 2, \ldots, 19$, while the spectral values in Fig. 9.16 are $X_{40}[k] = X(e^{j2\pi k/40})$ for $k = 0, 1, 2, \ldots, 39$. The frequency-sampling property tells us that they correspond to exactly the same frequency, i.e., $2\pi/20 = 2\pi(2)/40$.

One nagging question about Figs. 9.18 and 9.16 is the fact that with 20 spectral components we get a representation with only three nonzero components, but the $N = 40$ case has many additional nonzero spectral components. Why are these extra components needed? One answer is that the frequency samples for the $N = 20$ case happen to lie at the zeros of $X(e^{j\hat{\omega}})$, which is lucky for this particular signal. A better explanation is that additional spectral components are needed to represent the gap in the time domain when 20 zero samples are appended to the $L = 20$ samples of $x[n]$ to make a 40-point periodic signal.

9.5.3 Samples of the Frequency Response

One convenient application of FFT spectrum analysis is computing the frequency response of an FIR filter. If the impulse response $h[n]$ is of finite length L, the frequency response would have the form

$$H(e^{j\hat{\omega}}) = \sum_{n=0}^{L-1} h[n]e^{-j\hat{\omega}n} \tag{9.5.4}$$

If we compute the N-point DFT of $h[n]$ with zero-padding, then we are evaluating (9.5.4) at N equally spaced frequencies in the interval $0 \le \hat{\omega} < 2\pi$, so we obtain

$$H[k] = \sum_{n=0}^{L-1} h[n]e^{-j(2\pi/N)kn} = H(e^{j2\pi k/N}) \qquad k = 0, 1, 2, \ldots, N-1 \quad (9.5.5)$$

What do we have in (9.5.5)? Based on our discussion, we have an FFT spectrum of the finite-length sequence $h[n]$, but we also have samples of the frequency response (9.5.4) of the FIR system. If we use a large enough value for N, we can plot a smooth curve for the frequency response of the filter.

Example 9.5   Assume that the sequence of (9.5.3) is now the impulse response of a FIR filter, i.e.,

$$h[n] = \begin{cases} 0.5[1 - \cos(2\pi n/L)] & 0 \le n \le L - 1 \\ 0 & \text{otherwise} \end{cases} \quad (9.5.6)$$

A filter with this impulse response is sometimes called a *Hann filter*. Figure 9.21 shows plots of the magnitude of the frequency response for two Hann filters with lengths $L = 20$ and $L = 40$. The frequency responses were obtained using (9.5.5) with $N = 1024$. This figure illustrates that the Hann filter is a lowpass filter, because its passband is near $\hat{\omega} = 0$ where the frequency response is large, and its stopband covers the range $4\pi/L < \hat{\omega} < \pi$, where the frequency response values are relatively low. Do not be confused by the region $\pi < \hat{\omega} < 2\pi$, which is actually the negative frequency interval of the frequency response.

When we compare the two different lengths, we see that the frequency response becomes more concentrated around $\hat{\omega} = 0$ as we increase L. It can be shown that the first zero of $H(e^{j\hat{\omega}})$ occurs at $\hat{\omega} = 4\pi/L$ for the Hann filter. Thus, increasing L from 20 to 40 cuts the width of the passband in half (and also doubles its amplitude).   ◇

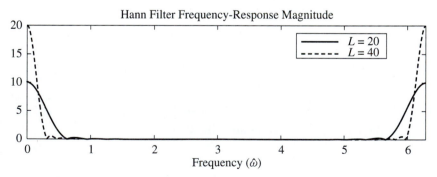

Figure 9.21   Frequency response of two Hann filters; length $L = 20$ (solid lines), and length $L = 40$ (dashed lines).

9.5.4 Spectrum Analysis of Continuing Nonperiodic Signals

When a signal is neither periodic nor finite-length, we can still use the DFT spectrum representation that was first derived for periodic signals and then adapted for use with finite-length signals. To see how this can be done, it is helpful to return to the filter-bank system of Fig. 9.10 and generalize that structure somewhat. Figure 9.22 shows a generalized version of the filter-bank system of Fig. 9.10, where the N-point running sum filters have been replaced by a more general LTI lowpass filter with impulse response $h[n]$.

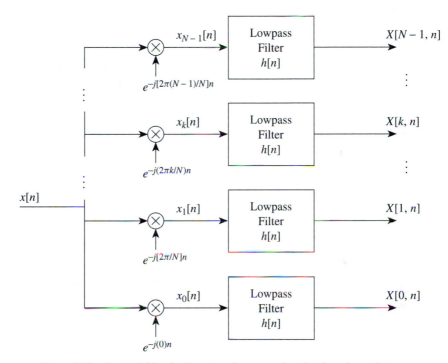

Figure 9.22 General filter-bank system for computing the time-dependent spectrum $X[k, n]$ for the frequencies $\hat{\omega} = 2\pi k/N, k = 0, 1, \ldots, N - 1$. Usually the lowpass filter is an FIR filter.

The filter-bank system of Fig. 9.22 is a set of modulators and filters whose collective outputs we would like to consider as the spectrum of the input signal. Each individual modulator-filter combination is called a *channel* of the filter bank, and the kth channel has an *analysis frequency* at $\hat{\omega}_k = 2\pi k/N$. The output of the lowpass filter block in the kth channel is

$$X[k, n] = \sum_{m=n-L+1}^{n} h[n - m]x[m]e^{-j(2\pi/N)km} \qquad k = 0, 1, 2, \ldots, N - 1 \quad (9.5.7)$$

if the impulse response $h[n]$ of each of the filters is of length L. For example, one possibility is to use the Hann filter of Section 9.5.1 for $h[n]$. A close look at (9.5.7) and Fig. 9.22 shows that $X[k, n]$ is really a function of two variables: frequency ($\hat{\omega} = 2\pi k/N$) and time (n). Examining (9.5.7) carefully, we see that for a particular frequency index $k = k_0$, the input signal $x[n]$ modulates the complex exponential $e^{-j(2\pi/N)k_0 n}$ to produce the signal

$$x_{k_0}[n] = x[n]e^{-j(2\pi/N)k_0 n}$$

which then is filtered by the lowpass filter $h[n]$ to produce the output $X[k_0, n]$. These steps are illustrated in Fig. 9.23 for the case $N = 100$ and $k_0 = 4$. The top panel shows 201 samples of a speech signal $x[m]$. Observe that the signal appears to be periodic, but on closer examination does not repeat exactly. The middle panel shows the real part of the complex exponential $e^{-j(2\pi/100)4m}$, for the fourth channel. The bottom panel of Fig. 9.23 shows the real part of the product $x_4[m] = x[m]e^{-j(2\pi/100)4m}$, which would be the input to the lowpass filter.

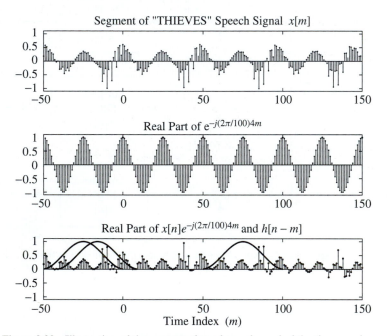

Figure 9.23 Illustration of the computation of one channel of the time-varying spectrum of a speech signal. The bottom panel also shows the sliding Hann window that performs the lowpass filtering.

Also shown in the bottom panel of Fig. 9.23 is the $L = 51$-point Hann filter impulse response $h[n - m]$ as a function of m for several different values of the parameter n, i.e., $n = 0, 10, 100$. Since the plot is a function of m, the impulse response $h[n - m]$ is actually flipped, but the Hann filter is symmetric so this is not noticeable. Also, although the impulse response of the Hann filter is discrete, it is shown as continuous to avoid confusion with the modulated discrete-time signal $x_4[n]$. The output $X[4, n]$ at each time n would be obtained by multiplying the samples of $x_4[n]$ that fall underneath $h[n-m]$ by the corresponding impulse response values and then summing the products as in (9.5.7). The final output of the fourth channel will be a function of time n.

SLIDING FFT SPECTRO-GRAM

As n increases, the Hann filter $h[n-m]$ moves to the right, enclosing a different subset of 51 points as it moves. Since only 51 points of the signal $x_4[m]$ are taken into the summation in (9.5.7) at a time, the function $h[n-m]$ is usually called a *sliding window* because only 51 points are "visible" to the summation at time n. The sliding window moves to the right, isolating 51-point segments of $x_4[m]$ as it moves. The minimum shift is one sample point, because $h[n+1-m]$ overlaps by 50 samples with $h[n-m]$.

Figure 9.24 shows the magnitude of the output versus n for three channels $(k = 0, 2, 4)$ when processing the speech signal from Fig. 9.23. Each of these outputs is very smooth because the channel's lowpass filter has a narrow passband. Since the output is so smooth, it is not necessary to compute the outputs for all n; instead, we could skip some values and still retain the basic shape of the output. Skipping outputs is equivalent to sliding the window by more than one sample.

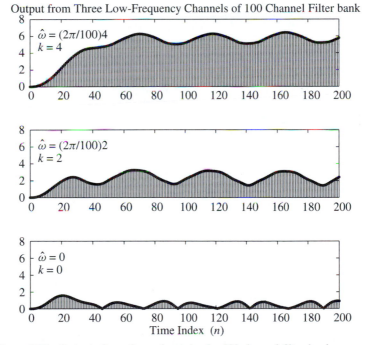

Figure 9.24 Outputs from three channels of a 100-channel filter bank for the speech input signal of Fig. 9.23. The analysis frequencies are $\hat{\omega} = 0$, $(2\pi/100)2$, and $(2\pi/100)4$, and the lowpass filter is a length–51 Hann filter.

We can run another experiment to show how the filter bank responds to different frequencies. A good test signal is a stepped-frequency sinusoid. We form an input signal containing three 50-point segments: the first one ($0 \le n \le 50$) has frequency $\hat{\omega} = 0$, the next $\hat{\omega} = 2\pi(4/100)$, and the final one $\hat{\omega} = 2\pi(8/100)$. In Fig. 9.25, the output signals from three channels ($k = 0, 4, 8$) of a 100-channel filter bank are shown. The analysis frequencies are $\hat{\omega} = 0, 2\pi(4/100), 2\pi(8/100)$, so the test signal should be analyzed perfectly as its frequency jumps from one channel to the next.

Notice how each channel in Fig. 9.25 responds best when the signal contains its frequency, but also notice that there is some cross-channel interference. This is due to the fact that the lowpass filters are not perfect; their stopband gain is small, but not zero. Therefore, neighboring channels will respond a little when there is a sinusoid at a nearby frequency. The channel responses are time-shifted because the channel lowpass filter is a length–51 Hann filter which has a delay of 25 samples.

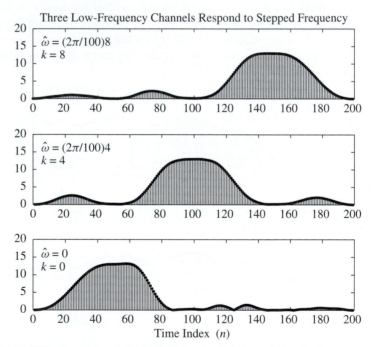

Figure 9.25   Response of three channels of a 100-channel filter bank to a frequency-stepped input signal. The signal's frequencies are $\hat{\omega} = 0, (2\pi/100)4, (2\pi/100)8$, starting at $n = 0, 50, 100$, respectively.

This filter bank structure can also provide some insight into the case where the signal is very long and the impulse response of the filters is also long. In our earlier discussion, we found that the ideal filter for measuring the spectral component at any arbitrary frequency would be an infinitely narrow lowpass filter such that $H(e^{j\hat{\omega}}) = 0$

at all frequencies except $\hat{\omega} = 0$. As was illustrated in Fig. 9.21, the Hann filter can approximate this condition as $L \to \infty$; i.e., as the impulse response becomes infinitely long. Thus, we can, in theory, compute the spectrum of a very long signal, but we would have to wait a very long time to collect all the samples of the signal, and then we would have to do a huge computation using (9.5.7) to get the desired result. Although this can be done for large (finite) lengths, it is neither necessary nor desirable to do so.

Instead, it is often more useful to compute a "local spectrum" that measures frequency content over a short time interval. The filter-bank structure does exactly that, as we saw in the stepped-frequency example of Fig. 9.25. In the next section, we will define a "time-dependent spectrum" by breaking up a long sequence into short finite-length segments and then using the FFT to determine the spectrum of each short segment. Finally, we will prove that the FFT method is equivalent to a filter bank.

9.6   THE SPECTROGRAM

We can interpret the filter bank of Fig. 9.22 and (9.5.7) in another way. If we fix the time index at $n = n_0$, then the set of values $X[k, n_0]$ is

$$X[k, n_0] = \sum_{m=n_0-L+1}^{n_0} h[n_0 - m]x[m]e^{-j(2\pi/N)km} \qquad k = 0, 1, 2, \ldots, N-1$$

$$(9.6.1)$$

When plotted versus k, $X[k, n_0]$ is what we want to call the "frequency spectrum at time n_0." Equation (9.6.1) can be made to look very similar to the DFT (9.3.11) if two observations are made. First of all, let $L = N$, so that the sum is taken over one period of the complex exponentials. Secondly, define a new function $w[n]$ as a flipped version of $h[n]$

$$w[n] = h[-n] \qquad\qquad (9.6.2)$$

This function $w[n]$ is called a *window*. With these two definitions, we get

$$X[k, n_0] = \sum_{m=n_0-N+1}^{n_0} (w[m - n_0]x[m])\, e^{-j(2\pi/N)km} \qquad k = 0, 1, 2, \ldots, N-1$$

$$(9.6.3)$$

In (9.6.3), $X[k, n_0]$ is just the DFT spectrum of the short segment of $x[m]$ that lies inside the sliding window. The sequence $w[m - n_0]x[m]$ is a finite-length sequence (since the window has length L) and therefore all the results of Section 9.5.1 can be applied to the interpretation of $X[k, n_0]$. The computations needed for (9.6.3) are the same as the DFT, so an N-point FFT of $w[m - n_0]x[m]$ can be used. Figure 9.26 shows a block diagram for using the FFT.

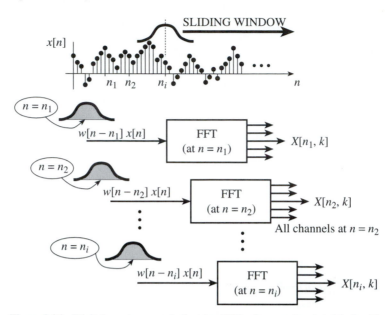

Figure 9.26 Digital spectrum analysis using FFTs of successive data blocks. The kth output of each FFT corresponds to the output signal from a single channel of the spectrum analyzer.

With either the sliding FFT or the filter-bank interpretation, we see that $X[k, n]$ is a two-dimensional sequence. The k dimension represents frequency (since $\hat{\omega}_k = 2\pi k/N$ is the analysis frequency of the kth channel), and the n dimension represents time. Since the result is a function of both time and frequency, we cannot plot *the* spectrum, because there is a different *local spectrum* for each time. To deal with this additional complexity, a three-dimensional graphical display is needed. This is done by plotting $|X[k, n]|$ (or $\log |X[k, n]|$) as a function of both k and n using perspective plots, contour plots, or gray-scale images. The preferred form is the *spectrogram*, which is a gray-scale image where the gray level at point (k, n) is proportional to $|X[k, n]|$ or $\log |X[k, n]|$; large values are black, and small ones white. Examples of the spectrogram can be seen in Figs. 9.27, 9.31, and 9.32. The horizontal axis is time and the vertical axis is frequency, starting with zero frequency at the bottom. Note that the filter bank in Fig. 9.10 was drawn so that the channel frequencies also increase from zero at the bottom to the highest frequency $2\pi(N-1)/N$ at the top. In the spectrogram image for a real signal, only the channels for $0 \le k \le N/2$ would be used.

9.6.1 Spectrograms in MATLAB

Since the spectrogram can be computed by doing many FFTs of windowed signal segments, MATLAB is an ideal environment for doing the calculation and displaying the image. The MATLAB command is

```
[B,F,T] = specgram(xx, NFFT, Fs, window, Noverlap)
```

where B is a two-dimensional array containing the complex spectrogram values, F is a vector of all the analysis frequencies, and T is a vector containing the times of the sliding window positions. The inputs are the signal xx, the FFT length NFFT, the sampling frequency Fs, the window coefficients window, and the number of points to overlap as the window slides Noverlap. Note that the window skip is NFFT - Noverlap. The overlap must be less than the window length, but choosing Noverlap equal to length(window)-1 would generate a lot of needless computation, because the window skip would be one. It is common to pick the overlap to be somewhere between 50 percent and 80 percent of the window length, depending on how smooth the final image needs to be. See help specgram in MATLAB for more details.[9]

The spectrogram image can be displayed by using

```
imagesc( T, F, abs(B) )
axis  xy, colormap(1-gray)
```

The color map of (1-gray) gives a negative gray scale that is useful for printing, but on a computer screen it is preferable to use color, e.g., colormap(jet). Finally, it may be advantageous to use a logarithmic amplitude scale in imagesc in order to see small amplitude components.

9.6.2 Spectrogram of a Sampled Periodic Signal

To illustrate the use of the spectrogram, we use the periodic signal (9.4.1) from Section 9.4 as a test signal. This signal consists of five harmonic frequencies that are multiples of a fundamental frequency at $f_0 = 100$ Hz. As discussed in Section 9.4, if the sampling rate is $f_s = 4000$ samples/sec, the sequence that results from sampling (9.4.1) has a period of 40 samples. If the window used in (9.6.3) has length $L = 80$ and the FFT length is also $N = 80$, we obtain

$$X[k, n] = \sum_{m=n-79}^{n} x[m]e^{-j(2\pi/80)km} \qquad k = 0, 1, 2, \ldots, 79 \qquad (9.6.4)$$

The resulting spectrogram is shown in Fig. 9.27. This gray-scale image shows five constant horizontal lines. These spectral components have different intensities, but there are no other spectral components. Why does the image look this way? From our discussion of periodic signals in Section 9.3.4, we know that $X[k, n]$ should not vary along the time dimension because the input is periodic and the running-sum filter impulse response spans exactly two complete periods of the signal. Furthermore,

---

[9] The function spectgr provided in the DSP First toolbox is equivalent to specgram, which is a part of the MATLAB Signal Processing toolbox.

from the discussion in Section 9.4, it follows that the only nonzero spectral components should be at $k = 4, 8, 10, 32, 34$.

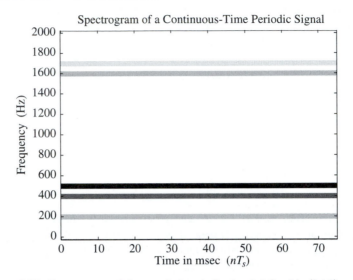

Figure 9.27   Spectrogram of the sampled periodic signal defined in (9.4.1); $f_s = 4000$ samples/sec and $f_0 = 100$ Hz. Only the second, fourth, fifth, sixteenth, and seventeenth harmonics are present. Channel filter is an 80-point running sum, whose length is exactly equal to two periods.

The frequency axis in Fig. 9.27 is calibrated in terms of the continuous-time cyclic frequency variable, f. In order to derive the relationship between f and the frequency index k, recall that $\hat{\omega} = 2\pi k/N$ and $\hat{\omega} = 2\pi(f/f_s)$ (from the sampling theorem). Equating these two formulas, we get

$$f = (k/N)f_s \tag{9.6.5}$$

which allows us to label the frequency axis in Hz. Applying (9.6.5) to the frequency indices $0 \le k \le N/2$, we get the continuous-time frequencies $0 \le f \le f_s/2$, which is the usual range of frequencies displayed for a real signal. The negative-frequency spectral components can be inferred from the symmetry of the magnitude spectrum $|X[k, n]|$ of a real signal. Therefore, in Fig. 9.27 there are only 41 discrete points on the vertical axis. The five frequency components are at $(k/80)4000$ for $k = 4, 8, 10, 32, 34$. The width of the spectral lines plotted at these values of k is due to MATLAB's image-plotting function, which zooms to show a larger image with many more than 41 pixels vertically.[10] Since we chose $L = 80$, no two spectral components are next to each other, so we can see the gray-scale amplitudes of the five spectral components.

---

[10] The gray scale uses white to represent zero amplitude, and black to represent the highest amplitude. This scale was chosen to make the picture mostly white for printing. On a computer monitor, it may be better to invert the scale or use color.

9.6.3 Resolution of the Spectrogram

In our earlier discussion leading to the spectrogram, two equivalent views have been presented: the filter-bank structure and the sliding window FFT. We note that calculation of the spectrum analysis requires that the filter length (L) be finite and that the number of analysis frequencies (N) be finite. The performance of the spectrogram analysis usually comes down to a statement about its resolution in either frequency or time. The key parameters that control resolution are $h[n]$ and its length L. In the filter-bank interpretation, $h[n]$ is the impulse response of the channel filters; in the FFT view, $h[n] = w[-n]$ is the shape of the window used prior to the FFT. We will use the filter-bank interpretation, so we refer to $h[n]$ as the "impulse response."

Resolution can be measured by determining the response of the system when processing closely spaced components. For example, to measure frequency resolution, we use an input signal that is the sum of two sinusoids whose frequencies are f_0 and f_1. The frequency resolution is the minimum separation $|f_0 - f_1|$, such that both sinusoids can be detected by the spectrum analyzer. Likewise, we can define time resolution by measuring how well we can detect the beginning or end of finite-length signals. For the purposes of this discussion, detection means that we can see them both in a spectrogram image.

RESOLUTION
of SPECTRO-
GRAM

In the filter-bank structure (Fig. 9.22), each channel uses an FIR lowpass filter whose frequency response determines the resolution. The filter bank works on the principle that each individual spectral component can be frequency shifted into the passband of the lowpass filter, while all the others will lie in the filter's stopband. Closely spaced frequency components demand that the passband of the LPF be narrower than the frequency separation, so we conclude that the narrower the passband of the LPF, the better the frequency resolution. The passband width is controlled primarily by the length of the impulse response, L. A secondary factor is the actual shape of the impulse response, $h[n]$. For example, the Hann filter (9.21) has a passband width that is approximately $2\pi(4/L)$, so we should make L large to improve the frequency resolution.

There is a down side to increasing L; namely, the time resolution will suffer. When L is large, short-duration time events or transitions will be lost when averaged together by a long impulse response. In Chapter 6, we experienced this phenomenon in the form of image blurring by lowpass filtering. The same sort of smoothing will happen at the endpoints of pulses passed through an LPF. Therefore, we have uncovered a basic constraint of Fourier spectrum analysis:

> *Good time resolution and good frequency resolution cannot be obtained simultaneously in the spectrogram.*

This statement is called the *uncertainty principle*; it is equivalent to the Heisenberg uncertainty principle in physics, which also relies on a Fourier relationship between two quantities—position and momentum.

9.6.3.1 Resolution Experiment   In order to test the resolving power of the spectrogram, we need a simple experiment. Our test will rely on analyzing a signal that is hard to "see" in both time and frequency. The test signal consists of two components. The first is a constant 960-Hz tone that persists over the entire data set; the second is a short signal whose frequency is 1000 Hz, and whose beginning and ending times are 200 and 400 msec. The ideal form of such a spectrogram is drawn in Fig. 9.28. The goal in this test is to simultaneously resolve the beginning and ending times of the second signal, as well as its frequency. In our discussion of the filter bank, we noted that the frequency resolution is dictated by the bandwidth of the lowpass filter in the channels. When a Hann filter is used, this bandwidth is approximately $4\pi/L$, so the frequency resolution is

$$\text{Frequency Resolution} \;\approx\; \frac{2}{L} f_s \quad \text{Hz} \tag{9.6.6}$$

The frequency-scaling relationship (9.6.5) was used to express the resolution in Hz.

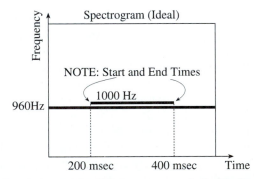

Figure 9.28   Resolution test for digital spectrum analysis. Idealized spectrum consisting of two sine waves with different durations and different frequencies.

The time resolution, on the other hand, is directly proportional to the length of the filter. The FIR filter performs a weighted average of all points within the filter length, so it causes blurring. If we have a signal that turns on at $t = t_0$, the Hann filter seems to blur over an interval from $t = t_0 - (L/4)T_s$ to $t = t_0 + (L/4)T_s$, so the uncertainty in the start time would be

$$\text{Time Resolution} \;\approx\; \frac{L}{2} T_s \quad \text{sec} \tag{9.6.7}$$

The trade-off between time and frequency resolution can be seen in Fig. 9.29(a) and (b) for the cases $L = 512$ and $L = 128$, respectively. The predicted resolutions from (9.6.6) and (9.6.7) are

Length	Frequency-Resolution	Time Resolution
512	15 Hz	0.064 sec
128	60 Hz	0.016 sec

The frequency resolution of the longer window ($L = 512$) is adequate to resolve the two signals, but it also causes a blurring of the signal's endpoints, which inhibits an accurate calculation of the starting and ending times, as is evident in Fig. 9.29(a). On the other hand, the shorter window ($L = 128$) in Fig. 9.29(b) does find the ends of the signal, but fails to resolve the two frequencies. This sort of time-frequency trade-off is always present in a spectrogram that relies on a filter-bank analysis with equally spaced filters, all having the same bandwidth.[11]

RESOLUTION OF SPEC-TROGRAM

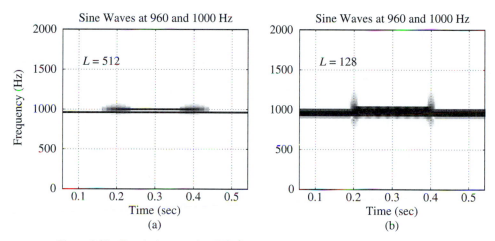

Figure 9.29   Resolution test for digital spectrum analysis using two different window lengths: (a) $L = 512$, which is 128 msec at $f_s = 4000$ Hz; (b) $L = 128$, or 32 msec. The FFT length N was set equal to L.

9.6.4 Spectrogram of a Musical Scale

One case where the spectrogram matches our intuition comes from the analysis of musical instrument sounds, as we discussed in Chapter 3. A musical score (Fig. 9.30) employs a notation that corresponds to the "time-frequency" image found in the spectrogram. Each note specifies the frequency of the tone to be played, the duration of the tone, and the time at which it is to be played (in relation to other notes). Therefore, we can say that musical notation is an idealized spectrogram, although it does not use gray-scale encoding to show amplitude.

---

[11] One approach to the problem posed by this resolution example is called *wavelet analysis*, which relies on filters with non-uniform spacing and variable bandwidth.

Figure 9.30   Musical score for Beethoven's *Für Elise*.

As a simple example of a musical passage, we can synthesize a scale using pure tones (i.e., sinusoids). Eight notes played in succession make up a scale. If the scale is C major, the notes are C, D, E, F, G, A, B, C, which have the frequencies given in the following table:

Middle C	D	E	F	G	A	B	C
262 Hz	294	330	349	392	440	494	523

The spectrogram of the synthetic scale is shown in Fig. 9.31. We can easily identify each note as it is played, although there is some blurring of the transitions between notes. This blurring is due to the fact that the window length of the spectrum analyzer is long enough to straddle two of the notes simultaneously. Shortening the window length would give sharper edges near the ends of each note, at the cost of fatter spectral lines for the tones.

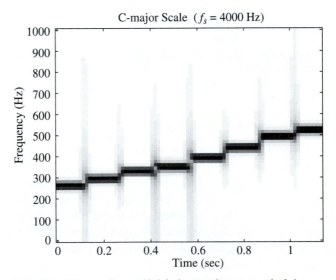

Figure 9.31   Spectrogram of an artificial piano scale composed of sine waves. Filter length of the channel LPFs was $L = 256$; sampling rate was $f_s = 4000$ Hz.

MUSIC GUI

A real piano plays notes that have a much more complex structure than the sine waves used for the C-major scale in Fig. 9.31. Most keys on a piano strike three strings when played, and the complex vibrations of these strings make the pleasing sound of the instrument. A spectrogram reveals the complex structure of the notes because the frequency spectrum is still fairly concentrated. Figure 9.32 shows the spectrogram for the opening passage of Beethoven's *Für Elise*. In this case, only one key is being played at any one time, but the spectrogram is certainly more complicated. In fact, we can identify the primary note being played and can also see second, and third, harmonics of that note, and sometimes even "undertones" at lower frequencies.

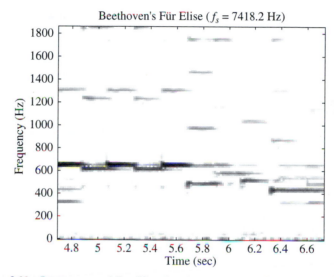

Beethoven's Für Elise ($f_s = 7418.2$ Hz)

Figure 9.32   Spectrogram of *Für Elise* played on a piano. Notice the harmonics at twice the actual note frequency owing to the complex sounds made by the piano. The window length was $L = 256$, the FFT length $N = 256$, and the sampling frequency $f_s = 7418.2$ Hz.

These two examples suggest that the spectrogram might be a useful tool for making an automatic music-writing program. If we could analyze the spectrogram to find its large peaks, then the program could "read" the spectrogram and determine the frequency and duration of the notes being played.

9.6.5 Spectrogram of a Speech Signal

As another example, Fig. 9.33 shows a speech signal, which has been sampled at a rate of $f_s = 8000$ samples/sec. The time-domain plot is given in a strip format consisting of five waveform lines, each representing 800 samples, or 100 msec in time. The beginning of the second line is the sample after the last sample in the first line, etc. The plot is drawn as a continuous waveform, because 800 individual samples per line would be very close together on this plotting scale.

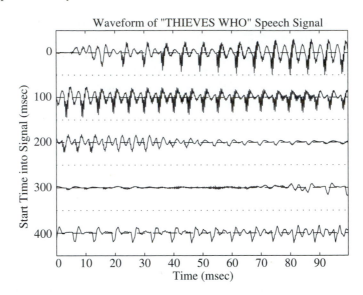

Figure 9.33   Waveform of a speech signal. Each line consists of 800 points.

We know that speech consists of a sequence of different sounds that alternate between voiced sounds (formed by vibrating vocal cords), such as vowels, and unvoiced sounds, such as "s", "sh" and "f". The waveform of Fig. 9.33 corresponds to the utterance "thieves who," so we can identify the voiced and unvoiced intervals. The vowel "ie" in "thieves" occupies the time region $0 \leq t \leq 200$ msec, while the unvoiced sound "s" can be found at $300 \leq t \leq 360$ msec. The vowel from "who" lies near the end, $390 \leq t \leq 500$ msec. The vowel signal is a loud, nearly periodic waveform; the unvoiced sound, on the other hand, is soft and seems to have a random structure. These sounds are the major events in the signal, but overall the waveform changes slowly with time, being relatively unchanged over intervals of 20 to 80 msec. Hence the spectral properties of the signal will also change slowly, and the spectrogram provides an invaluable tool for visualizing the changing character of the speech signal.

The spectrogram of the speech signal of Fig. 9.33 is shown in Fig. 9.34 for the case where $L = 250$. This image is a plot of $|X[k, n]|$ for (9.5.7) where $h[n]$ is a 250-point Hann filter impulse response (9.5.3) and the FFT length is $N = 400$. In order for such a plot to be useful, we must understand why it looks the way that it does, and be able to interpret the various features of the image in terms of the time waveform and the concept of a time-varying frequency spectrum. Notice that in the first three lines of the waveform plot in Fig. 9.33, the vowel waveform is composed of pulses that occur somewhat evenly spaced in time. Indeed, if we limit our view to an interval of length 20 to 30 msec, we see that the waveform generally appears to be almost periodic over that interval. In computing the time-varying spectrum $|X[k, n]|$, the window length is $L = 250$, which corresponds to a time interval of $250/8000 = 31.25$ msec. As the window moves along the waveform, different segments of length 31.25 msec are analyzed. As the pulses change shape and their spacing changes with time, so do the

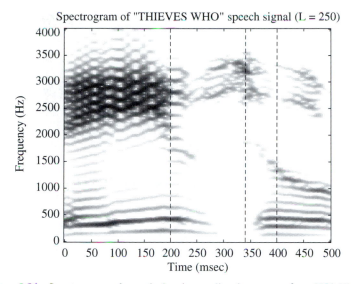

Spectrogram of "THIEVES WHO" speech signal (L = 250)

Figure 9.34   Spectrogram of speech signal; sampling frequency $f_s = 8000$ Hz, window length $L = 250$ (31.25 msec). Spectrum slices at times $nT_s = 200, 340$, and 400 msec are shown individually in Fig. 9.35.

spectral characteristics of the segments within the sliding window. We see this clearly in Fig. 9.34. The dark bars represent equally spaced frequency components whose spacing in the vertical (frequency) direction depends on the "fundamental frequency" at the corresponding time. The bars vary together; their spacing increases when the "fundamental frequency" increases (longer period), and vice versa. Both vowel regions, $0 \leq t \leq 200$ msec and $390 \leq t \leq 500$ msec, exhibit this regular structure, although the spacing of the bars is much closer in the last 110-msec interval.

Returning to Fig. 9.33, notice that in the time interval 300 to 360 msec (the fourth line) the time waveform of the unvoiced sound decreases in amplitude and the waveform is no longer periodic. Correspondingly, in the spectrogram of Fig. 9.34, during the time interval 300 to 360 msec, the image fades, the regular structure disappears, and most of the spectral content lies at high frequency near 3000 Hz.

In the sliding-window interpretation of the spectrogram, the image in Fig. 9.34 is just a sequence of spectral slices stacked side-by-side with the shade of gray representing the amplitude. In Fig. 9.34, the dotted vertical lines show the location of three "spectral slices" that have been extracted and plotted in Fig. 9.35. The two spectra at $t = 200$ and 400 msec have the general character of what we would expect for periodic signals, i.e., highly concentrated at equally spaced frequencies. Note that the peaks in the top panel ($t = 200$) are more widely spaced than those in the bottom panel ($t = 400$) . Examination of the time waveform at 200 and 400 msec, respectively, shows that the fundamental period is significantly shorter at 200 msec than it is at 400 msec. Thus, the spectral peaks should be farther apart at 200 msec than they are at 400 msec. The spectrum in the middle panel ($t = 340$) is typical of unvoiced sounds which have their energy concentrated at relatively high frequencies.

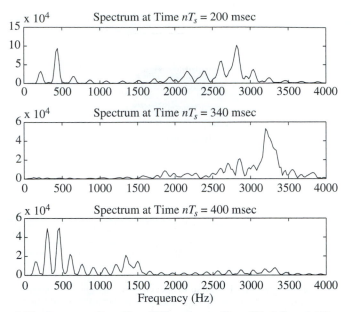

Figure 9.35   Spectrum slices ($L = 250$) at times $nT_s = 200$, 340, and 400 msec taken from Fig. 9.34.

An important point in our interpretation of the spectrogram of the speech signal is that, over the time interval of the window, the waveform "looks periodic." In other words, given only the waveform within the window, we cannot tell anything about the waveform outside the window. It could continue periodically, it could be zero outside the window, or it could change its character as the speech signal does. Thus, the length and shape of the window are important factors in the spectrum analysis of signals like speech that continue in time.

Suppose that the window is shorter than the local period of the signal. In this case, there is not enough signal within the window to measure the local period. Indeed, in this case, it would be more appropriate to think of the signal within the window interval as a finite-length signal. Thus, we would expect a significantly different spectrogram if the window is short. This is demonstrated in Fig. 9.36, which shows the spectrogram of the speech waveform of Fig. 9.33 for the case where $h[n]$ is the Hann filter impulse response with $L = 50$ and the frequency spectrum is again evaluated with an FFT of length $N = 400$. Notice that the fine detail in the vertical direction in Fig. 9.34 is no longer present in Fig. 9.36. The thin wavy bars have been replaced by broader bars. In other words, the fine detail of the spectrum is no longer *resolved*. This point is made clear in the spectral slices taken at 200, 340, and 400 msec, as shown in Fig. 9.37. The frequency peaks in Fig. 9.37 are much wider than the corresponding ones in Fig. 9.35. Also note that the image of Fig. 9.36 seems to consist of vertical slices that alternate between dark and light. This is due to the fact that the window length is so short that sometimes it covers the high-amplitude portion of one of the pulses and later covers the low-amplitude part. Thus, the spectrum appears to fade in and out along the time dimension.

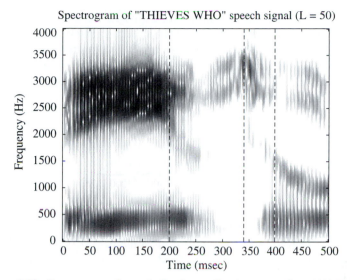

Figure 9.36   Spectrogram of speech signal; sampling frequency $f_s = 8000$ Hz, window length $L = 50$ (6.25 msec). Spectrum slices at times $nT_s = 200, 340,$ and 400 msec are shown individually in Fig. 9.37.

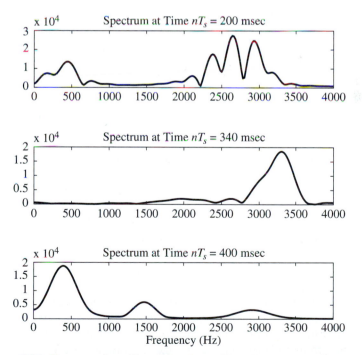

Figure 9.37   Spectrum slices ($L = 50$) at times $nT_s = 200, 340,$ and 400 msec taken from Fig. 9.36.

To conclude this discussion, we answer the question, "What would the spectrum of the entire 4000 samples of the speech signal be like?" Since an FFT length of $N = 4000$ is not really a problem for modern computers, we can simply use the entire sequence as the input and evaluate (9.3.11) with $N = 4000$. The result is shown in Fig. 9.38. In this case, we have a single computed spectrum. Observe that Fig. 9.38 shows 2001 distinct discrete-time frequencies $f_k = 2k$ for $k = 0, 1, \ldots, 2000$, corresponding to the continuous-time frequencies in the range $0 \le f \le 4000$ Hz. Notice that Fig. 9.38 is similar in some respects to the spectral slices in Figs. 9.35 and 9.37, but it has much more detail and does not have the regularly spaced peaks that characterize a periodic signal. This is because the signal is *not* periodic over the entire 4000-sample segment. The large broad peaks give an indication of the concentration of frequencies in the signal, but that is about all that can be inferred from this plot. Thus we have demonstrated that we can have either a long-term or a short-term spectrum for the speech signal. The choice will be determined by many factors, but it is clear that the short-term, time-dependent spectrum portrays a great deal about the speech signal that would be hidden in the long-term spectrum.

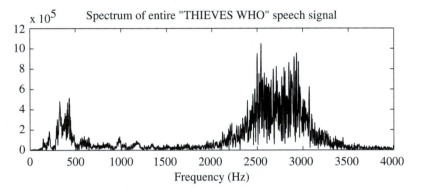

Figure 9.38   Spectrum of the entire sequence of 4000 samples in Fig. 9.22.

9.7 FILTERED SPEECH

As a final example, we show that lowpass filtering applied to the speech signal will alter the spectrogram by removing all high-frequency components. For the lowpass filter we use the LTI system whose impulse response is plotted in the top panel of Fig. 9.39 and whose frequency response (magnitude) is shown in the bottom panel of the figure. If the speech signal of Fig. 9.33 is the input to this filter, we would expect that "the high frequencies would be removed by the filter." Figure 9.40 shows the output signal when the input is the speech signal of Fig. 9.33. A careful comparison of the two waveforms shows that (1) the output is delayed relative to the input, and (2) the output is "smoother" than the input. Both of these results are expected, since the filter has a delay of 25 samples, and high frequencies are required to make rapid changes in the waveform. Clearly, the filter removes the high frequencies.

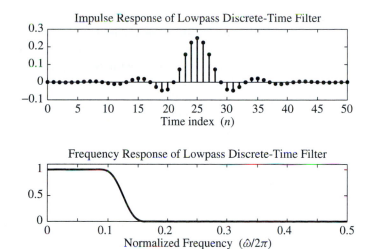

Figure 9.39 Impulse response and frequency response of a lowpass discrete-time filter.

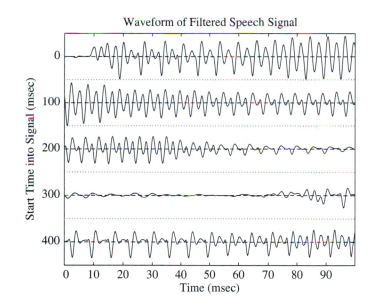

Figure 9.40 Waveform of lowpass filtered speech signal.

This is further demonstrated by the spectrogram in Fig. 9.41, where all components above 1000 Hz have been removed from the spectrum, while below 1000 Hz the spectrograms of Figs. 9.34 and 9.41 appear to be identical. This is entirely consistent with our understanding of lowpass filtering, since the "cut-off frequency" of the discrete-time filter is approximately $\hat{\omega}_c = \pi/4$, which corresponds to continuous-time frequency $\omega_c = \hat{\omega}_c f_s = (\pi/4)8000 = 2\pi(1000)$. The detailed spectral slices at times $t = 200$ and 400 msec shown in Fig. 9.42 also confirm our conclusions. Note that the solid lines show the local spectrum at those times, and the dotted lines are a plot of the magnitude of the frequency response (scale on right-hand side) of the discrete-time filter plotted as a function of corresponding continuous-time cyclic frequency. The spectral slice at $t = 340$ msec corresponding to the middle panel in Fig. 9.35 is not shown since it is effectively zero because the speech signal has virtually no energy below 1000 Hz for the unvoiced sound.

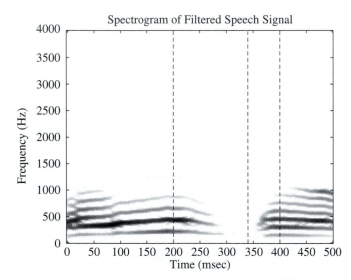

Figure 9.41   Spectrogram of lowpass filtered speech signal. Window length $L = 250$; sampling frequency $f_s = 8000$ Hz.

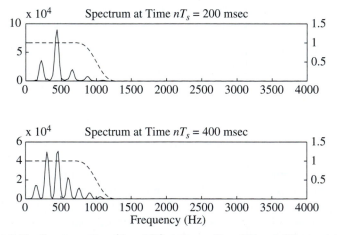

Figure 9.42   Spectrum slices ($L = 250$) at times $nT_s = 200$ and 400 msec taken from Fig. 9.41.

9.8   THE FAST FOURIER TRANSFORM (FFT)

The material in this section is optional reading. It is included for completeness, since the FFT is the most important algorithm and computer program for doing spectrum analysis.

9.8.1 Derivation of the FFT

In Section 9.3.7 we discussed the FFT as an efficient algorithm for computing the DFT. In this section, we will give the basic divide-and-conquer method that leads to the FFT. From this derivation, it should be possible to write an FFT program that runs in time proportional to $(N/2) \log_2 N$ time. We need to assume that N is a power of two, so that the decomposition can be carried out recursively. Such algorithms are called *radix–2* algorithms.

The DFT summation (9.3.11) and the IDFT summation (9.3.10) are essentially the same, except for a minus sign in the exponent of the DFT and a factor of $1/N$ in the inverse DFT. Therefore, we will concentrate on the DFT calculation, knowing that a program written for the DFT could be modified to do the IDFT by changing the sign of the complex exponentials and multiplying the final values by $1/N$. The DFT summation can be broken into two sets, one sum over the even-indexed points of $x[n]$ and another sum over the odd-indexed points.

$$X[k] = \mathrm{DFT}_N\{x[n]\} \tag{9.8.1}$$

$$= \sum_{n=0}^{N-1} x[n] e^{-j(2\pi/N)kn} \tag{9.8.2}$$

$$= x[0]e^{-j0} + x[2]e^{-j(2\pi/N)2k} + \ldots + x[N-2]e^{-j(2\pi/N)k(N-2)}$$
$$+ \; x[1]e^{-j(2\pi/N)k} + x[3]e^{-j(2\pi/N)3k} + \ldots + x[N-1]e^{-j(2\pi/N)k(N-1)}$$

$$\tag{9.8.3}$$

$$X[k] = \sum_{\ell=0}^{N/2-1} x[2\ell]e^{-j(2\pi/N)k(2\ell)} + \sum_{\ell=0}^{N/2-1} x[2\ell+1]e^{-j(2\pi/N)k(2\ell+1)} \tag{9.8.4}$$

At this point, two clever steps are needed: First, the exponent in the second sum must be broken into the product of two exponents, so we can factor out the one that does not depend on ℓ; second, the factor of two in the exponents (2ℓ) can be associated with the N in the denominator of $2\pi/N$.

$$X[k] = \sum_{\ell=0}^{N/2-1} x[2\ell]e^{-j(2\pi/N)k(2\ell)} + e^{-j(2\pi/N)k} \sum_{\ell=0}^{N/2-1} x[2\ell+1]e^{-j(2\pi/N)k(2\ell)}$$

$$X[k] = \sum_{\ell=0}^{N/2-1} x[2\ell]e^{-j(2\pi/(N/2))k\ell} + e^{-j(2\pi/N)k} \sum_{\ell=0}^{N/2-1} x[2\ell+1]e^{-j(2\pi/(N/2))k\ell}$$

Now we have the correct form. Each of the summations is a DFT of length $N/2$, so we can write:

$$X[k] = \text{DFT}_{N/2}\{x[2\ell]\} + e^{-j(2\pi/N)k}\, \text{DFT}_{N/2}\{x[2\ell+1]\} \qquad (9.8.5)$$

The formula (9.8.5) for reconstructing $X[k]$ from the two smaller DFTs has one hidden feature: It must evaluated for $k = 0, 1, 2, \ldots, N-1$. The $N/2$-point DFTs give output vectors that contain $N/2$ elements, e.g., the DFT of the odd-indexed points would be

$$X_{N/2}^o[k] = \text{DFT}_{N/2}\{x[2\ell+1]\} \qquad k = 0, 1, 2, \ldots, N/2-1$$

so we need an extra bit of information to calculate $X[k]$, for $k \geq N/2$. It is easy to verify that

$$X_{N/2}^o[k + N/2] = X_{N/2}^o[k]$$

and likewise for the DFT of the even-indexed points, so we need merely to periodically extend the results of the $N/2$-point DFTs before doing the sum in (9.8.5). This requires no additional computation.

The decomposition in (9.8.5) is enough to specify the entire FFT algorithm: Compute two smaller DFTs and then multiply the outputs of the DFT over the odd indices by the exponential factor $e^{-j(2\pi/N)k}$. Refer to Fig. 9.43, where three levels of the recursive decomposition can be seen. If a recursive structure is adopted, the two $N/2$ DFTs can be decomposed into four $N/4$-point DFTs, and those into eight $N/8$-point DFTs, etc. If N is a power of two, this decomposition will continue ($\log_2 N - 1$) times and then eventually reach the point where the DFT lengths are equal to two. For two-point DFTs, the computation is trivial:

$$X_2[0] = x_2[0] + x_2[1]$$

$$X_2[1] = x_2[0] + e^{-j2\pi/2} x_2[1] = x_2[0] - x_2[1]$$

The two outputs of the two-point DFT are the sum and the difference of the inputs.

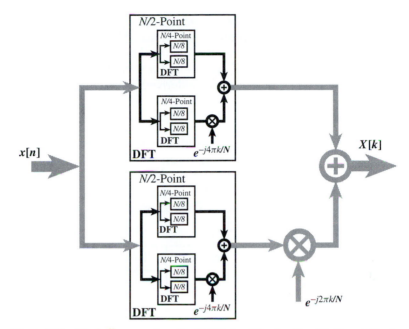

Figure 9.43   Block diagram of a radix–2 FFT algorithm for $N = 2^v$. The width of the lines is proportional to the amount of data being processed. For example, each $N/4$-point DFT must transform a data vector containing $N/4$ elements.

9.8.1.1   FFT Operation Count

The foregoing derivation is a bit sketchy, but the basic idea for writing an FFT program using the two-point DFT and the complex exponential as basic operators has been covered. However, the important point about the FFT is not how to write the program, but rather the number of operations needed to complete the calculation. When it was first published, the FFT made a huge impact on how people thought about problems, because it made the frequency domain accessible numerically. Spectrum analysis became a routine calculation, even for very long signals. Operations such as filtering, which seem to be natural for the time domain, could be done more efficiently in the frequency domain for very long FIR filters.

The number of operations needed to compute the FFT can be expressed in a simple formula. We have said enough about the structure of the algorithm to count the number of operations. The count goes as follows: the N point DFT can be done with two $N/2$ point DFTs followed by N complex multiplications and N complex additions, as we can see in (9.8.5).[12] Thus, we have

$$\mu_c(N) = 2\mu_c(N/2) + N \qquad \alpha_c(N) = 2\alpha_c(N/2) + N$$

---

[12] Actually, the number of complex multiplications can be reduced to $N/2$, because $e^{-j2\pi(N/2)/N} = -1$.

where $\mu_c(N)$ is the number of complex multiplications for a length-N DFT, and $\alpha_c(N)$ is the number of complex additions. This equation can be evaluated successively for $N = 2, 4, 8, \ldots$, because we know that $\mu_c(2) = 0$ and $\alpha_c(2) = 2$. Table 9.1 lists the number of operations for some transform lengths that are powers of two. The formula for each can be derived by matching the table:

$$\mu_c(N) = N(\log_2 N - 1) \qquad \alpha_c(N) = N \log_2 N$$

Since complex number operations ultimately must be done as multiplies and adds between real numbers, it is useful to convert the number of operations to real adds and real multiplies. Each complex addition requires two real additions, but each complex multiplication is equivalent to 4 real multiplies and 2 real adds. Therefore, we can put two more columns in Table 9.1 with these counts.

N	$\mu_c(N)$	$\alpha_c(N)$	$\mu_r(N)$	$\alpha_r(N)$	$4N^2$
2	0	2	0	4	16
4	4	8	16	16	64
8	16	24	64	48	256
16	48	64	192	128	1024
32	128	160	512	320	4096
64	320	384	1280	768	16384
128	768	896	3072	1792	65536
256	1792	2048	7168	4096	262144
⋮	⋮	⋮	⋮	⋮	⋮

Table 9.1: Number of operations for the radix–2 FFT when N is a power of two. Notice how much smaller $\mu_c(N)$ is than $4N^2$.

The bottom line for operation counts is that the total count is something proportional to $N \log_2 N$. The exact formulas from Table 9.1 are

$$\mu_r(N) = 4N(\log_2 N - 1) \qquad \alpha_r(N) = 2N(\log_2 N - 1) + 2N = 2N \log_2 N$$

for the number of real multiplications and additions, respectively. Even these counts are a bit high because certain symmetries in the complex exponentials can be exploited to further reduce the computations.

9.9 SUMMARY AND LINKS

In this chapter, we have tried to establish a basic understanding of spectrum analysis and an appreciation for the concept of time-frequency representations of a discrete-time signal. We began with simple examples and showed, through the fundamental concept of filtering, how we can measure the complex amplitudes of complex exponential components that make up the signal. We then generalized our spectrum-analysis approach to periodic signals and derived what is widely known as the DFT. We examined the DFT in the context of finite-length signals and in the context of continuing signals such as speech or music. Finally, we defined the spectrogram and gave two interpretations that provide understanding of its properties: the filter-bank structure of Fig. 9.22 and the sliding window FFT of Fig. 9.26. Several examples were shown of spectrograms for speech and music to illustrate the resolution trade-off inherent in spectrum analysis.

The example of Section 9.7 is ample evidence that the concept of frequency spectrum has wide-ranging validity. It means that we can think of the effect of an LTI system on *any* signal as being represented by its frequency response. We can say things like "The signal has a bandwidth of 4 kHz," or "The high frequencies were emphasized by the filter." Certainly this should not be a surprising conclusion, since even without this semiformal discussion, most of us, engineers or not, are comfortable with thinking about frequency responses of hi-fi sets, loudspeakers, headphones, and even our ears. What makes the frequency response such a useful concept is that any signal can be thought of as a sum of (perhaps infinitely many) complex exponential signals. This chapter has tried to demonstrate this with many examples and some equations. There is much more to learn about Fourier analysis and the concept of the frequency spectrum, but the essence of this infinitely interesting subject is contained here.

The demonstrations and projects for this chapter have a lot in common with those from Chapter 3, so the reader should consult those earlier demonstrations. One new laboratory project is available in Appendix C and on the CD-ROM: Lab C.11 requires students to develop a music-analysis program that will write the music for a song from a recording. Basically, the spectrogram is used to get a time-frequency representation of the music, and then a peak-picking algorithm and editing program must be written to find the spectral peaks that correspond to actual notes. This project can be considered as the inverse to Lab C.3. In addition, Lab C.7 involves some practical systems that work with sinusoidal signals, such as a touch-tone phone. Write-ups of the labs can also be found on the CD-ROM.

The CD-ROM also contains three demonstrations related to the spectrogram:

SPECTRO-
GRAMS AND
SOUNDS:
WIDEBAND
FM

1. Computation of the spectrogram is illustrated by a movie showing the sliding-window FFT moving through a signal.

2. Spectrograms of chirp signals show how the window length of the FFT affects the spectrogram image that you see and hear. For the chirp case, the rate of change of the frequency affects the spectrogram image depending on how fast the frequency changes within the window duration. It also affects the sound your hear, because the human auditory system can be modeled as a filter bank with variable (but known) bandwidths.

MUSIC GUI

3. A MATLAB graphical user interface that presents the connection between musical notation and the spectrogram.

Finally, the reader is reminded of the large number of solved homework problems on the CD-ROM that are available for review and practice.

PROBLEMS

9.1 Suppose that a discrete-time signal $x[n]$ is a sum of complex exponential signals

$$x[n] = 3 + 2e^{j0.2\pi n} + 2e^{-j0.2\pi n} - 7je^{j0.7\pi n} + 7je^{-j0.7\pi n}$$

  (a) Make a plot of the spectrum for $x[n]$ using only positive frequencies.

  (b) Suppose that $x_b[n] = x[n]e^{j0.4\pi n}$. Make a plot of the spectrum for $x_b[n]$ using only positive frequencies.

  (c) Suppose that $x_c[n] = (-1)^n x[n]$. Make a plot of the spectrum for $x_c[n]$ using only positive frequencies.

9.2 Determine the 10-point DFT of the following:

  (a) $x_0[n] = \begin{cases} 1 & n = 0 \\ 0 & n = 1, 2, \ldots, 9 \end{cases}$

  (b) $x_1[n] = 1 \qquad$ for $n = 0, 1, 2, \ldots, 9$

  (c) $x_2[n] = \begin{cases} 1 & n = 4 \\ 0 & n \neq 4 \end{cases}$

  (d) $x_3[n] = e^{j2\pi n/5} \qquad$ for $n = 0, 1, 2, \ldots, 9$

9.3 Determine the 10-point inverse DFT (IDFT) of the following:

  (a) $X_a[k] = \begin{cases} 1 & k = 0 \\ 0 & k = 1, 2, \ldots, 9 \end{cases}$

  (b) $X_b[k] = 1 \qquad$ for $k = 0, 1, 2, \ldots, 9$

  (c) $X_c[k] = \begin{cases} 1 & k = 3, 7 \\ 0 & k = 0, 1, 2, 4, 5, 6, 8, 9 \end{cases}$

  (d) $X_d[k] = \cos(2\pi k/5) \qquad$ for $k = 0, 1, 2, \ldots, 9$

9.4 Determine the 12-point DFT of the following:

  (a) $y_0[n] = \begin{cases} 1 & n = 0, 1, 2, 3 \\ 0 & n = 4, 5, \ldots, 11 \end{cases}$

  (b) $y_1[n] = \begin{cases} 1 & n = 0, 2, 4, 6, 8, 10 \\ 0 & n = 1, 3, 5, 7, 9, 11 \end{cases}$

  (c) $y_2[n] = \begin{cases} e^{j2\pi n/5} & n = 0, 1, 2, 3, 4 \\ 0 & n = 5, 6, 7, 8, 9, 10, 11 \end{cases}$

9.5 Prove the *orthogonality property* of the complex exponential $e^{j2\pi kn/N}$

$$\sum_{n=0}^{N-1} e^{j2\pi \ell n/N} e^{-j2\pi kn/N} = \begin{cases} N & \ell = k \bmod N \\ 0 & \ell \neq k \bmod N \end{cases}$$

where the notation $\ell = k \bmod N$ means ℓ can be equal to k, $k \pm N$, $k \pm 2N$, etc. This property is true only when the complex exponentials have the same period N.

9.6 Suppose that $x[n]$ has a spectrum with components at $\hat{\omega}_1 = 0.25\pi$, $\hat{\omega}_2 = 0.5\pi$, and $\hat{\omega}_3 = 0.75\pi$, and a DC value of $X_0 = 3$. Use the 8-point running-sum FIR filter with coefficients $\{b_k\} = \{1, 1, 1, 1, 1, 1, 1, 1, \}$ to process this signal. Show that the output is $y[n] = 24$ for all n. Use values from a plot of $|H(e^{j\hat{\omega}})|$ versus $\hat{\omega}$ to simplify your calculations.

9.7 Determine the response of a 12-point running-sum filter to the following input signals:

   (a) $x[n] = \delta[n]$

   (b) $x[n] = e^{j\pi n/4}$ for all n

   (c) $x[n] = \cos(\pi n/4)$ for $n \geq 0$

   (d) $x[n] = e^{j\pi n/4}$ for $0 \leq n < 20$

   Exploit your knowledge of the frequency response $H(e^{j\hat{\omega}})$ to simplify the calculations as much as possible.

9.8 Suppose that continuous-time signal $x(t)$ consists of several sinusoidal sections:

$$x(t) = \begin{cases} \cos(2\pi(600)t) & 0 \leq t < 0.5 \\ \sin(2\pi(1100)t) & 0.3 \leq t < 0.7 \\ \cos(2\pi(500)t) & 0.4 \leq t < 1.2 \\ \cos(2\pi(700)t - \pi/4) & 0.4 \leq t < 0.45 \\ \sin(2\pi(800)t) & 0.35 \leq t < 1.0 \end{cases}$$

   (a) If the signal is sampled at $f_s = 8000$ Hz, make a sketch of the *ideal* spectrogram that corresponds to the signal definition.

   (b) If the signal is sampled at $f_s = 8000$ Hz, make a sketch of the *actual* spectrogram that would be obtained with an FFT length of $N = 256$ and a Hann window length of $L = 256$. Make approximations in order to do the sketch without actually calculating the spectrogram in MATLAB.

9.9 The C-major scale was analyzed in the spectrogram of Fig. 9.31 with a window length of $L = 256$. Make a sketch of the spectrogram that would be obtained if the window length were $L = 100$. Explain how your sketch would differ from Fig. 9.31.

9.10 Assume that a speech signal has been sampled at 8000 Hz and then analyzed with MATLAB's specgram function using the following parameters: Hann window with length $L = 100$, FFT length of $N = 256$, overlap of 80 points. Determine the resolution of the resulting spectrogram image.

   (a) Determine the frequency resolution (in Hz)

   (b) Determine the time resolution (in sec)

9.11 The resolution of a filter-bank system is determined by the frequency response of the lowpass filter used in the channels. Suppose that we sample a continuous-time signal at $f_s = 10{,}000$ Hz and we want to have frequency resolution of 250 Hz.

   (a) If we use a Hann filter for the channel LPF, what filter length L would be needed? Estimate a minimum value for L.

   (b) Design a Hann FIR filter that will satisfy the resolution requirement. Demonstrate that this lowpass filter will be sufficient by plotting its frequency response (magnitude).

A

Complex Numbers

The basic manipulations of complex numbers are reviewed in this appendix. The algebraic rules for combining complex numbers are reviewed, and then a geometric viewpoint is taken to visualize these operations by drawing vector diagrams. This geometric view is a key to understanding how complex numbers can be used to represent signals. Specifically, the following three significant ideas about complex numbers are treated in this appendix:

- *Simple algebraic rules:* Operations on complex numbers ($z = x + jy$) follow exactly the same rules as real numbers, with j^2 replaced everywhere by -1.[1]
- *Elimination of trigonometry:* Euler's formula for the complex exponential $z = re^{j\theta} = r\cos\theta + jr\sin\theta$ provides a connection between trigonometric identities and simple algebraic operations on complex numbers.
- *Representation by vectors:* A vector drawn from the origin to a point (x, y) in a two-dimensional plane is equivalent to $z = x + jy$. The algebraic rules for z are, in effect, the basic rules of vector operations. More important, however, is the *visualization* gained from the vector diagrams.

The first two ideas concern the algebraic nature of $z = x + jy$, the other its role as a *representer* of signals. Skill in algebraic manipulations is important, but the use of complex numbers in representation is more important in the long run. Complex numbers in electrical engineering are used as a convenience because when they stand for the sinusoidal signals they can simplify manipulations of the signals. Thus a sinusoidal problem (such as the solution to a differential equation) is converted into a complex number problem that can be (1) solved by the simple rules of algebra

---

[1] Mathematicians and physicists use the symbol i for $\sqrt{-1}$; electrical engineers prefer to reserve the symbol i for current in electric circuits.

and (2) visualized through vector geometry. The key to all this is the higher-order thinking that permits abstraction of the problem into the world of complex numbers. Ultimately, we are led to the notion of a "transform" such as the Fourier or Laplace transform to reduce many other sophisticated problems to algebra. If you were skillful at high-school algebra, your study of signals and systems should be productive. If not, a careful study of this appendix is advised.

Once such insight is gained, it still will be necessary to return occasionally to the lower-level drudgery of calculations. When you have to manipulate complex numbers, a calculator will be most useful, especially one with built-in complex arithmetic capability. It is worthwhile to learn how to use this feature on your calculator. However, it is also important to do some calculations by hand, so that you will *understand* what your calculator is doing!

Finally, it's too bad that complex numbers are called "complex." Most students, therefore, think of them as complicated. However, their elegant mathematical properties usually simplify calculations quite a bit.

A.1   INTRODUCTION

A complex number system is an extension of the real number system. Complex numbers are necessary to solve equations such as

$$z^2 = -1 \qquad (A.1.1)$$

The symbol j is introduced to stand for $\sqrt{-1}$, so the previous equation (A.1.1) has the two solutions $z = \pm j$. More generally, complex numbers are needed to solve for the two roots of a quadratic equation

$$az^2 + bz + c = 0$$

which, according to the quadratic formula, has the two solutions:

$$z = \frac{-b \pm \sqrt{b^2 - 4ac}}{2a}$$

Whenever the discriminant $(b^2 - 4ac)$ is negative, the solution must be expressed as a complex number. For example, the roots of

$$z^2 + 6z + 25 = 0$$

are $z = -3 \pm j4$, because $\sqrt{b^2 - 4ac} = \sqrt{36 - 4(25)} = \sqrt{-64} = \pm j8$.

A.2   NOTATION FOR COMPLEX NUMBERS

Several different mathematical notations can be used to represent complex numbers. The two basic types are polar form and rectangular form. Converting between them quickly and easily is an important skill.

A.2.1 Rectangular Form

In *rectangular form*, all of the following notations define the same complex number.

$$z = (x, y)$$
$$= x + jy$$
$$= \Re\{z\} + j\Im\{z\}$$

The ordered pair (x, y) can be interpreted as a point in the two-dimensional plane.[2]
A complex number can also be drawn as a vector whose tail is at the origin and whose head is at the point (x, y), in which case x is the horizontal coordinate of the vector and y the vertical coordinate. See Fig. A.1 for some examples. The complex number $z = 2 + j5$ is represented by the point $(2, 5)$, which lies in the first quadrant; likewise, $z = 4 - j3$ lies in the fourth quadrant at the location $(4, -3)$.

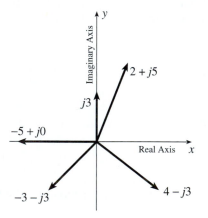

Figure A.1   Complex numbers plotted as vectors in the two-dimensional "complex plane." Each $z = x + jy$ is represented by a vector from the origin to the point with coordinates (x, y) in the complex plane.

Since the complex number notation $z = x + jy$ represents the point (x, y) in the two-dimensional plane, the number j represents $(0, 1)$, which is drawn as a vertical vector from the origin up to $(0, 1)$, as in Fig. A.2. Thus multiplying a real number, such as 5, by j changes it from pointing along the horizontal axis to pointing vertically, i.e., $j(5 + j0) = 0 + j5$.

---

[2] This is also the notation used on some calculators when entering complex numbers.

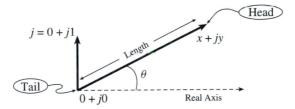

Figure A.2   Complex numbers represented as a vector from the origin to (x, y).

Rectangular form is also referred to as *Cartesian form*. The horizontal coordinate x is called the *real part*, and the vertical coordinate y the *imaginary part*. The operators $\Re e\{z\}$ and $\Im m\{z\}$ are provided to extract the real and imaginary parts of $z = x + jy$:

$$x = \Re e\{z\}$$

$$y = \Im m\{z\}$$

A.2.2 Polar Form

In *polar form*, the vector is defined by its length (r) and its direction (θ) as in Figs. A.2 and A.4. Therefore, we use the following descriptive notation sometimes:

$$z \longleftrightarrow r\angle\theta$$

Some examples are shown in Fig. A.3 where the direction θ is given in degrees. The angle is measured from the positive x-axis and may be either positive (counterclockwise) or negative (clockwise). However, we generally specify the *principal value* of the angle so that $-180° < \theta \le 180°$. This requires that integer multiples of $360°$ be subtracted from or added to the angle until the result is between $-180°$ and $+180°$. Thus the vector $3\angle-80°$ is the principal value of $3\angle280°$.

Z-DRILL

A.2.3 Conversion: Rectangular and Polar

Both the polar and rectangular forms are commonly used to represent complex numbers. The prevalence of the polar form, for sinusoidal signal representation, makes it necessary to convert quickly and accurately between the two representations. Referring to Fig. A.4, we see that the x and y coordinates of the vector are given by

$$x = r\cos\theta$$
$$y = r\sin\theta$$

(A.2.1)

Therefore, a valid formula for z is

$$z = r\cos\theta + jr\sin\theta$$

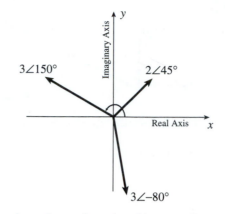

Figure A.3   Several complex numbers plotted in terms of length (r) and direction (θ) of their vector representation. The angle is always measured with respect to the positive real axis; its units are usually radians, but are shown as degrees in this case.

Example A.1   In Fig. A.3, the three complex numbers are

$$2\angle 45° \quad \longleftrightarrow \quad z = \sqrt{2} + j\sqrt{2}$$

$$3\angle 150° \quad \longleftrightarrow \quad z = -\frac{3\sqrt{3}}{2} + j\frac{3}{2}$$

$$3\angle -80° \quad \longleftrightarrow \quad z = 0.521 - j2.954$$

$\diamond$

The conversion from (x, y) to $r\angle\theta$ is a bit trickier. From Fig. A.4, the formulas are

$$r = \sqrt{x^2 + y^2} \qquad \text{(Length)} \qquad\qquad \text{(A.2.2)}$$

$$\theta = \arctan(y/x) \qquad \text{(Direction)} \qquad\qquad \text{(A.2.3)}$$

The arctangent must give a four-quadrant answer, and the direction θ is usually given in radians rather than degrees.

Exercise A.1. At this point, the reader should convert to polar form the five complex numbers shown in Fig. A.1. The answers, in a random order, are: $3\angle 90°$, $5\angle -36.87°$, $4.243\angle 225°$, $5.385\angle 68.2°$, and $5\angle 180°$.

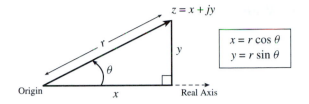

Figure A.4   Basic triangle for relating (x, y) to r and θ.

In Section A.3, we will introduce two other polar notations:

$$z = re^{j\theta}$$

$$= |z|e^{j \arg z}$$

where $|z| = r = \sqrt{x^2 + y^2}$ is called the *magnitude* of the vector and $\arg z = \theta = \arctan(y/x)$ is its *phase* in radians (not degrees). This exponential notation, which relies on Euler's formula, has the advantage that when it is used in algebraic expressions, the standard laws of exponents apply.

A.2.4 Difficulty in Second or Third Quadrant

The formula (A.2.3) for the angle θ as the $\arctan(y/x)$ must be used with care, especially when the real part is negative (see Fig. A.5). For example, the complex number $z = -3 + j4$ would require that we evaluate $\arctan(-4/3)$ to get the angle; the same calculation would be needed if $z = 3 - j4$. The arctangent of $-4/3$ is -0.95 rad, or about $-53°$, which is the correct angle for $z = 3 - j4$. However, for $z = -3 + j4$, the vector lies in the second quadrant and the angle must satisfy $90° \leq \theta \leq 180°$. In this case, the correct angle is $\pi - 0.95 = 2.2$ rad, or about $180° - 53° = 127°$.

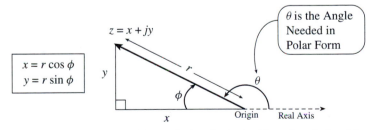

Figure A.5   In the second quadrant, the interior angle ϕ is easily calculated from x and y, but is not the correct angle for the polar form, which requires the exterior angle θ with respect to the positive real axis.

A.3 EULER'S FORMULA

The conversion from polar to rectangular form (A.2.1) suggests the following formula:

$$e^{j\theta} = \cos\theta + j\sin\theta \qquad (A.3.1)$$

Equation (A.3.1) defines the *complex exponential* $e^{j\theta}$, which is equivalent to $1\angle\theta$, i.e., a vector of length 1 at angle θ. A proof of Euler's formula based on power series is outlined in Problem 2.4 of Chapter 2. Figure A.6 shows $r\angle\theta$ which is equivalent to $re^{j\theta}$.

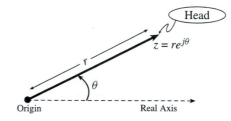

Figure A.6   Polar form for a complex number expressed in exponential notation.

The amazing discovery was that the laws of exponents apply to $e^{j\theta}$. Euler's formula (A.3.1) is so important that it must be instantly recalled; likewise, the inverse Euler formulas (A.3.2) and (A.3.3) should also be committed to memory.

Example A.2   Here are some examples:

$$(90°):\quad e^{j\pi/2} = \cos(\pi/2) + j\sin(\pi/2) = 0 + j1 = j \quad\longleftrightarrow\quad 1\angle\pi/2$$

$$(180°):\quad e^{j\pi} = \cos(\pi) + j\sin(\pi) = -1 + j0 = -1 \quad\longleftrightarrow\quad 1\angle\pi$$

$$(45°):\quad e^{j\pi/4} = \cos(\pi/4) + j\sin(\pi/4) = \tfrac{1}{\sqrt{2}} + j\tfrac{1}{\sqrt{2}} \quad\longleftrightarrow\quad 1\angle\pi/4$$

$$(60°):\quad e^{j\pi/3} = \cos(\pi/3) + j\sin(\pi/3) = \tfrac{1}{2} + j\tfrac{1}{2}\sqrt{3} \quad\longleftrightarrow\quad 1\angle\pi/3$$

$\diamond$

Example A.3   Referring back to Fig. A.3, the three complex numbers can be rewritten as:

$$2\angle 45° \quad\longleftrightarrow\quad z = 2e^{j\pi/4}$$

$$3\angle 150° \quad\longleftrightarrow\quad z = 3e^{j5\pi/6}$$

$$3\angle -80° \quad\longleftrightarrow\quad z = 3e^{-j4\pi/9} = 3e^{-j1.396}$$

$\diamond$

Numbers like -1.396 rad are difficult to visualize, because we are more used to thinking of angles in degrees. It may be helpful to express the angles used in the exponents as a fraction of π rad, i.e., $-1.396 = -(1.396/\pi)\pi = -0.444\pi$ rad. This is a good habit to adopt, because it simplifies the conversion between degrees and radians. If θ is given in radians, the conversion is:

$$\theta \times \left(\frac{180}{\pi}\right) = \text{direction in degrees}$$

A.3.1 Inverse Euler Formulas

Euler's formula (A.3.1) can be solved separately for the cosine and sine parts. The result will be called the *inverse Euler relations*

$$\cos\theta = \frac{e^{j\theta} + e^{-j\theta}}{2} \qquad (A.3.2)$$

$$\sin\theta = \frac{e^{j\theta} - e^{-j\theta}}{2j} \qquad (A.3.3)$$

Recalling that $\cos(-\theta) = \cos\theta$ and $\sin(-\theta) = -\sin\theta$, the proof of the $\sin\theta$ version is as follows:

$$e^{-j\theta} = \cos(-\theta) + j\sin(-\theta)$$

$$= \cos\theta - j\sin\theta$$

$$e^{+j\theta} = \cos\theta + j\sin\theta$$

$$\implies e^{j\theta} - e^{-j\theta} = 2j\sin\theta$$

$$\implies \sin\theta = \frac{e^{j\theta} - e^{-j\theta}}{2j}$$

A.4   ALGEBRAIC RULES FOR COMPLEX NUMBERS

The basic arithmetic operators for complex numbers follow the usual rules of algebra as long as the symbol j is treated as a special token that satisfies $j^2 = -1$. In rectangular form, all of these rules are relatively straightforward. The five fundamental rules are the following:

$$(\text{Addition}) \quad z_1 + z_2 = (x_1 + jy_1) + (x_2 + jy_2)$$
$$= (x_1 + x_2) + j(y_1 + y_2)$$

$$(\text{Subtraction}) \quad z_1 - z_2 = (x_1 + jy_1) - (x_2 + jy_2)$$
$$= (x_1 - x_2) + j(y_1 - y_2)$$

$$(\text{Multiplication}) \quad z_1 \times z_2 = (x_1 + jy_1) \times (x_2 + jy_2)$$
$$z_1 z_2 = x_1 x_2 + j^2 y_1 y_2 + jx_1 y_2 + jx_2 y_1$$
$$= (x_1 x_2 - y_1 y_2) + j(x_1 y_2 + x_2 y_1)$$

$$(\text{Conjugate}) \quad z_1^* = (x_1 + jy_1)^*$$
$$= x_1 - jy_1$$

$$(\text{Division}) \quad z_1 \div z_2 = (x_1 + jy_1)/(x_2 + jy_2)$$
$$\frac{z_1}{z_2} = \frac{z_1 z_2^*}{z_2 z_2^*} = \frac{z_1 z_2^*}{|z_2|^2}$$
$$= \frac{(x_1 + jy_1)(x_2 - jy_2)}{x_2^2 + y_2^2}$$
$$= \frac{(x_1 x_2 + y_1 y_2) + j(x_2 y_1 - x_1 y_2)}{x_2^2 + y_2^2}$$

Addition and subtraction are straightforward because we need only add or subtract the real and imaginary parts. On the other hand, addition (or subtraction) in polar form cannot be carried out directly on r and θ; instead, an intermediate conversion to rectangular form is required. In contrast, the operations of multiplication and division, which are rather messy in rectangular form, reduce to simple manipulations in polar form. For multiplication, multiply the magnitudes and add the angles; for division, divide the magnitudes and subtract the angles. The conjugate in polar form requires only a change of sign of the angle.

$$(\text{Multiplication}) \quad z_1 \times z_2 = r_1 e^{j\theta_1} \times r_2 e^{j\theta_2}$$
$$= (r_1 r_2) e^{j(\theta_1 + \theta_2)}$$

$$(\text{Conjugate}) \quad z_1^* = (r_1 e^{j\theta_1})^*$$
$$= r_1 e^{-j\theta_1}$$

$$(\text{Division}) \quad z_1 \div z_2 = \frac{r_1 e^{j\theta_1}}{r_2 e^{j\theta_2}}$$
$$= \frac{r_1}{r_2} e^{j(\theta_1 - \theta_2)}$$

Exercise A.2. The inverse or reciprocal of a complex number z is the number z^{-1} such that $z^{-1}z = 1$. A common mistake with the inverse is to invert $z = x + jy$ by taking the inverse of x and y separately. To show that this is wrong, take the specific case where $z = 4 + j3$ and $w = \frac{1}{4} + j\frac{1}{3}$. Show that w is not the inverse of z, because $wz \neq 1$. Determine the correct inverse of z.

Polar form presents difficulties when adding (or subtracting) two complex numbers and expressing the final answer in polar form. An intermediate conversion to rectangular form must be done. Here is the recipe for adding complex numbers in polar form.

1. Starting in polar form, we have:

$$z_3 = z_1 \pm z_2 = r_1 e^{j\theta_1} \pm r_2 e^{j\theta_2}$$

2. Convert both z_1 and z_2 to Cartesian form:

$$z_3 = (r_1 \cos\theta_1 + j\,r_1 \sin\theta_1) \pm (r_2 \cos\theta_2 + j\,r_2 \sin\theta_2)$$

3. Perform the addition in Cartesian form:

$$z_3 = (r_1 \cos\theta_1 \pm r_2 \cos\theta_2) + j\,(r_1 \sin\theta_1 \pm r_2 \sin\theta_2)$$

4. Identify the real and imaginary parts of z_3:

$$x_3 = \Re e\{z_3\} = r_1 \cos\theta_1 \pm r_2 \cos\theta_2$$
$$y_3 = \Im m\{z_3\} = r_1 \sin\theta_1 \pm r_2 \sin\theta_2$$

5. Convert back to polar form using (A.2.2) and (A.2.3):

$$z_3 = x_3 + jy_3 \qquad \longleftrightarrow \qquad z_3 = r_3 e^{j\theta_3}$$

If you have a calculator that converts between polar and rectangular form, learn how to use it; it will save many hours of hand calculation and also be more accurate. Most "scientific" calculators even have the capability to use both notations, so the conversion is transparent to the user.

Example A.4   Here is a numerical example of adding two complex numbers given in polar form:

$$z_3 = 7e^{j4\pi/7} + 5e^{-j5\pi/11}$$

$$z_3 = (-1.558 + j6.824) + (0.712 - j4.949)$$

$$z_3 = -0.846 + j1.875$$

$$z_3 = 2.057e^{j1.995} = 2.057e^{j0.635\pi} = 2.057\angle 114.3°$$

Remember: When the angle appears in the exponent, its units must be in radians.   ◇

A.4.1 Exercises

To practice computations for complex numbers, try the following exercises.

Exercise A.3. Add and multiply the following, then plot the results:

$$z_4 = 5e^{j4\pi/5} + 7e^{-j5\pi/7}$$

$$z_5 = 5e^{j4\pi/5} \times 7e^{-j5\pi/7}$$

The answers are $z_4 = -8.41 - j2.534 = 8.783e^{-j0.907\pi}$ and $z_5 = 35e^{j3\pi/35}$.

Exercise A.4. For the conjugate the simple rule is to change the sign of all the j terms. Work the following:

$$(3 - j4)^* = ?$$

$$(j(1 - j))^* = ?$$

$$(e^{j\pi/2})^* = ?$$

Exercise A.5. Prove that the following identities are true:

$$\Re e\{z\} = (z + z^*)/2$$

$$\Im m\{z\} = (z - z^*)/2j$$

$$|z|^2 = zz^*$$

Z-DRILL

More drill problems can be generated by using the MATLAB program `zdrill.m`, which presents a GUI that asks questions for each of the complex operations, and also plots the vectors that represent the solutions. The `zdrill` has both a novice and an advanced level.

A.5   GEOMETRIC VIEWS OF COMPLEX OPERATIONS

It is important to develop an ability to visualize complex number operations. This is done by plotting the vectors that represent the numbers in the (x, y) plane, where $x = \Re e\{z\}$ and $y = \Im m\{z\}$. The key to this is to recall that, as shown in Fig. A.2, the complex number $z = x + jy$ is a vector whose tail is at the origin and whose head is at (x, y).

A.5.1 Geometric View of Addition

For complex addition, $z_3 = z_1 + z_2$, both z_1 and z_2 are viewed as vectors with their tails at the origin. The sum z_3 is the result of vector addition, and is constructed as follows (see Fig. A.7):

1. Draw a copy of z_1 with its tail at the head of z_2. Call this displaced vector $\hat{z}_1$.
2. Draw the vector from the origin to the head of $\hat{z}_1$. This vector is the sum z_3.

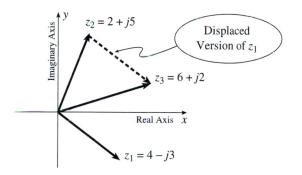

Figure A.7   Graphical construction of complex number addition $z_3 = z_1 + z_2$.

This method of addition can be generalized to many vectors. **Figure A.8** shows the result of adding the four complex numbers:

$$(1 + j) + (-1 + j) + (-1 - j) + (1 - j)$$

where the answer happens to be zero.

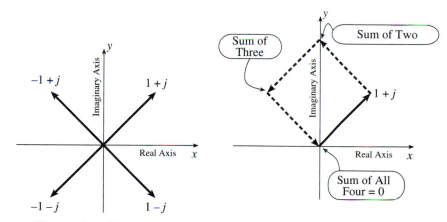

Figure A.8   Adding the four vectors $\{1 + j, -1 + j, -1 - j, 1 - j\}$ by using the "head-to-tail" graphical method.

A.5.2 Geometric View of Subtraction

The visualization of subtraction requires that we construct a triangle with z_1 and z_2 as sides; then the vector $z_2 - z_1$ is the third side, if displaced so that its tail is at the head of z_1, as in Fig. A.9. Notice that $z_2 - z_1$ and $z_1 - z_2$ could both be the third side of the triangle; they have the same length, just opposite direction. There are three comments to be made about subtraction:

1. Since $z_3 = z_2 + (-z_1)$, we could add $(-z_1)$ to z_2 to get the answer. This sum is also shown in Fig. A.9 and is equivalent to the visualization of addition, as in Fig. A.7.

2. Since $z_2 = z_1 + z_3$, a displaced version of the difference vector would be drawn with its tail at the head of z_1 and its head of z_2.

3. Thus a triangle is defined by the three points z_1, z_2, and the origin. The sides of the triangle are z_1, z_2, and $z_2 - z_1$.

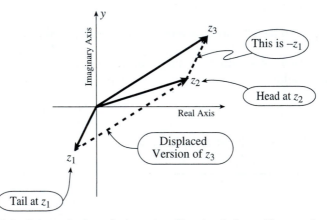

Figure A.9   Geometric view of subtraction. The triangle formed by z_1 and z_2 has a displaced version of $z_3 = z_2 - z_1$ as its third side.

Exercise A.6. Prove the triangle inequality:

$$|z_2 - z_1| \leq |z_1| + |z_2|$$

Use either an algebraic method by squaring both sides, or a geometric argument based on the intuitive idea that "the shortest distance between two points is a straight line."

A.5.3 Geometric View of Multiplication

It is best to view multiplication in terms of polar form where we multiply the magnitudes, and add the angles. In order to draw the product vector z_3, we must decide whether or not $|z_1|$ and/or $|z_2|$ are greater than 1. In Fig. A.10, it is assumed that both are larger than 1.

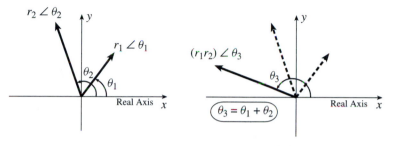

Figure A.10   Geometric view of complex multiplication $z_3 = z_1 z_2$.

A special case occurs when $|z_2| = 1$, because then there is no scaling, and the multiplication by $z_2 = e^{j\theta_2}$ becomes a rotation. Figure A.11 shows the case where $z_2 = j$, which gives a rotation by $\pi/2$ or $90°$, because $j = e^{j\pi/2}$.

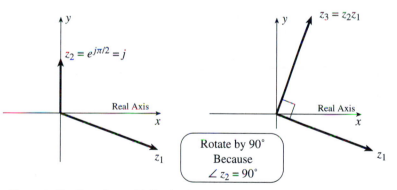

Figure A.11   Complex multiplication becomes a rotation when $|z_2| = 1$. The case where $z_2 = j$ gives a rotation by $90°$.

A.5.4 Geometric View of Division

Division is very similar to the visualization of multiplication, except that we must now subtract the angles and divide the magnitudes (Fig. A.12).

$$z_3 = \frac{z_1}{z_2} = \frac{r_1}{r_2} e^{j(\theta_1 - \theta_2)}$$

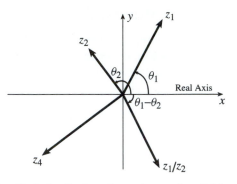

Figure A.12   Graphical visualization of complex number division $z_3 = z_1/z_2$. Notice that the angles subtract.

Exercise A.7. Given two complex numbers z_4 and z_2, as in Fig. A.12, where the angle between them is $90°$ and the magnitude of z_4 is twice that of z_2. Evaluate z_4/z_2.

A.5.5 Geometric View of Inverse, z^{-1}

This is a special case of division where $z_1 = 1$, so we just negate the angle and take the reciprocal of the magnitude.

$$z^{-1} = \frac{1}{z} = \frac{1}{r}e^{-j\theta}$$

Refer to Fig. A.13 for examples of the inverse.

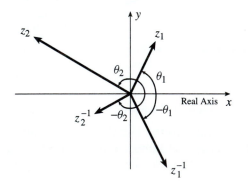

Figure A.13   Graphical construction of complex number inverse $1/z$. For the vectors shown, $|z_1| < 1$ and $|z_2| > 1$.

A.5.6 Geometric View of Conjugate z^*

In this case, we negate the angle, which has the effect of flipping the vector about the horizontal axis. The length of the vector remains the same.

$$z^* = x - jy = re^{-j\theta}$$

Figure A.14 shows two examples of the geometric interpretation of complex conjugation.

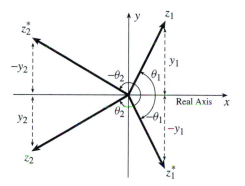

Figure A.14   Graphical construction of the complex conjugate z^*; only the imaginary part changes sign. The vectors flip about the real axis: z_1 flips down, and z_2 flips up.

Exercise A.8. Prove the following fact:

$$\frac{1}{z^*} = \frac{1}{r}e^{j\theta}$$

Plot an example for $z = 1 + j$; also plot $1/z$ and z^*.

A.6 POWERS AND ROOTS

Integer powers of a complex number can be defined in the following manner:

$$z^N = \left(re^{j\theta}\right)^N = r^N e^{jN\theta}$$

In other words, the rules of exponents still apply, so the angle θ is multiplied by N and the magnitude is raised to the Nth power. Figure A.15 shows a sequence of these:

$$\{z^\ell\} = \{z^0,\ z^1,\ z^2,\ z^3, \dots \}$$

where the angle steps by a constant amount; in this case, exactly $\pi/6$ rad. The magnitude of z is less than one, so the successive powers spiral in toward the origin. If $|z| > 1$, the points would spiral outward; if $|z| = 1$, all the powers z^N would lie on the *unit circle* (a circle of radius 1). One famous identity is DeMoivre's formula:

$$(\cos\theta + j\sin\theta)^N = \cos N\theta + j\sin N\theta$$

The proof of this seemingly difficult trigonometric identity is actually trivial if we invoke Euler's formula (A.3.1) for $e^{jN\theta}$.

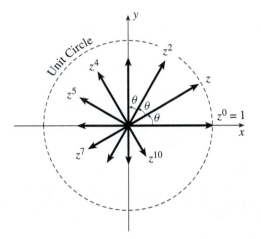

Figure A.15   A sequence of powers z^ℓ for $\ell = 0, 1, 2, \ldots, 10$. Since $|z| = 0.9 < 1$, the vectors spiral in toward the origin. The angular change between successive powers is constant, $\theta = \arg z = \pi/6$.

Exercise A.9. Let z^{N-1}, z^N, and z^{N+1} be three consecutive members of sequence such as is shown in Fig. A.15. If $z = 0.98e^{-j\pi/6}$ and $N = 11$, plot the three numbers.

A.6.1 Roots of Unity

In a surprising number of cases, the following equation must be solved,

$$z^N = 1 \tag{A.6.1}$$

where N is an integer. One solution is $z = 1$, but there are many others, because (A.6.1) is equivalent to finding all the roots of the Nth-degree polynomial $z^N - 1$, which must have N roots. It turns out that all the solutions are given by

$$z = e^{j2\pi\ell/N} \qquad \text{for } \ell = 0, 1, 2, \ldots N-1$$

which are called the *N*th *roots of unity*. As shown in Fig. A.16, these N solutions are numbers equally spaced around the unit circle. The angular spacing between them is $2\pi/N$.

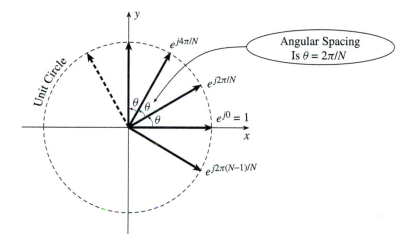

Figure A.16   Graphical display of the *N*th roots of unity ($N = 12$). These are the solutions of $z^N = 1$. Notice that there are only N distinct roots.

A.6.1.1   Procedure for Finding Multiple Roots   Once we know that the solutions to (A.6.1) are the *N*th roots of unity, we can describe a structured approach to solving equations with multiple roots. A slightly more general situation is

$$z^N = c$$

where c is a complex constant, $c = |c|e^{j\phi}$.

1. Write z^N as $r^N e^{jN\theta}$.

2. Write c as $|c|e^{j\phi}e^{j2\pi\ell}$ where ℓ is an integer. Note that when $c = 1$ we would write the number 1 as

$$1 = e^{j2\pi\ell} \qquad \text{for } \ell = 0, \pm1, \pm2, \ldots$$

3. Equate the two sides and solve for the magnitudes and angles separately.

$$r^N e^{jN\theta} = |c|e^{j\phi}e^{j2\pi\ell}$$

4. The magnitude is the positive *N*th root of a positive number $|c|$:

$$r = |c|^{1/N}$$

5. The angle contains the interesting information, because there are N different solutions:

$$N\theta = \phi + 2\pi\ell \qquad \ell = 0, 1, \ldots N - 1$$

$$\theta = \frac{\phi + 2\pi\ell}{N}$$

$$\theta = \frac{\phi}{N} + \frac{2\pi\ell}{N}$$

6. Thus the N different solutions all have the same magnitude, but their angles are equally spaced with a difference of $2\pi/N$ between each one.

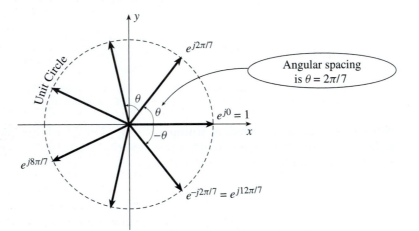

Figure A.17   Graphical display of the 7th roots of unity. Notice that the sequence $e^{j2\pi\ell/7}$ repeats with a period equal to 7.

Example A.5   Solve $z^7 = 1$, using the procedure above.

$$r^7 e^{j7\theta} = e^{j2\pi\ell}$$

$$\implies r = 1$$

$$\implies 7\theta = 2\pi\ell$$

$$\theta = \frac{2\pi}{7}\ell \qquad \ell = 0, 1, 2, 3, 4, 5, 6$$

Therefore, these solutions are equally spaced around the unit circle, as shown in Fig. A.17. In this case, the solutions are called the *seventh roots of unity*.   ◇

Exercise A.10. Solve the following equation:

$$z^5 = -1$$

Use the fact that $-1 = e^{j\pi}$. Plot all the solutions.

A.7   SUMMARY AND LINKS

This appendix has presented a brief review of complex numbers and their visualization as vectors in the two-dimensional complex plane. Although this material should have been seen before by most students during high-school algebra, our intense use of complex notation demands much greater familiarity. The first two labs in Appendix C deal with various aspects of complex numbers, and also introduce MATLAB. In Lab C.1, we also have included a number of MATLAB functions for plotting vectors from complex numbers (zvect, zcat) and for changing between Cartesian and polar forms (zprint).

Z-DRILL

In addition to the labs, we have written a MATLAB GUI (graphical user interface) that will generate drill problems for each of the complex operations studied here: addition, subtraction, multiplication, division, inverse, and conjugate. A screen shot of the GUI is shown in Fig. A.18. This program can be installed from the CD, along with the other dspfirst software for MATLAB. The readme.txt file on the CD-ROM describes the installation process for different platforms.

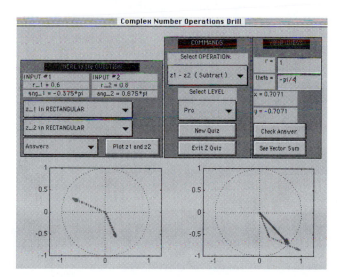

Figure A.18   The MATLAB GUI zdrill for practicing complex number operations.

PROBLEMS

A.1   Convert the following to polar form:

   (a) $z = 0 + j2$

   (b) $z = (-1, 1)$

   (c) $z = -3 - j4$

   (d) $z = (0, -1)$

A.2   Convert the following to rectangular form:

   (a) $z = \sqrt{2}\,e^{j(3\pi/4)}$

   (b) $z = 1.6\,\angle\,(\pi/6)$

   (c) $z = 3e^{-j(\pi/2)}$

   (d) $z = 7\,\angle\,(7\pi)$

A.3   Evaluate the following by reducing the answer to rectangular form:

   (a) j^3

   (b) $e^{j(\pi + 2\pi m)}$ (m an integer)

   (c) j^{2n} (n an integer)

   (d) $j^{1/2}$ (find two answers)

A.4   Simplify the following complex-valued expressions:

   (a) $3e^{j2\pi/3} - 4e^{-j\pi/6}$

   (b) $(\sqrt{2} - j2)^8$

   (c) $(\sqrt{2} - j2)^{-1}$

   (d) $(\sqrt{2} - j2)^{1/2}$

   (e) $\Im m\{je^{-j\pi/3}\}$

   Give the answers in both Cartesian and polar form.

A.5   Evaluate each example and give the answer in both rectangular and polar form (in all cases, assume that $z_1 = -4 + j3$ and $z_2 = 1 - j$):

   (a) z_1^*

   (b) z_2^2

   (c) $z_1 + z_2^*$

   (d) jz_2

   (e) $z_1^{-1} = 1/z_1$

   (f) z_1/z_2

   (g) e^{z_2}

   (h) $z_1 z_1^*$

   (i) $z_1 z_2$

A.6   Simplify the following complex-valued sum:

$$z = e^{j9\pi/3} + e^{-j5\pi/8} + e^{j13\pi/8}$$

   Give the numerical answer for z in polar form. Draw a vector diagram for the three vectors and their sum (z).

A.7   Simplify the following complex-valued expressions. Give your answers in polar form. Reduce the answers to a simple numerical form.

   (a) For $z = -3 + j4$, evaluate $1/z$.

   (b) For $z = -2 + j2$, evaluate z^5.

   (c) For $z = -5 + j13$, evaluate $|z|^2$.

   (d) For $z = -2 + j5$, evaluate $\Re e\{ze^{-j\pi/2}\}$

A.8   Solve the following equation for z:

$$z^4 = j$$

   Be sure to find all possible answers, and express your answer(s) in polar form.

A.9   Let $z_0 = e^{j2\pi/N}$. Prove that $z_0^{N-1} = 1/z_0$.

A.10   Evaluate $(-j)^{1/2}$ and plot the result(s).

B

Programming in MATLAB

MATLAB will be used extensively in the laboratory exercises of Appendix C. This appendix provides an overview of MATLAB and some of its capabilities. We focus on programming issues since a wealth of information is already available on syntax and basic commands.[1] MATLAB has an extensive on-line help system, which can be used to answer any questions not answered in this brief presentation. In fact, an ideal way to read this appendix would be to have MATLAB running so that help can be used whenever necessary, and the examples can be run and modified.

MATLAB (short for Matrix Laboratory) is an environment for numerical analysis and computing. It originated as an interface to the collections of numerical routines from the LINPACK and EISPACK projects, but it is now a commercial product of The Mathworks, Inc. MATLAB has evolved into a powerful programming environment containing many built-in functions for doing signal processing, linear algebra, and other mathematical calculations. The language has also been extended by means of toolboxes containing additional functions for MATLAB. For example, the CD-ROM that accompanies this book contains a toolbox of functions needed for the laboratory exercises. The toolboxes are installed as separate directories within the MATLAB directory. Please follow the instructions on the CD-ROM to install the *DSP First* toolbox before doing the laboratory exercises.

DSP First
Toolbox

Since MATLAB is extensible, users find it convenient to write new functions whenever the built-in functions fail to do a vital task. The programming necessary to create new functions and scripts is not too difficult if the user has some experience with C, PASCAL, or FORTRAN. This appendix gives a brief overview of MATLAB for the purpose of programming.

---

[1] One useful reference book is D. Hanselman and B. Littlefield, *Mastering MATLAB: A Comprehensive Tutorial and Reference*, Prentice Hall, Upper Saddle River, NJ, 1996.

B.1 MATLAB HELP

MATLAB provides an on-line help system accessible by using the `help` command. For example, to get information about the function `filter`, enter the following at the command prompt:

```
>> help filter
```

The command prompt is indicated by `>>` in the command window. The `help` command will return text information in the command window. Help is also available for categories; for example, `help punct` summarizes punctuation as used in MATLAB's syntax. In more recent versions of MATLAB, the help system has been given a Web-browser interface. In version 5, the commands `helpdesk` and `helpwin` bring up this interface.

A useful command for getting started is `intro`, which covers the basic concepts in the MATLAB language. Also, there are many demonstration programs that illustrate the various capabilities of MATLAB; these can be started with the command `demo`.

Finally, if you are searching for other tutorials, some are freely available on the Web.

When unsure about a command, use `help`.

B.2 MATRIX OPERATIONS AND VARIABLES

The basic variable type in MATLAB is a matrix.[2] To declare a variable, simply assign it a value at the MATLAB prompt. For example,

```
>> M = [1 2 6; 5 2 1]
   M =
         1   2   6
         5   2   1
```

When the definition of a matrix involves a long formula or many entries, then a very long MATLAB command can be broken onto two (or more) lines by placing an ellipses (...) at the end of the line to be continued. For example,

```
P = [ 1, 2, 4, 6, 8 ] + [ pi, 4, exp(1), 0, -1 ] + ...
    [ cos(0.1*pi), sin(pi/3), tan(3), atan(2), sqrt(pi) ];
```

If an expression is followed by a semicolon (;), then the result is not echoed to the screen. This is very useful when dealing with large matrices.

---

[2] It is the only type in version 4, but version 5 offers many other data types found in conventional programming languages. This appendix will not discuss these version 5 extensions.

The size of the matrix can always be extracted with the `size` operator:

```
>> Msize = size(M)
   Msize =
          2   3
```

Therefore, it becomes unnecessary to assign separate variables to track the number of rows and number of columns. Two special types of matrix variables are worthy of mention: *scalars* and *vectors*. A scalar is a matrix with only one element; its size is 1×1. A vector is a matrix that has only one row or column. In the *DSP First* laboratory exercises, signals will often be stored as vectors.

Individual elements of a matrix variable may be accessed by giving the row index and the column index, for example:

```
>> M13 = M(1,3)
   M13 =
          6
```

Submatrices may be accessed in a similar manner by using the colon operator as explained in Section B.2.1.

B.2.1 The Colon Operator

The colon operator (`:`) is useful for creating index arrays, creating vectors of evenly spaced values, and accessing submatrices. Use `help colon` for a detailed description of its capabilities.

The colon notation is based on the idea that an index range can be generated by giving a start, a skip, and then the end. Therefore, a regularly spaced vector of numbers is obtained by means of:

$$iii = start:skip:end$$

Without the `skip` parameter, the default increment is 1. This sort of counting is similar to the notation used in FORTRAN DO loops. However, MATLAB takes it one step further by combining it with matrix indexing. For a 9×8 matrix A, A(2,3) is the scalar element located at the 2nd row and 3rd column of A, so a 4×3 submatrix can be extracted with A(2:5,1:3). The colon also serves as a wild card; i.e., A(2,:) is the 2nd row. Indexing backwards flips a vector, e.g., x(9:-1:1) for a length–9 vector. Finally, it is sometimes necessary to work with a list of all the values in a matrix, so A(:) gives a 72×1 column vector that is merely the columns of A concatenated together. This is an example of *reshaping* the matrix. More general reshaping of the matrix A can be accomplished with the reshape(A,M,N) function. For example, the 9×8 matrix A can be reshaped into a 12×6 matrix with: Anew = reshape(A,12,6).

B.2.2 Matrix and Array Operations

The default operations in MATLAB are matrix operations. Thus A*B means matrix multiplication, which is defined and reviewed next.

B.2.2.1 A Review of Matrix Multiplication   The operation of matrix multiplication AB can be carried out only if the two matrices have compatible dimensions, i.e., the number of columns in A must equal the number of rows in B. For example, a 5×8 matrix can multiply a 8×3 matrix to give a result AB that is 5×3. In general, if A is $m \times n$, then B must be $n \times p$, and the product AB would be $m \times p$. Usually matrix multiplication is *not* commutative, i.e., $AB \neq BA$. If $p \neq m$, then the product BA cannot be defined, but even when BA is defined, we find that the commutative property applies only in special cases.

Each element in the product matrix is calculated with an inner product. To generate the first element in the product matrix, $C = AB$, simply take the first row of A and multiply *point by point* with the first column of B, and then sum. For example, if

$$A = \begin{bmatrix} a_{1,1} & a_{1,2} & a_{1,3} \\ a_{2,1} & a_{2,2} & a_{2,3} \end{bmatrix} \quad \text{and} \quad B = \begin{bmatrix} b_{1,1} & b_{1,2} \\ b_{2,1} & b_{2,2} \\ b_{3,1} & b_{3,2} \end{bmatrix}$$

then the first element of $C = AB$ is:

$$c_{1,1} = a_{1,1}b_{1,1} + a_{1,2}b_{2,1} + a_{1,3}b_{3,1}$$

which is, in fact, the inner product between the first row of A and the first column of B. Likewise, $c_{2,1}$ is found by taking the inner product between the *second* row of A and the *first* column of B, and so on for $c_{1,2}$ and $c_{2,2}$. The final result would be:

$$\begin{aligned} C &= \begin{bmatrix} c_{1,1} & c_{1,2} \\ c_{2,1} & c_{2,2} \end{bmatrix} \\ &= \begin{bmatrix} a_{1,1}b_{1,1} + a_{1,2}b_{2,1} + a_{1,3}b_{3,1} & a_{1,1}b_{1,2} + a_{1,2}b_{2,2} + a_{1,3}b_{3,2} \\ a_{2,1}b_{1,1} + a_{2,2}b_{2,1} + a_{2,3}b_{3,1} & a_{2,1}b_{1,2} + a_{2,2}b_{2,2} + a_{2,3}b_{3,2} \end{bmatrix} \end{aligned} \quad \text{(B.2.1)}$$

Some special cases of matrix multiplication are the *outer product* and the *inner product*. In the *outer product*, a column vector multiplies a row vector to give a matrix. If we let one of the vectors be all 1s, then we can get a repeating result:

$$\begin{bmatrix} a_1 \\ a_2 \\ a_3 \end{bmatrix} \begin{bmatrix} 1 & 1 & 1 & 1 \end{bmatrix} = \begin{bmatrix} a_1 & a_1 & a_1 & a_1 \\ a_2 & a_2 & a_2 & a_2 \\ a_3 & a_3 & a_3 & a_3 \end{bmatrix}$$

With all 1s in the row vector, we end up repeating the column vector four times.

For the *inner product*, a row vector multiplies a column vector, so the result is a scalar. If we let one of the vectors be all 1s, then we will sum the elements in the other vector:

$$\begin{bmatrix} a_1 & a_2 & a_3 & a_4 \end{bmatrix} \begin{bmatrix} 1 \\ 1 \\ 1 \\ 1 \end{bmatrix} = a_1 + a_2 + a_3 + a_4$$

B.2.2.2   Pointwise Array Operations   If we want to do a pointwise multiplication between two arrays, some confusion can arise. In the pointwise case, we want to multiply the matrices together element-by-element, so they must have exactly the same size in both dimensions. For example, two 5×8 matrices can be multiplied pointwise, although we cannot do matrix multiplication between two 5×8 matrices. To obtain pointwise multiplication in MATLAB, we use the "point-star" operator A .* B. For example, if A and B are both 3×2, then

$$
D \;=\; A \;.* \; B \;=\; \begin{bmatrix} d_{1,1} & d_{1,2} \\ d_{2,1} & d_{2,2} \\ d_{3,1} & d_{3,2} \end{bmatrix} \;=\; \begin{bmatrix} a_{1,1}\,b_{1,1} & a_{1,2}\,b_{1,2} \\ a_{2,1}\,b_{2,1} & a_{2,2}\,b_{2,2} \\ a_{3,1}\,b_{3,1} & a_{3,2}\,b_{3,2} \end{bmatrix}
$$

where $d_{i,j} = a_{i,j}\,b_{i,j}$. We will refer to this type of multiplication as *array multiplication*. Notice that array multiplication is commutative because we would get the same result if we computed D = B.*A.

A general rule in MATLAB is that when "point" is used with another arithmetic operator, it modifies that operator's usual matrix definition to a pointwise one. Thus we have ./ for pointwise division and .^ for pointwise exponentiation. For example, xx = (0.9) .^ (0:49) generates a vector whose values are equal to $(0.9)^n$, for $n = 0, 1, 2, \ldots, 49$.

B.3   PLOTS AND GRAPHICS

MATLAB is capable of producing two-dimensional x-y plots and three-dimensional plots, displaying images, and even creating and playing movies. The two most common plotting functions that will be used in the *DSP First* laboratory exercises are plot and stem. The calling syntax for both plot and stem takes two vectors, one for the x-axis points, and the other for the y-axis.[3] The invocation plot(x,y) produces a connected plot with straight lines between the data points

$$\{(x(1),y(1)),\ (x(2),y(2)),\ \ldots,\ (x(N),y(N))\}$$

as shown in the top panel of Fig. B.1. The same call with stem(x,y) produces the "lollipop" presentation of the same data in the bottom panel of Fig. B.1. MATLAB has numerous plotting options that can be studied by using help plotxy, help plotxyz, or help graphics in version 4; or by using help graph2d, help graph3d, or help specgraph in version 5.

---

[3] If only one argument is given, plot(y) uses the single argument as the y-axis, and uses 1:length(y) for the x-axis.

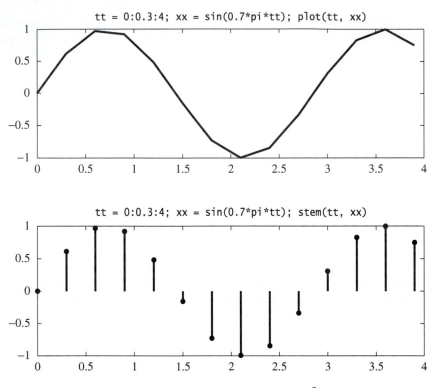

Figure B.1   Example of two different plotting formats, plot and stem.

B.3.1 Figure Windows

Whenever MATLAB makes a plot, it writes the graphics to a *figure window*. You can have multiple figure windows open, but only one of the them is considered the *active* window. Any plot command executed in the command window will direct its graphical output to the active window. The command figure(n) will pop up a new figure window that can be refered to by the number n, or makes it active if it already exists. Control over many of the window attributes (size, location, color, etc.) is also possible with the figure command, which does initialization of the plot window.

B.3.2 Multiple Plots

Multiple plots per window can be done with the subplot function. This function does not do the actual plotting; it merely divides the window into tiles. To set up a 3×2 tiling of the figure window, use subplot(3,2,tile_number). For example, subplot(3,2,3) will direct the next plot to the third tile, which is in the second row, left side. The graphs in Fig. B.1 were done with subplot(2,1,1) and subplot(2,1,2).

B.3.3 Printing and Saving Graphics

Plots and graphics may be printed to a printer or saved to a file using the print command. To send the current figure window to the default printer, simply type print without arguments. To save the plot to a file, a device format and filename must be specified. The device format specifies which language will be used to store the graphics commands. For example, a useful format for including the file in a document is encapsulated PostScript (EPS), which can be produced as follows:

```
>> print -deps myplot.eps
```

The postscript format is also convenient when the plots are kept for printing at a later time. For a complete list of available file formats, supported printers, and other options, see help print.

B.4   PROGRAMMING CONSTRUCTS

MATLAB supports the paradigm of "functional programming" in which it is possible to nest a sequence of function calls. Consider the following equation, which can be implemented with one line of MATLAB code.

$$\sum_{n=1}^{L} \log(|x_n|)$$

Here is the MATLAB equivalent:

```
sum( log( abs(x) ) )
```

where x is a vector containing the elements x_n. This example illustrates MATLAB in its most efficient form, where individual functions are combined to get the output. Writing efficient MATLAB code requires a programming style that generates small functions that are vectorized. Loops should be avoided. The primary way to avoid loops is to use calls to toolbox functions as much as possible.

B.4.1 MATLAB Built-in Functions

Many MATLAB functions operate on arrays just as easily as they operate on scalars. For example, if x is an array, then cos(x) returns an array of the same size as x containing the cosine of each element of x.

$$
\cos(\mathsf{x}) = \begin{bmatrix} \cos(x_{1,1}) & \cos(x_{1,2}) & \cdots & \cos(x_{1,n}) \\ \cos(x_{2,1}) & \cos(x_{2,2}) & \cdots & \cos(x_{2,n}) \\ \vdots & \vdots & \vdots & \vdots \\ \cos(x_{m,1}) & \cos(x_{m,2}) & \cdots & \cos(x_{m,n}) \end{bmatrix}
$$

Notice that no loop is needed, even though cos(x) does apply the cosine function to every array element. Most transcendental functions follow this pointwise rule. In some cases, it is crucial to make this distinction, such as the matrix exponential (expm) versus the pointwise exponential (exp):

$$
\exp(\mathsf{A}) = \begin{bmatrix} \exp(a_{1,1}) & \exp(a_{1,2}) & \cdots & \exp(a_{1,n}) \\ \exp(a_{2,1}) & \exp(a_{2,2}) & \cdots & \exp(a_{2,n}) \\ \vdots & \vdots & \vdots & \vdots \\ \exp(a_{m,1}) & \exp(a_{m,2}) & \cdots & \exp(a_{m,n}) \end{bmatrix}
$$

B.4.2 Program Flow

Program flow can be controlled in MATLAB using if statements, while loops, and for loops. In MATLAB version 5, there is also a switch statement. These are similar to any high-level language. Descriptions and examples for each of these program constructs can be viewed by using the MATLAB help command.

B.5 MATLAB SCRIPTS

Any expression that can be entered at the MATLAB prompt can also be stored in a text file and executed as a script. The text file can be created with any plain ASCII editor such as notepad on a PC, emacs or vi on UNIX, and the built-in MATLAB editor on a Macintosh or Windows platform. The file extension must be .m, and the script is executed in MATLAB simply by typing the filename (with or without the extension). These programs are usually called M-files. Here is an example:

```
tt = 0:0.3:4;
xx = sin(0.7*pi*tt);
subplot(2,1,1)
plot( tt, xx )
title('tt = 0:0.3:4; xx = sin(0.7*pi*tt); plot( tt, xx)')
subplot(2,1,2)
stem( tt, xx )
title('tt = 0:0.3:4; xx = sin(0.7*pi*tt); stem( tt, xx)')
```

If these commands are saved in a file named `plotstem.m`, then typing `plotstem` at the command prompt will run the file, and all eight commands will be executed as if they had been typed in at the command prompt. The result is the two plots that were shown in Fig. B.1.

B.6   WRITING A MATLAB FUNCTION

You can write your own functions and add them to the MATLAB environment. These functions are another type of M-file, and are created as an ASCII file with a text editor. The first word in the M-file must be the keyword `function` to tell MATLAB that this file is to be treated as a function with arguments. On the same line as the word `function` is the calling template that specifies the input and output arguments of the function. The filename for the M-file must end in `.m`, and the filename will become the name of the new command for MATLAB. For example, consider the following file, which extracts the last L elements from a vector:

```
function  y = foo( x, L )
%FOO      get last L points of x
%   usage:
%              y = foo( x, L )
%   where:
%           x = input vector
%           L = number of points to get
%           y = output vector
N = length(x);
if( L > N )
   error('input vector too short')
end
y = x((N-L+1):N);
```

If this file is called `foo.m`, the operation may be invoked from the MATLAB command line by typing

```
aa = foo( (1:2:37), 7 );
```

The output will be the last seven elements of the vector (1:2:37), i.e.,

```
aa = [ 25 27 29 31 33 35 37 ]
```

B.6.1 Creating A Clip Function

Most functions can be written according to a standard format. Consider a clip *function* M-file that takes two input arguments (a signal vector and a scalar threshold) and returns an output signal vector. You can use an editor to create an ASCII file clip.m that contains the following statements:

These lines of comments at the beginning of the function will be the response to help clip.

First step is to figure out matrix dimensions of x.

Input could be row or column vector.

Since x is local, we can change it without affecting the workspace.

Create output vector.

```
function  y = clip( x, Limit )
%CLIP       saturate mag of x[n] at Limit
%      when |x[n]| > Limit, make |x[n]| = Limit
%
%   usage:  y = clip( x, Limit )
%
%       x   - input signal vector
%   Limit   - limiting value
%       y   - output vector after clipping

[nrows ncols] = size(x);

if( ncols ~= 1 & nrows ~= 1 )        %-- NEITHER
    error('CLIP: input not a vector')
end
Lx = max([nrows ncols]);          %-- Length

for n=1:Lx                %-- Loop over entire vector
    if( abs(x(n)) > Limit )
        x(n) = sign(x(n))*Limit;     %-- saturate
    end              Preserve the sign of x(n)
end
y = x;                   %-- copy to output vector
```

Figure B.2   Illustration of a MATLAB function.

We can break down the M-file clip.m into four elements:

1. *Definition of Input–Output:* Function M-files must have the word function as the very first item in the file. The information that follows function on the same line is a declaration of how the function is to be called and what arguments are to be passed. The name of the function should match the name of the M-file; if there is a conflict, it is the name of the M-file on the disk that is known to the MATLAB command environment.

Input arguments are listed inside the parentheses following the function name. Each input is a matrix. The output argument (also a matrix) is on the left side of the equals sign. Multiple output arguments are also possible if square brackets surround the list of output arguments; e.g., notice how the size(x) function returns the number of rows and number of columns into separate output variables. Finally, observe that there is no explicit command for returning the outputs; instead, MATLAB returns whatever value is contained in the output matrix when the function completes. For clip the last line of the function assigns the clipped vector to y, so that the clipped vector is returned. MATLAB does have a command called return, but it just exits the function, it does not take an argument.

The essential difference between the function M-file and the script M-file is dummy variables versus permanent variables. MATLAB uses "call by value" so that the function makes local copies of its arguments. These local variables disappear after the function completes. For example, the following statement creates a clipped vector wwclipped from the input vector ww.

```
>> wwclipped = clip(ww, 0.9999);
```

The arrays ww and wwclipped are permanent variables in the MATLAB workspace. The temporary arrays created inside clip (i.e., y, nrows, ncols, Lx and i) exist only while clip runs; then they are deleted. Furthermore, these variable names are local to clip.m, so the name x may also be used in the workspace as a permanent name. These ideas should be familiar to anyone experienced with a high-level computer language like C, FORTRAN, or PASCAL.

2. *Self-Documentation:* A line beginning with the % sign is a comment line. The first group of these in a function is used by MATLAB's help facility to make M-files automatically self-documenting. That is, you can now type help clip and the comment lines from *your* M-file will appear on the screen as help information. The format suggested in clip.m follows the convention of giving the function name, its calling sequence, a brief explanation, and then definitions of the input and output arguments.

3. *Size and Error Checking:* The function should determine the size of each vector or matrix that it will operate on. This information does not have to be passed as a separate input argument, but can be extracted with the size function. In the case of the clip function, we want to restrict the function to operating on vectors, but we would like to permit either a row $(1 \times L)$ or a column $(L \times 1)$. Therefore, one of the variables nrows or ncols must be equal to 1; if not, we terminate the function with the bail-out function error, which prints a message to the command line and quits the function.

4. *Actual Function Operations:* In the case of the clip function, the actual clipping is done by a for loop, which examines each element of the x vector for its size compared to the threshold Limit. In the case of negative numbers, the clipped value must be set to -Limit, hence the multiplication by sign(x(n)). This

assumes that Limit is passed in as a positive number, a fact that might also be tested in the error-checking phase.

This particular implementation of clip is very inefficient owing to the for loop. In Section B.7.1, we will show how to vectorize this program for speed.

B.6.2 Debugging a MATLAB M-file

Since MATLAB is an interactive environment, debugging can be done by examining variables in the workspace. MATLAB versions 4 and 5 contain a symbolic debugger with support for break points. Since different functions can use the same variable names, it is important to keep track of the local context when examining variables. Several useful debugging commands are listed here, others can be found with help debug.

dbstop is used to set a breakpoint in an M-file. It can also be used to give you a prompt when an error occurs by typing dbstop if error before executing the M-file. This allows you to examine variables within functions and also the calling workspace (by typing dbup).

dbstep incrementally steps through your M-file, returning you to a prompt after each line is executed.

dbcont causes normal program execution to resume, or, if there was an error, returns you to the MATLAB command prompt.

dbquit quits the debug mode and returns you to the MATLAB command prompt.

keyboard can be inserted into the M-file to cause program execution to pause, giving you a MATLAB prompt of the form K> to indicate that it is not the command-line prompt.

B.7 PROGRAMMING TIPS

This section presents a few programming tips that should improve the speed of your MATLAB programs. For more ideas and tips, list some of the function M-files in the toolboxes of MATLAB by using the type command. For example,

```
type angle
type conv
type trapz
```

Copying the style of other programmers is always an efficient way to improve your own knowledge of a computer language. In the following hints, we discuss some of the most important points involved in writing good MATLAB code. These comments will become increasingly useful as you develop more experience in MATLAB.

B.7.1 Avoiding Loops

Since MATLAB is an interpreted language, certain common programming habits are intrinsically inefficient. The primary one is the use of for loops to perform simple operations over an entire matrix or vector. *Whenever possible*, you should try to find a vector function (or the nested composition of a few vector functions) that will accomplish the desired result, rather than writing a loop. For example, if the operation is summing all the elements in a matrix, the difference between calling sum and writing a loop that looks like FORTRAN code can be astounding; the loop is unbelievably slow owing to the interpreted nature of MATLAB. Consider the following three methods for matrix summation:

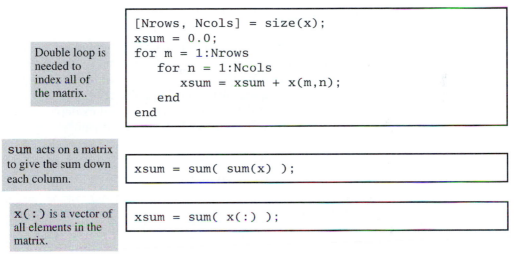

Double loop is needed to index all of the matrix.

```
[Nrows, Ncols] = size(x);
xsum = 0.0;
for m = 1:Nrows
    for n = 1:Ncols
        xsum = xsum + x(m,n);
    end
end
```

sum acts on a matrix to give the sum down each column.

```
xsum = sum( sum(x) );
```

x(:) is a vector of all elements in the matrix.

```
xsum = sum( x(:) );
```

Figure B.3   Three ways to add all the elements of a matrix.

The first method is the MATLAB equivalent of conventional programming. The last two methods rely on the built-in function sum, which has different characteristics depending on whether its argument is a matrix or a vector (called "operator overloading"). When acting on a matrix, sum returns a row vector containing the column sums; when acting on a row (or column) vector, the sum is a scalar. To get the third (and most efficient) method, the matrix x is converted to a column vector with the colon operator. Then one call to sum will suffice.

B.7.2 Repeating Rows or Columns

Often it is necessary to form a matrix from a vector by replicating the vector in the rows or columns of the matrix. If the matrix is to have all the same values, then functions such as ones(M,N) and zeros(M,N) can be used. But to replicate a column vector x to create a matrix that has identical columns, a loop can be avoided by using the outer-product matrix multiply operation discussed in Section B.2.2. The following MATLAB code fragment will do the job for eleven columns:

```
x = (12:-2:0)';        % prime indicates conjugate transpose
X = x * ones(1,11)
```

If x is a length L column vector, then the matrix X formed by the outer product is $L \times 11$. In this example, $L = 7$. Note that MATLAB is case-sensitive, so the variables x and X are different. We have used capital X to indicate a matrix, as would be done in mathematics.

B.7.3 Vectorizing Logical Operations

It is also possible to vectorize programs that contain if, else conditionals. The clip function (Fig. B.2) offers an excellent opportunity to demonstrate this type of vectorization. The for loop in that function contains a logical test and might not seem like a candidate for vector operations. However, the relational and logical operators in MATLAB, such as greater than, apply to matrices. For example, a greater than test applied to a 3×3 matrix returns a 3×3 matrix of ones and zeros.

```
>> x = [ 1  2 -3; 3 -2  1; 4  0  -1]    %-- create a test matrix
   x = [ 1  2 -3
         3 -2  1
         4  0 -1 ]
>> mx = x > 0          %-- check the greater than condition
   mx = [ 1  1  0
          1  0  1
          1  0  0 ]
>> y = mx .* x         %-- pointwise multiply by masking matrix
   y = [ 1  2  0
         3  0  1
         4  0  0 ]
```

The zeros mark where the condition was false; the ones denote where the condition was true. Thus, when we do the pointwise multiply of x by the masking matrix mx, we get a result that has all negative elements set to zero. Note that these last two statements process the entire matrix without ever using a for loop.

Since the saturation done in clip.m requires that we change the large values in x, we can implement the entire for loop with three array multiplications. This leads to a vectorized saturation operator that works for matrices as well as vectors:

```
y = Limit*(x > Limit) - Limit*(x < -Limit) + x.*(abs(x) <= Limit);
```

Three different masking matrices are needed to represent the three cases of positive saturation, negative saturation, and no action. The additions correspond to the logical OR of these cases. The number of arithmetic operations needed to carry out this statement is $3N$ multiplications and $2N$ additions, where N is the total number of elements in x. This is actually more work than the loop in clip.m if we counted only arithmetic operations. However, the cost of code interpretation is high. This vectorized statement is interpreted only once, whereas the three statements inside the for loop must be reinterpreted N times. If the two implementations are timed with etime, the vectorized version will be much faster for long vectors.

B.7.4 Creating an Impulse

Another simple example is given by the following trick for creating an impulse signal vector:

```
nn = [-10:25];
impulse = (nn==0);
```

This result may be plotted with stem(nn, impulse). In a sense, this code fragment is perfect because it captures the essence of the mathematical formula that defines the impulse as existing only when $n = 0$.

$$\delta[n] = \begin{cases} 1 & n = 0 \\ 0 & n \neq 0 \end{cases}$$

B.7.5 The Find Function

An alternative to masking is to use the find function. This is not necessarily more efficient; it just gives a different approach. The find function will determine the list of indices in a vector where a condition is true. For example, find(x > Limit) will return the list of indices where the vector is greater than the Limit value. Thus we can do saturation as follows:

```
y = x;
jkl = find(y > Limit);
y( jkl ) =  Limit*ones(size(jkl));
jkl = find(y < -Limit);
y( jkl ) = -Limit*ones(size(jkl));
```

The ones function is needed to create a vector on the right-hand side that is the same size as the number of elements in jkl. In version 5, this would be unnecessary, since a scalar assigned to a vector is assigned to each element of the vector.

B.7.6 Seek to Vectorize

The dictum to "avoid for loops" is not always an easy path to follow, because it means the algorithm must be cast in a vector form. If matrix-vector notation is incorporated into MATLAB programs, the resulting code will run much faster. Even loops with logical tests can be "vectorized" if masks are created for all possible conditions. Thus, a reasonable goal is the following:

Eliminate all for *loops.*

B.7.7 Programming Style

If there were a proverb to summarize good programming style, it would probably read something like:

May your functions be short and your variable names long.

–Anon

This is certainly true for MATLAB. Each function should have a single purpose. This will lead to short, simple modules that can be linked together by functional composition to produce more complex operations. Avoid the temptation to build super functions with many options and a plethora of outputs.

MATLAB supports long variable names (up to 32 characters). Take advantage of this feature to give variables descriptive names. In this way, the number of comments littering the code can be drastically reduced. Comments should be limited to help information and the documentation of tricks used in the code.

C

Laboratory Projects

This appendix contains eleven computer-based laboratories. These correspond roughly to each of the different chapters, although in some cases the labs use concepts from several chapters. The following table summarizes the lab material and the chapters involved in each lab.

Lab	Subject	Cross-Reference
1	Introduction to MATLAB	Appendix B
2	Introduction to Complex Exponentials	Chapter 2, Appendix A
3	Synthesis of Sinusoidal Signals	Chapter 3
4	AM and FM Sinusoidal Signals	Chapter 3
5	FIR Filtering of Sinusoidal Waveforms	Chapter 5
6	Filtering Sampled Waveforms	Chapters 5 and 6
7	Everyday Sinusoidal Signals	Chapters 5 and 6
8	Filtering and Edge Detection of Images	Chapters 5 and 6
9	Sampling and Zooming of Images	Chapters 4, 5, 6, and 7
10	The z-, n-, and $\hat{\omega}$-Domains	Chapters 7 and 8
11	Extracting Frequencies of Musical Tones	Chapter 9

The structure of the labs is as follows:

Overview: Each lab starts with a brief review of the relevant theory to be studied and implemented.

Warm-up: The warm-up section consists of a few simple exercises that introduce the MATLAB functionality needed for the implementation of that lab. The warm-

Instructor
Verification

up exercises in each lab should be completed during a supervised lab time, so that students can ask questions of an expert. In our own use of these experiments, we have implemented a system where the laboratory instructor must verify the appropriate steps by initialing an **Instructor Verification** sheet for a few key steps. One example of this verification sheet is included at the end of Lab C.1; the rest are included on the CD-ROM.

Exercises: The bulk of the work in each lab consists of exercises that require some MATLAB programming and plotting. All of the exercises are designed to illustrate the theoretical ideas presented in the text. Furthermore, we have included numerous processing examples that involve real signals such as speech, music, and images.

Projects: In some cases, the labs require implementations that are so large and complicated that it is no longer fair to call them exercises. A better description would be a project. In the music synthesis lab, the touch-tone decoding lab, or the music writing lab, the problem statement is more like that of a design project, so that the students are given some flexibility in creating an implementation that satisfies a general objective. In addition, many individual parts must be completed to make the whole project function correctly.

DSP-First
Toolbox for
MATLAB

These laboratory exercises can be done with either version 4 or version 5 of MATLAB. The student version will suffice in most cases, but not for the processing of long signals or large images. Thus, some of the real signals cannot be manipulated in the student version. We have also provided a package of M-files containing functions developed for this book.

Before doing any of the laboratory exercises, it is necessary to install the DSP First Toolbox. Refer to the instructions on the DSP First CD-ROM.

C.1   LABORATORY: INTRODUCTION TO MATLAB

The Warm-up section of each lab should be completed during a supervised lab session, and the laboratory instructor should verify the appropriate steps by initialing the **Instructor Verification** line. An Instructor Verification sheet can be found at the end of this lab.

C.1.1 Overview and Goals

MATLAB will be used extensively in all the succeeding labs. The goals of this first lab are to gain familiarity with MATLAB and to build some basic skills in the MATLAB language. Read the material in Appendix B on *Using MATLAB* for a more complete overview. If you desire an in-depth presentation that covers most of the language, consult the MATLAB reference manual, which is also available on-line with a Web-browser interface.

There are several specific objectives in this lab:

1. Learn to use the help system to study basic MATLAB commands and syntax.
2. Learn to write functions and M-files in MATLAB.
3. Learn some advanced programming techniques for MATLAB, e.g., vectorization.

DSP-First
Toolbox for
MATLAB

C.1.2 Warm-up

Each lab will start with a warm-up section like this one. The warm-up usually consists of a few simple exercises to introduce MATLAB commands. Since this is the first lab, make sure that you have installed the *DSP First Toolbox* for MATLAB. Refer to the instructions on the *DSP First* CD-ROM. When you start MATLAB on your computer, the path should include a directory called `dspfirst`.

C.1.2.1 Basic Commands   The following exercises provide an orientation to MATLAB.

1. View the MATLAB introduction by typing `intro` at the MATLAB prompt. This short introduction will demonstrate some of the basics of using MATLAB.
2. Explore the MATLAB `help` capability. Type each of the following lines to read about these commands:

```
help
help plot
help colon
help ops
help punct
help zeros
help ones
lookfor filter      %<--- keyword search
```

If the lines scroll past the bottom of the screen, it is possible to force MATLAB to display only one screenful of information at a time by issuing the command more on.

3. Use MATLAB as a calculator. Try the following:

```
pi*pi - 10
sin(pi/4)
ans ^ 2     %<--- 'ans' holds the last result
```

4. Variable names can store values and matrices in MATLAB. Try the following:

```
xx = sin( pi/5 );
cos( pi/5 )                 %<--- assigned to what?
yy = sqrt( 1 - xx*xx )
ans
```

5. Complex numbers are natural in MATLAB.[1] Notice that the names of some basic operations are unexpected, e.g., abs for magnitude. Try the following:

```
zz = 3 + 4i
conj(zz)
abs(zz)
angle(zz)
real(zz)
imag(zz)
help zprint       %<--- requires DSP First Toolbox
exp( sqrt(-1)*pi )
exp( j*[ pi/4 -pi/4 ] )
```

6. Plotting is easy in MATLAB, for both real and complex numbers. The basic plot command will plot a vector yy versus a vector xx. Try the following:

```
xx = [-3 -1 0 1 3 ];
yy = xx.*xx - 3*xx;
plot( xx, yy )
zz = xx + yy*sqrt(-1);
plot( zz )       %<--- complex values can be plotted
```

Drop the semicolons, if you want to display the values in the xx, yy, and zz vectors. Use help arith to learn how the operation xx.*xx works; compare to matrix multiply.

---

[1] Refer to Appendix A for a review of complex numbers.

> *When unsure about a command, use* `help`.

C.1.2.2   MATLAB Array Indexing

1. Be sure to understand the colon notation. In particular, explain what the following MATLAB code will produce:

```
jkl =  2 : 4 : 17
jkl = 99 : -1 : 88
ttt =  2 : (1/9) : 4
tpi = pi * [ 2 : (-1/9) : 0 ]
```

2. Extracting or inserting numbers in a vector is very easy to do. Consider the following definition:

```
xx = [ ones(1,4), [2:2:11], zeros(1,3) ]
xx(3:7)
length(xx)
xx(2:2:length(xx))
```

Explain the result echoed from the last three lines of this code.

3. In item 2, the vector xx contains 12 elements. Observe the result of the following assignment:

```
xx(3:7) = pi*(1:5)
```

Now write a statement that will replace the odd-indexed elements of xx with the constant −77 (i.e., xx(1), xx(3), etc). *Use vector indexing and vector replacement.*

Instructor Verification (separate page)

C.1.2.3   MATLAB Script Files

1. Experiment with vectors in MATLAB. Think of a vector as a list of numbers. Try the following:

```
kset = -3:11;
kset
cos( pi*kset/4 )       %<---comment: compute cosines
```

Explain how the last example computes the different values of cosine without a loop. The text following the % is a comment; it may be omitted. If you remove the semicolon at the end of the first statement, all the elements of kset will be echoed to the screen.

2. *Vectorization* is an essential programming skill in MATLAB. Loops can be written in MATLAB, but they are not the most efficient way to get things done. It's better to *avoid loops and use the vector notation instead.* For example, this code uses a loop to compute values of the sine function. Rewrite this computation without using the loop (as in item 1).

```
xx = [ ];    %<--- initialize the xx vector to a null
for k=0:7
    xx(k+1) = sin( k*pi/4 )    %<--- xx(0) would fail
end
xx
```

3. Use the built-in MATLAB editor (on Windows or a Mac), or an external one such as emacs (on UNIX), to create a script file called funky.m containing the following lines:

```
tt = -2 : 0.05 : 3;
xx = sin( 2*pi*0.789*tt );
plot( tt, xx ), grid on    %<--- plot a sinusoid
title('TEST PLOT of SINUSOID')
xlabel('TIME (sec)')
```

4. Run your script from MATLAB. To run the file funky created in item 3, try

```
funky           %<---will run the commands in the file
type funky      %<---will type out the contents of
                %    funky.m to the screen
which funky     %<---will show directory containing funky.m
```

5. Add three lines of code to your script, so that it will plot a cosine on top of the sine. Use the hold function to add a plot of

```
0.5*cos( 2*pi*0.789*tt )
```

to the plot created in item 3. See help hold in MATLAB.

Instructor Verification (separate page)

C.1.2.4   MATLAB Demos   There are many demonstration files in MATLAB. Run the MATLAB demos from a menu by typing demo, and explore some of the different demos of basic MATLAB commands and plots.

Reminder: *When unsure about a command, use* help.

C.1.2.5   MATLAB Sound

1. Run the MATLAB sound demo by typing xpsound at the MATLAB prompt. If you are unable to hear the sounds in the MATLAB demo, then check the sound hardware on your machine. Since there are so many variations in the types of sound hardware on different computers, this may require consultation with an expert in system configuration or MATLAB installation.

2. Now generate a tone (i.e., a sinusoid) in MATLAB and listen to it with the sound command. The frequency of your tone should be 2 kHz, and the duration should be 1 sec. The following lines of code should be saved in a file called mysound.m and run from the command line.

```
dur = 1.0;
fs = 8000;
tt = 0 : (1/fs) : dur;
xx = sin( 2*pi*2000*tt );
sound( xx, fs )
```

The sound hardware will convert the vector of numbers xx into a sound wave-form at a certain rate, called the sampling rate. In this case, the sampling rate is 8000 samples/sec, but other values might be used depending on the capability of the sound hardware. What is the length of the vector xx? Read the on-line help for sound (or soundsc) to get more information on using this command.

Instructor Verification (separate page)

C.1.2.6   Functions   The following warm-up exercises illustrate how to write functions in MATLAB. Although the questions below contain minor errors, they do exemplify the correct structure and syntax for writing functions.

1. Find the mistake in the following function:

```
function  xx = cosgen(f,dur)
%COSGEN   Function to generate a cosine wave
%   usage:
%       xx = cosgen(f,dur)
%        f = desired frequency
%      dur = duration of the waveform in seconds
%
tt = [0:1/(20*f):dur];   % gives 20 samples per period
xx = cos(2*pi*f*tt);
```

2. Find the mistake in the following function:

```
function  [sum,prod] = sumprod(x1,x2)
%SUMPROD  Function to add and multiply two complex numbers
%  usage:
%          [sum,prod] = sumprod(x1,x2)
%              x1 = a complex number
%              x2 = another complex number
%            sum = sum of x1 and x2
%          prod = product of x1 and x2
%
sum = z1+z2;
prod = z1*z2;
```

3. Explain how the following lines of MATLAB code work:

```
yy = ones(7,1) * rand(1,4);
```

```
xx = randn(1,3);
yy = xx(ones(6,1),:);
```

4. Write a function that performs the same task as the following without using a loop. Consult Section B.2.2.1 for some clever solutions.

```
function Z = expand(xx,ncol)
%EXPAND  Function to generate a matrix Z with identical
%          columns equal to an input vector xx
%  usage:
%          Z = expand(xx,ncol)
%          xx = the input vector containing one column for Z
%        ncol = the number of desired columns
%
xx = xx(:); %-- makes the input vector xx into a column vector
Z = zeros(length(xx),ncol);
for i=1:ncol
    Z(:,i) = xx;
end
```

C.1.2.7   Vectorization

1. Explain the results obtained from the following lines of Matlab code:

```
A = randn(6,3);
A = A .* (A > 0);
```

2. Write a new function that performs the same task as the following function without using a for loop. Use the idea in item 1, and also consult Section B.7.1. In addition, the Matlab logical operators are summarized with help relop.

```
function  Z = replacez(A)
%REPLACEZ  Function that replaces the negative elements
%              of a matrix with the number 77
%   usage:
%          Z = replacez(A)
%          A = input matrix whose negative elements are to
%              be replaced with 77
%
[M,N] = size(A);
for i=1:M
   for j=1:N
      if A(i,j) < 0
         Z(i,j) = 77;
      else
         Z(i,j) = A(i,j);
      end
   end
end
```

C.1.3 Exercises: Using Matlab

The following exercise can be completed on your own time. Results from each part should be included in a brief lab report write-up.

C.1.3.1   Manipulating Sinusoids with MATLAB   Generate two 3000-Hz sinusoids with different amplitudes and phases.

$$x_1(t) = A_1 \cos(2\pi (3000)t + \phi_1) \qquad\qquad x_2(t) = A_2 \cos(2\pi (3000)t + \phi_2)$$

1. Select the value of the amplitudes as follows: Let $A_1 = 13$, and use your age for A_2. For the phases, use the last two digits of your telephone number for ϕ_1 (in degrees), and take $\phi_2 = -30°$. *When doing computations in MATLAB, be sure to convert degrees to radians.*

2. Make a plot of both signals over a range of t that will exhibit approximately 3 cycles. Be sure that the plot starts at a negative time, so that it will include $t = 0$, *and be sure that your have at least 20 samples per period of the wave.*

3. Verify that the phase of the two signals $x_1(t)$ and $x_2(t)$ is correct at $t = 0$; also verify that each one has the correct maximum amplitude.

4. Use subplot(3,1,1) and subplot(3,1,2) to make a three-panel subplot that puts both of these plots on the same window. See help subplot.

5. Create a third sinusoid as the sum: $x_3(t) = x_1(t) + x_2(t)$. In MATLAB, this amounts to summing the vectors that hold the samples of each sinusoid. Make a plot of $x_3(t)$ over the same range of time as used in the previous two plots. Include this as the third panel in the window by using subplot(3,1,3).

6. Measure the magnitude and phase of $x_3(t)$ directly from the plot. In your lab report, explain how the magnitude and phase were measured by making annotations on each of the plots.

C.1.4 Lab Review Questions

In general, the lab write-up should indicate an enhanced understanding of the topics treated by the laboratory assignment. Here are a few questions to answer in order to assess your understanding of this lab's objective—a working knowledge of the basics of MATLAB. If you do not know the answers to these questions, go back to the lab and try to figure them out in MATLAB (remember the commands help and lookfor).

1. You saw how easy it is for MATLAB to generate and manipulate vectors (i.e., one-dimensional arrays of numbers). For example, consider the following:

```
yy = 0:10;
yy = zeros(1,25);
yy = 1:0.25:5;
```

   (a) How would you modify one of these lines of MATLAB code to create a vector that runs from 0 to 10 in steps of 0.5?

   (b) How would you modify one of the lines in the code to create a vector of one hundred 100s?

2. You also learned that MATLAB has no problem handling complex numbers. Consider the following line of code:

$$yy = 3+5j;$$

   (a) How do you get MATLAB to return the magnitude of the complex number yy?

   (b) How do you get MATLAB to return the phase of the complex number yy? What are the units of the answer?

3. In Section C.1.2.3, you learned that multiple lines of MATLAB code can be stored in a file with a .m extension. MATLAB then executes the code in the order that it appears in the file. Consider the following file, named example.m:

```
f = 200;
tt = [0:1/(20*f):1];
z = exp(j*2*pi*f*tt);
subplot(2,1,1)
plot(real(z))
title('Real part of exp(j*2*pi*200*tt)')
subplot(2,1,2)
plot(imag(z))
title('Imaginary part of exp(j*2*pi*200*tt)')
```

   (a) How do you execute the file from the MATLAB prompt?

   (b) Suppose the file were named example.dog. Would it run? How could you change it to make it work in MATLAB ?

   (c) Assuming that the M-file runs, what do you expect the plots to look like? If you're not sure, type in the code and run it.

Lab 1

Instructor Verification Sheet

Staple this page to the end of your lab report.[1]

Instructor
Verification
Sheet

Name:_____   Date:_____

Part C.1.2.2   Vector replacement using the colon operator:

Instructor Verification_____

Part C.1.2.3   Run the modified function funky from a file:

Instructor Verification_____

Part C.1.2.5   Use sound to play a 2 kHz tone in MATLAB:

Instructor Verification_____

Part C.1.2.7   Modify replacez using vector logicals:

Instructor Verification_____

---

[2] This page will appear only once, to give an example of the verification sheet. For all the labs, an Instructor Verification Sheet can be obtained from the CD-ROM and printed out as needed.

C.2 LABORATORY: INTRODUCTION TO COMPLEX EXPONENTIALS

The goal of this laboratory is to gain familiarity with complex numbers and their use in representing sinusoidal signals as complex exponentials.

C.2.1 Overview

Manipulating sinusoidal functions using complex exponentials turns trigonometric problems into simple arithmetic and algebra. In this lab, we first review the complex exponential signal and the phasor addition property needed for adding cosine waves. Then we will use MATLAB to make plots of phasor diagrams that show the vector addition needed when combining sinusoids. Consult Appendix A for a review of complex numbers.

DSP First
MATLAB
Toolbox

C.2.1.1   Complex Numbers in MATLAB   MATLAB can be used to compute complex-valued formulas and also to display the results as vector or "phasor" diagrams. For this purpose, several new functions have been written and are available on the *DSP First* CD-ROM. Be sure that this toolbox has been installed by doing help on the new M-files: zvect, zcat, ucplot, zcoords, and zprint. Each of these functions can plot (or print) several complex numbers at once, if the input is formed into a vector of complex numbers. The following example will plot five vectors all on one graph:

```
zvect( [ 1+j, j, 3-4*j, exp(j*pi), exp(2i*pi/3) ] )
```

Here are some of MATLAB's complex number operators:

conj	Complex conjugate
abs	Magnitude
angle	Angle (or phase) in radians
real	Real part
imag	Imaginary part
i,j	predefined as $\sqrt{-1}$
x = 3 + 4i	i suffix defines imaginary constant
exp(j*theta)	Function for the complex exponential $e^{j\theta}$

Each of these functions takes a vector (or matrix) as its input argument and operates on each element of the vector.

Z-DRILL

Finally, there is a complex-numbers drill program called `zdrill` that generates complex number problems and tests your answers. *Please spend some time working with this drill, since it is very useful in helping you to get a feel for complex arithmetic.*

Reminder: *When unsure about a command, use* `help`.

C.2.1.2   Sinusoid Addition Using Complex Exponentials   Recall that sinusoids may be expressed in the form:

$$x(t) = A\cos(2\pi f_0 t + \phi) = \Re\left\{Ae^{j\phi}e^{j2\pi f_0 t}\right\} \tag{C.2.1}$$

Consider the sum of cosine waves given by (C.2.2):

$$x(t) = \sum_{k=1}^{N} A_k \cos(2\pi f_k t + \phi_k) \tag{C.2.2}$$

where each cosine wave in the sum might have a different frequency, f_k.

When all the frequencies are identical, $f_k = f_0$, this sum reduces to a single cosine. It is difficult to simplify using trigonometric identities, but it reduces to an algebraic sum of complex numbers when solved using complex exponentials. This is the *phasor addition rule* presented in Section 2.6.2. A summary of the phasor addition rule using the complex exponential representation of the cosines (C.2.1) is:

$$x(t) = \Re\left\{\sum_{k=1}^{N} X_k e^{j2\pi f_0 t}\right\} \tag{C.2.3}$$

$$= \Re\left\{\left(\sum_{k=1}^{N} X_k\right) e^{j2\pi f_0 t}\right\} \tag{C.2.4}$$

$$= \Re\left\{X_s e^{j2\pi f_0 t}\right\} \tag{C.2.5}$$

$$= A_s \cos(2\pi f_0 t + \phi_s) \tag{C.2.6}$$

where

$$X_k = A_k e^{j\phi_k} \tag{C.2.7}$$

and

$$X_s = \sum_{k=1}^{N} X_k = A_s e^{j\phi_s} \tag{C.2.8}$$

We see that the sum signal $x(t)$ is a single sinusoid, and it is periodic with period $T_0 = 1/f_0$.

C.2.1.3   Harmonic Sinusoids   There is an important extension where $x(t)$ is the sum of N cosine waves whose frequencies (f_k) are all multiples of one basic frequency f_0.

$$f_k = k f_0 \qquad \text{(Harmonic Frequencies)}$$

The sum of N cosine waves given by (C.2.2) becomes

$$x(t) = \sum_{k=1}^{N} A_k \cos(2\pi k f_0 t + \phi_k) = \Re e \left\{ \sum_{k=1}^{N} X_k e^{j 2\pi k f_0 t} \right\} \qquad \text{(C.2.9)}$$

This particular signal $x(t)$ is also periodic with period $T_0 = 1/f_0$. The frequency f_0 is called the *fundamental frequency*, and T_0 is called the *fundamental period*.

C.2.2   Warm-up

Instructor
Verification
Sheet

The instructor verification sheet can be found on the CD-ROM and printed out as needed.

C.2.2.1   Complex Numbers   To exercise your understanding of complex numbers, do the following:

1. Define $z_1 = -1 + j0.3$ and $z_2 = 0.8 + j0.7$. Enter these in MATLAB and plot them with zvect, and print them with zprint.

2. Compute the conjugate z^* and the inverse $1/z$ for both z_1 and z_2, and plot the results. In MATLAB, see help conj. Display the results numerically with zprint.

3. Compute $z_1 + z_2$ and plot the sum. Use zcat to show the sum as vectors head-to-tail. Use zprint to display the results numerically.

4. Compute $z_1 z_2$ and z_1/z_2 and plot. Use the zvect plot function to show how the angles of z_1 and z_2 determine the angles of the product and quotient. Use zprint to display the results numerically.

5. Work a few problems on the complex-number drill program. To start the program, simply type dspfirst and select the *Complex Number Drill*.

   Instructor Verification (separate page)

C.2.2.2   Sinusoidal Synthesis with an M-File   Write an M-file that will synthesize a waveform in the form of (C.2.2). Write the function without using loops. Take advantage of the fact that matrix–vector multiplication computes a sum of products. For example,

$$\mathtt{c = A*b} \qquad \Longrightarrow \qquad c_n = \sum_{k=1}^{L} a_{nk} b_k \qquad \text{(C.2.10)}$$

where c_n represents the nth element of the vector **c**, a_{nk} is the element in the nth row and kth column of the matrix **A**, b_k is the kth element of the column vector **b**, and L is the number of columns in **A**. The first few statements of the M-file should look like

```
function          xx = sumcos(f, X, fs, dur)
%SUMCOS   Function to synthesize a sum of cosine waves
%  usage:
%        xx = sumcos(f, X, fs, dur)
%         f = vector of frequencies
%                 (these could be negative or positive)
%         X = vector of complex exponentials: Amp*e^(j*phase)
%        fs = the sampling rate in Hz
%       dur = total time duration of signal
%
%    Note: f and X must be the same length.
%              X(1) corresponds to frequency f(1),
%              X(2) corresponds to frequency f(2), etc.
```

The MATLAB syntax length(f) returns the number of elements in the vector f, so we do not need a separate input argument for the number of frequencies. On the other hand, the programmer should provide error-checking to make sure that the lengths of f and X are the same. It is possible (although not required) to complete this function in a single line. For some hints, refer to the review of matrix multiplication in section B.2.2.1.

In order to use this M-file to synthesize periodic waveforms, you would simply choose the entries in the frequency vector to be integer multiples of the desired fundamental frequency. Try the following tests and plot the results.

```
xx = sumcos([20],    [1],    200, 0.25);
xx = sumcos([20 40], [1 1/2], 200, 0.25);
xx = sumcos([20 40 60 80], [1 -1 1 -1], 200, 0.25);
```

Instructor Verification (separate page)

C.2.3 Exercises: Complex Exponentials

C.2.3.1 Representation of Sinusoids with Complex Exponentials   In MATLAB, consult help on exp, real, and imag.

1. Generate the signal $x(t) = A e^{j(\omega_0 t + \phi)}$ for $A = 3$, $\phi = -0.4\pi$, and $\omega_0 = 2\pi(1250)$. Take a range for t that will cover two or three periods.

2. Plot the real part of $x(t)$ versus t and the imaginary part versus t. Use subplot(2,1,1) and subplot(2,1,2) to put both plots in the same window.

3. Verify that the real and imaginary parts are sinusoids, and that they have the correct frequency, phase, and amplitude.

C.2.3.2   Verify Addition of Sinusoids Using Complex Exponentials   Generate four sinusoids with the following amplitudes and phases:

$$x_1(t) = 5\cos(2\pi(15)t + 0.5\pi)$$

$$x_2(t) = 5\cos(2\pi(15)t - 0.25\pi)$$

$$x_3(t) = 5\cos(2\pi(15)t + 0.4\pi)$$

$$x_4(t) = 5\cos(2\pi(15)t - 0.9\pi)$$

1. Make a plot of all four signals over a range of t that will exhibit approximately three cycles. Be sure that the plot includes negative time so that the phase at $t = 0$ can be measured. *In order to get a smooth plot, be sure to have at least 20 samples per period of the wave.*[3]

2. Verify that the phase of all four signals is correct at $t = 0$, and also verify that each one has the correct maximum amplitude. Use subplot(3,2,i), i=1,2,3,4 to make a 6-panel subplot that puts all of these plots on the same page.

3. Create the sum sinusoid by using : $x_5(t) = x_1(t) + x_2(t) + x_3(t) + x_4(t)$. Make a plot of $x_5(t)$ over the same range of time as used in the last plot. Include this as the lower left panel in the plot by using subplot(3,1,3).

4. Measure the magnitude and phase of $x_5(t)$ directly from the plot. In your lab report, include this plot with sufficient annotation to show how the magnitude and phase were measured.

5. Now, do some complex arithmetic. Create the complex amplitudes corresponding to the sinusoids $x_i(t)$:

$$z_i = A_i e^{j\phi_i} \qquad i = 1, 2, 3, 4, 5$$

Give the numerical values of z_i in polar *and* Cartesian form.

6. Verify that $z_5 = z_1 + z_2 + z_3 + z_4$. Show a plot of these five complex numbers as vectors. Use the MATLAB functions zvect, zcat and zprint discussed in the Warm-up.

7. Relate the magnitude and phase of z_5 to the plot of $x_5(t)$.

Reminder: *When unsure about a command, use* help.

---

[3] If you have already studied sampling in Chapter 4, then you will realize that this requirement of 20 samples per period amounts to considerable oversampling.

C.2.4 Periodic Waveforms

Each of the following waveforms can be synthesized with a simple call to the function sumcos. Plot a short section of the signal to observe its characteristic shape.

Note: It is important to have a sampling rate that is at least *twice as high as the highest frequency component in your signal.* This is an important fact about sampling that is discussed in Chapter 4, but, for this lab, simply choose a number for f_s that gives a smooth plot. (This may require a bit of experimentation.)

1. Try your sumcos M-file with the fundamental $f_0 = 25$ Hz, $f_k = kf_0$, and

$$X_k = \begin{cases} \frac{j4}{k\pi} & k \text{ an odd integer} \\ 0 & k \text{ an even integer} \end{cases} \qquad (C.2.11)$$

Specify the duration to get three periods of the waveform.

Make plots for three different cases: $N = 5, 10,$ and 25 (where N is the number of cosines). Use a 3-panel subplot to show all three signals together. Explain how the period of the synthesized waveform is related to the fundamental frequency.

Explain what happens as $N \to \infty$. To what waveshape do the plots converge? Although the waveshape is converging to a simple form, it is not perfect. Describe any unusual features in the converging waveform as $N \to \infty$.

2. It is informative to listen to these signals for different numbers of coefficients. Repeat the synthesis from item 1 with $f_0 = 1$ kHz and listen to the cases where $N = 1, 2, 3, 4, 5,$ and 10. You need about 1 second of the signal to hear differences. When using sound(x,fs), the sampling frequency should be very high to avoid aliasing effects (discussed in Chapter 4).

3. Now try the coefficients

$$X_k = \frac{j(-1)^k}{\pi k} \qquad k = 1, 2, 3, \ldots \qquad (C.2.12)$$

Choose the fundamental frequency to be $f_0 = 25$ Hz. Compute the signal for three cases: $N = 5, 10,$ and 25, and plot all three functions together with a 3-panel subplot. What waveshape is approximated with this sum of cosines as $N \to \infty$? Explain how the period of the synthesized waveshape is related to the fundamental frequency.

C.3   LABORATORY: SYNTHESIS OF SINUSOIDAL SIGNALS

This lab includes a project on music synthesis with sinusoids. One of several candidate pieces may be selected when doing the synthesis program, or some other piece can be selected if sheet music is available. The project requires an extensive programming effort and should be documented with a complete lab report. A good report should include the following items: a cover sheet, commented MATLAB code, explanations of your approach, conclusions, and any additional tweaks that you implemented for the synthesis. Since the project must be evaluated by listening to the quality of the synthesized piece, the criteria for judging a good piece are given at the end of this lab description. In addition, it may be convenient to place the final piece on a Web site so that it can be accessed remotely by a lab instructor, who can then evaluate its quality.

C.3.1   Overview

The properties of sinusoidal waveforms of the form

$$x(t) = A \cos(\omega_0 t + \phi) \tag{C.3.1}$$

are considered in detail in Chapters 2 and 3. In this lab, we will put them to use. We will synthesize waveforms composed of sums of sinusoidal signals, sample them, and then reconstruct them for listening. We will use combinations of the basic sinusoid (C.3.1) to synthesize the following signals:

MUSIC
SYNTHESIS

1. Sine waves at a specific frequency played through a D-to-A converter.

2. Sinusoids that create a synthesized version of the piece *Für Elise*.

3. Any other piece can be used for the synthesis project. This lab write-up and the CD-ROM include information about four alternative pieces: *Jesu, Joy of Man's Desiring*; *Minuet in G*; *Beethoven's Fifth Symphony*; and *Twinkle, Twinkle, Little Star*.

The primary objective of the lab is to establish the connection between musical notes, their frequencies, and sinusoids. A secondary objective is the challenge of trying to add other features to the synthesis in order to improve the subjective quality for listening. Students who take this challenge will be motivated to learn more about the spectral representation of signals, a topic that underlies this entire book.

C.3.2   Warm-up: Music Synthesis

Instructor
Verification
Sheet

A copy of the instructor verification sheet (if needed) can be found on the CD-ROM and printed out as needed.

In this lab, sine waves and music signals will be created with the intention of playing them out through a speaker. Therefore, it is necessary to take into account the fact that a conversion is needed from the digital samples, which are numbers stored in the computer memory, to the actual voltage waveform that will be amplified for the speakers. The layout of a piano keyboard will also be explored, to obtain a formula that gives the frequency for each key.

C.3.2.1 D-to-A Conversion   The digital-to-analog conversion process has a number of aspects, but in its simplest form the only thing we need to worry about at this point is that the time spacing (T_s) between the signal samples must correspond to the rate of the D-to-A hardware that will be used. From MATLAB, the sound output is done by the sound(x,fs) function, which supports variable sampling rate if the hardware on the machine has such capability. A convenient choice for the D-to-A conversion rate is 8000 samples per second, for which $T_s = 1/8000$ sec. Another common choice is 11,025 Hz, which is one-quarter of the rate used for audio CDs. Both of these rates satisfy the requirement of sampling fast enough, as explained in Section C.3.2.2. In fact, most piano notes have relatively low frequencies, so an even lower sampling rate may be used. In some cases, it also will be necessary to scale the vector x so that it lies between ± 1.[4]

C.3.2.2 Theory of Sampling   Even though Chapter 4 treats sampling in detail, we provide a quick summary of essential facts here. The idealized process of sampling a signal and the subsequent reconstruction of the signal from its samples is depicted in Fig. C.1. This figure shows a continuous-time input signal $x(t)$, which is sampled by the continuous-to-discrete (C-to-D) converter to produce a sequence of samples $x[n] = x(nT_s)$, where n is the integer sample index and T_s is the sampling period. The sampling rate is $f_s = 1/T_s$. As described in Chapter 4, the ideal discrete-to-continuous (D-to-C) converter takes the input samples and interpolates a smooth curve between them. The *sampling theorem* tells us that if the input is a sum of sine waves, then the output $y(t)$ will be equal to the input $x(t)$ if the sampling rate is more than twice the highest frequency f_{max} in the input, i.e., $f_s > 2f_{max}$.

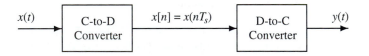

Figure C.1   Sampling and reconstruction of a continuous-time signal.

Most computers have a built-in analog-to-digital (A-to-D) converter and a digital-to-analog (D-to-A) converter (usually on the sound card). These hardware systems are physical realizations of the idealized concepts of C-to-D and D-to-C

---

[4] In MATLAB version 5, there is a function, soundsc(x,fs), which performs that scaling.

converters, respectively, and for purposes of this lab we will assume that they are perfect realizations.

1. The ideal C-to-D converter will be implemented in MATLAB by taking the formula for the continuous-time signal and evaluating it at the sample times, nT_s. This assumes perfect knowledge of the input signal, but for sinusoidal signals we have a mathematical equation for the continuous-time signal.

   To begin, compute a vector x1 of samples of a sinusoidal signal with $A = 100$, $\omega_0 = 2\pi(1100)$, and $\phi = 0$. Use a sampling rate of 8000 samples/sec, and compute a total number of samples equivalent to 2 seconds' time duration. You may find it helpful to recall that the MATLAB statement tt=(0:0.01:3); would create a vector of numbers from 0 through 3 with increments of 0.01. Therefore, it is necessary only to determine the time increment needed to obtain 8000 samples in one second.[5]

   Using sound( ), play the resulting vector through the D-to-A converter of your computer, assuming that the hardware can support the $f_s = 8000$ Hz rate (or $f_s = 11{,}025$ Hz). Listen to the output.

2. Now compute a vector x2 of samples (again with duration 2 secs) of the sinusoidal signal for the case $A = 100$, $\omega_0 = 2\pi(1650)$, and $\phi = \pi/3$. Listen to the signal reconstructed from these samples. How does it compare to the signal in item 1? Put both signals together in a new vector defined with the following MATLAB statement (assuming that both x1 and x2 are row vectors):

   ```
   xx = [x1 zeros(1,2000) x2];
   ```

   Listen to this signal. Explain what you heard.

3. Now send the vector xx to the D-to-A converter again, but double the sampling rate in sound( ) to 16,000 samples/sec. Do not recompute the samples in xx; just tell the D-to-A converter that the sampling rate is 16,000 samples/sec. Describe what you heard. Observe how the *duration* and *pitch* of the signal changed. Explain.

   Instructor Verification (separate page)

C.3.2.3 Piano Keyboard Section C.3.3 of this lab will consist of synthesizing the notes of a well-known piece of music.[6] Since we will use sinusoidal tones to represent piano notes, a quick introduction to the frequency layout of the piano keyboard is

---

[5] Another popular rate is 11,025 samples/sec, which is one-fourth of the rate used in audio CD players.

[6] If you have little or no experience reading music, don't be intimidated. Only a little knowledge is needed to carry out this lab. On the other hand, the experience of working in an application area where you must quickly acquire knowledge is a valuable one. Many real-world engineering problems have this flavor, especially in signal processing, which has such a broad applicability in such diverse areas as geophysics, medicine, radar, speech, and the like.

needed (see Fig. C.2). On a piano, the keyboard is divided into octaves, the notes in each octave being twice the frequency of the notes in the next lower octave. For example, the reference note is the A above middle C, which is usually called A-440 (or A_4) because its frequency is 440 Hz. Each octave contains 12 notes (5 black keys and 7 white keys) and the ratio between the frequencies of the notes is constant between successive notes. Thus, this ratio must be $2^{1/12}$. Since middle C is 9 keys below A-440, its frequency is approximately 261.6 Hz. Consult Section 3.5 in Chapter 3 for more details.

MUSIC GUI

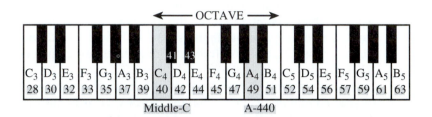

Figure C.2 Layout of a piano keyboard. Key numbers are shaded. The notation C_4 means the C key in the fourth octave.

Musical notation shows which notes are to be played, and their relative timing (half notes last twice as long as quarter notes, which, in turn, last twice as long as eighth notes). Figure C.3 shows how the keys on the piano correspond to notes drawn in musical notation.

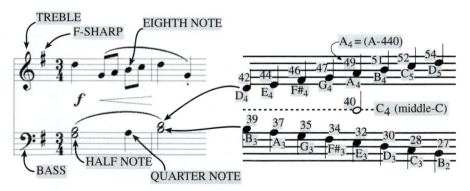

Figure C.3 Musical notation is a time-frequency diagram where vertical position indicates the frequency of the note to be played.

Another interesting relationship is the ratio of fifths and fourths as used in a chord. Strictly speaking, the fifth note should be 1.5 times the frequency of the base note. For middle C, the fifth is G, but the frequency of G is about 392 Hz, which

MUSIC GUI

is not exactly 1.5 times 261.6. It is very close, but the slight detuning introduced by the ratio $2^{1/12}$ gives a better sound to the piano overall. This innovation in tuning is called "equally tempered," and was introduced in Germany in the 1760s and made famous by J. S. Bach in *The Well-Tempered Clavichord*.

You can use the ratio $2^{1/12}$ to calculate the frequency of notes anywhere on the piano keyboard. For example, the E flat above middle C (key number 43) is 6 keys below A-440, so its frequency should be $f = 440 \times 2^{-6/12} = 440/\sqrt{2} \approx 311$ Hz.

1. Generate a sinusoid of 2 seconds' duration to represent the note E_5 above A-440 (key number 56). Choose the appropriate values for T_s and fs. Remember that fs should be at least twice as high as the frequency of the sinusoid you are generating. Also, T_s and fs must "match" in order for the note played out of the D-to-A converter to sound correct.

2. Now write an M-file to produce a desired note for a given duration. Your M-file should be in the form of a function called note.m. You may want to call the sumcos function that you wrote for Lab C.2. Your function should have the following form:

```
function tone = note(keynum,dur)
% NOTE   Produce a sinusoidal waveform corresponding to a
%           given piano key number
%
% usage:   tone = note (keynum, dur)
%
%            tone = the output sinusoidal waveform
%          keynum = the piano keyboard number of the desired note
%             dur = the duration (in seconds) of the output note
%
fs = 8000;   %-- use 11025 Hz on PC/Mac, 8000 on UNIX
tt = 0:(1/fs):dur;
freq =
tone =
```

For the freq = line, use the formulas based on $2^{1/12}$ to determine the frequency for a sinusoid in terms of its key number. You should start from a reference note (middle C or A-440 is recommended) and solve for the frequency based on this reference. For the tone = line, generate the actual sinusoid at the proper frequency and duration.

3. The following is an incomplete M-file that will play scales:

```
%--- play_scale.m
%---
keys =   [ 40  42  44  45  47  49  51  52 ];
%--- NOTES: C   D   E   F   G   A   B   C
%  key #40 is middle-C
%
dur  = 0.25 * ones(1,length(keys));
fs   = 8000;        %-- use 11025 Hz on PC/Mac, 8000 on UNIX
xx   = zeros(1,sum(dur)*fs+1);
n1 = 1;
for kk = 1:length(keys)
    keynum = keys(kk);
    tone =                         %<=== FILL IN THIS LINE
    n2 = n1 + length(tone) - 1;
    xx(n1:n2) = xx(n1:n2) + tone;
    n1 = n2;
end
sound( xx, fs )
```

For the tone = line, generate the actual sinusoid for keynum by making a call to the function note( ) written previously. Note that the code in play_scale.m allocates a vector of zeros large enough to hold the entire scale, then adds each note into its proper place in the vector xx.

Instructor Verification (separate page)

C.3.3 Lab: Synthesis of Musical Notes

The audible range of musical notes consists of well-defined frequencies assigned to each note in a musical score. Five different pieces are given here, but you need only choose one for your synthesis program. Before starting the project, make sure that you have a working knowledge of the relationship between a musical score, key number, and frequency. In the process of actually synthesizing the music, follow these steps:

1. Determine a sampling frequency that will be used to play out the sound through the D-to-A system of the computer. This will dictate the time T_s between samples of the sinusoids.

2. Determine the total time duration needed for each note.

3. Determine the frequency (in Hz) for each note (utilize the note.m function written in the warm-up and key numbers from Fig. C.2).

4. Synthesize the waveform as a combination of sinusoids, and play it out through the computer's built-in speaker or headphones using sound( ).

5. A chord can be synthesized by adding the sinusoids for each note in the chord. This will be a vector addition of the sinusoidal values for each note. Likewise, if you have more than one melody line playing at the same time, you can produce separate signal vectors for each melody (treble and bass) and then combine them by adding the signal vectors.

6. Make a plot of a few periods of two or three of the sinusoids to illustrate that you have the correct signals for each note.

C.3.3.1   Spectrogram of the Music   Musical notation describes how a song is composed of different frequencies and when they should be played. This representation can be considered to be a *time-frequency* representation of the signal that synthesizes the song. In MATLAB, we can can compute a time-frequency representation from the signal itself. This is called the spectrogram, and is implemented with the MATLAB function specgram.[7] To aid your understanding of music and its connection to frequency content, a MATLAB GUI is available so that you can visualize the spectrogram along with musical notation. This GUI also has the capability to synthesize music from a list of notes, but these notes are given in "standard" musical notation, not key number. For more information, consult the help on musicgui.m which runs only in MATLAB version 5.

MUSIC GUI

C.3.3.2   *Für Elise*   *Für Elise* is a well-known piece written by Beethoven. You can listen to a recording of the part that you will synthesize by following the links on the *DSP First* CD-ROM. The first few measures are shown in Fig. C.4, and more of the piece can be found on the CD-ROM, where an entire page of the music for *Für Elise* is reproduced.

Fur Elise

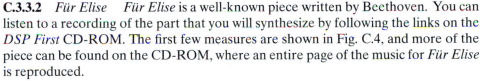

Figure C.4   The first few measures of Beethoven's Für Elise.

Determine the notes that are played in *Für Elise*, by mapping each note to a key number (Fig. C.2) and then synthesize sine waves to re-create the piece. Use either the short form (Fig. C.4) or the long form found on the CD. Use sine waves sampled at 8000 samples/sec (for UNIX) or 11,025 samples/sec (for other platforms). Listen to the example on the CD, and then estimate the time duration needed for each note. If you define a fixed time duration for a quarter note, say T_q then all the other durations will be defined: an eighth note is $\frac{1}{2}T_q$, a sixteenth note is $\frac{1}{4}T_q$, a half note $2T_q$, and so on. After defining the time duration for all notes, you still

---

[7] The *DSP First Toolbox* contains a function spectgr.m that will compute the spectrogram of a signal.

may need to make adjustments in the timing to improve the subjective quality of the synthesized song. In addition, adding short pauses between notes usually improves the music because it imitates the natural transition that a musician must make from one note to the next.

After you finish the project, assess the quality of your synthesized result. Suggest some other features that could be incorporated into your program if you had more time to work on it.

C.3.3.3   Musical Tweaks   The musical passage is likely to sound artificial, because it is created from pure sinusoids. Therefore, you may want to try to improve the quality of the sound by incorporating some modifications. For example, you could multiply each pure tone signal by an envelope $E(t)$ so that it would fade in and out.

$$x(t) = E(t) \cos(2\pi f_0 t + \phi) \qquad \qquad \text{(C.3.2)}$$

If an envelope is used, it should "fade in" quickly and fade out more slowly. An envelope such as a half-cycle of a sine wave $\sin(\pi t/\text{dur})$ is *not* good because it does not turn on quickly enough, so simultaneous notes of different durations no longer appear to begin at the same time. A standard way to define the envelope function is to divide $E(t)$ into four sections: attack (A), delay (D), sustain (S), and release (R). Together, these are called ADSR. The attack is a quickly rising front edge, the delay is a small short-duration drop, the sustain is more or less constant, and the release drops quickly back to zero. Figure C.5 shows a linear approximation to the ADSR profile.

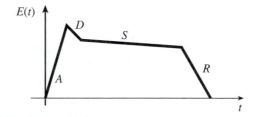

Figure C.5   ADSR profile for an envelope function $E(t)$.

Some other issues that affect the quality of your synthesis include relative timing of the notes, correct durations for tempo, rests (pauses) in the appropriate places, relative amplitudes to emphasize certain notes and make others soft, and harmonics. True piano sounds contain several frequency components, such as second and third harmonics. Since we have been studying harmonics, this modification would be simple, but be careful to make the amplitudes of the harmonics smaller than the fundamental frequency component. Experiment to hear what sounds best.

C.3.3.4   Programming Tips   You may want to modify your note function to accept additional parameters describing amplitude, duration, etc. You will also want to change it to add an envelope or harmonics. Chords are created on a computer by simply adding the signal vectors of several notes.

For testing, we have provided a MATLAB script that initializes vectors containing the note values and durations for *Für Elise*. This will save you from typing it all in yourself, but you are free to modify the duration values or anything else. This script called `fenotes.m` contains both the bass and treble, and it is available on the *DSP First* CD-ROM.

C.3.3.5   Alternative Piece: *Jesu, Joy of Man's Desiring*   Follow the project description given in Section C.3.3.2, but use the piece, *Jesu, Joy of Man's Desiring* written by Bach. The first few measures are shown in Fig. C.6, and you can listen to the part that you will synthesize by following the links on the *DSP First* CD-ROM. More of the song can be found on the CD-ROM, where an entire page of the music is reproduced.

Figure C.6   The first few measures of *Jesu, Joy of Man's Desiring*.

C.3.3.6   Alternative Piece: *Minuet in G*   Follow the project description given in Section C.3.3.2, but use the piece *Minuet in G* written by Bach. The first few measures are shown in Fig. C.7, and you can listen to the part that you will synthesize by following the links on the *DSP First* CD-ROM. More of the song can be found on the CD-ROM, where an entire page of the music is reproduced.

Figure C.7   The first few measures of *Minuet in G*.

C.3.3.7   Alternative Piece: *Beethoven's Fifth Symphony*   Follow the project description given in Section C.3.3.2, but use the theme from *Beethoven's Fifth Symphony*. The first few measures are shown in Fig. C.8, and you can listen to the part that you will synthesize by following the links on the *DSP First* CD-ROM. More of the song can be found on the CD-ROM, where an entire page of the music is reproduced.

Figure C.8   The first few measures of the theme from *Beethoven's Fifth*.

C.3.3.8   Alternative Piece: *Twinkle, Twinkle, Little Star*   Follow the project description given in Section C.3.3.2, but use the piece *Twinkle, Twinkle, Little Star* written by Mozart. The first few measures are shown in Fig. C.9, and you can listen to the part that you will synthesize by following the links on the *DSP First* CD-ROM. More of the song can be found on the CD-ROM, where an entire page of the music is reproduced.

Twinkle, Twinkle, Little Star

Figure C.9   The first few measures of *Twinkle, Twinkle, Little Star*.

C.3.4 Sound Evaluation Criteria

Here are some guidelines for evaluating the music synthesis projects
     Does the file play notes?   All Notes____   Most____   Treble only____

Overall Impression:

Excellent: Enjoyable sound, good use of extra features such as harmonics, envelopes, etc.

Good: Bass and treble clefs synthesized and in sync, few errors, one or two special features.

Average: Basic sinusoidal synthesis, including the bass, with only a few errors.

Poor: No bass notes, or treble and bass not synchronized, many wrong notes.

C.4   LABORATORY: AM AND FM SINUSOIDAL SIGNALS

The objective of this lab is to introduce more complicated signals that are related to the basic sinusoid. These signals that implement frequency modulation (FM) and amplitude modulation (AM) are widely used in communication systems such as radio and television, but they also can be used to create interesting sounds that mimic musical instruments. A number of demonstrations on the CD-ROM provide examples of these signals for many different conditions.

C.4.1 Overview

The properties of sinusoidal waveforms of the form

$$x(t) = A\cos(2\pi f_0 t + \phi) = \Re e\left\{Ae^{j\phi}e^{j2\pi f_0 t}\right\} \qquad \text{(C.4.1)}$$

are considered in detail in Chapters 2 and 3, and in Lab C.3. In this lab, we will continue to investigate sinusoidal waveforms, but for more complicated signals composed of sums of sinusoidal signals, or sinusoids with changing frequency.

C.4.1.1   Amplitude Modulation   If we add several sinusoids, each with a different frequency (f_k), we can express the result as:

$$x(t) = \sum_{k=1}^{N} A_k \cos(2\pi f_k t + \phi_k) = \Re e\left\{\sum_{k=1}^{N} X_k\, e^{j2\pi f_k t}\right\} \qquad \text{(C.4.2)}$$

where $X_k = A_k e^{j\phi_k}$ is the complex exponential amplitude. The choice of f_k will determine the nature of the signal. For amplitude modulation we pick two or three frequencies very close together; see Chapter 3.

C.4.1.2   Frequency Modulated Signals   We will also consider signals for which the frequency varies as a function of time. In the constant-frequency sinusoid (C.4.1), the argument of the cosine is also the exponent of the complex exponential, so the phase of this signal is the exponent $(2\pi f_0 t + \phi)$. This phase function changes *linearly* versus time, and its time derivative is $2\pi f_0$, which equals the constant frequency of the cosine in rad/sec.

FM Synthesis

A generalization is possible if we adopt the following notation for the class of signals with time-varying phase:

$$x(t) = A\cos(\psi(t)) = \Re e\left\{Ae^{j\psi(t)}\right\} \qquad \text{(C.4.3)}$$

The time derivative of the phase from (C.4.3) gives a radian frequency

$$\omega_i(t) = \frac{d}{dt}\psi(t) \qquad \text{(rad/sec)}$$

but we prefer units of hertz, so we divide by 2π to define the *instantaneous frequency*

$$f_i(t) = \frac{1}{2\pi} \frac{d}{dt} \psi(t) \quad \text{(Hz)} \quad\quad \text{(C.4.4)}$$

Spectrograms
& Sounds:
Wideband FM

C.4.1.3   Chirp, or Linearly Swept Frequency   A *chirp* signal is a sinusoid whose frequency changes linearly from some low value to a high one. The formula for such a signal may be defined by creating a complex exponential signal with quadratic phase by defining $\psi(t)$ in (C.4.3) as

$$\psi(t) = 2\pi \mu t^2 + 2\pi f_0 t + \phi$$

The derivative of $\psi(t)$ yields an instantaneous frequency (C.4.4) that changes *linearly* versus time.

$$f_i(t) = 2\mu t + f_0$$

(Note that if $\mu = 0$, we have the case of the constant-frequency sinusoid.) The slope of $f_i(t)$ is equal to 2μ, and its intercept is equal to f_0. If the signal starts at $t = 0$, then f_0 is also the starting frequency. The frequency variation produced by the time-varying phase is called *frequency modulation,* and signals of this class are called FM signals. Finally, since the linear variation of the frequency can produce an audible sound similar to a siren or a bird chirp, the linear FM signals are also called "chirps."

C.4.1.4   Advanced Topic: Spectrograms   It is often useful to think of signals in terms of their spectra. A signal's spectrum is a representation of the frequencies present in the signal. For a constant-frequency sinusoid as in (C.4.1), the spectrum consists of two complex exponential components: one at f_0, the other at $-f_0$. For more complicated signals, the spectra may be more complicated, and, in the case of FM, the spectrum is considered to be time-varying. One way to represent the time-varying spectrum of a signal is the *spectrogram* (see Section 3.5 in Chapter 3). A spectrogram is found by estimating the frequency content in short sections of the signal. The magnitude of the spectrum over individual sections is plotted as intensity or color on a two-dimensional plot versus frequency and time.

Sounds &
Spectrograms

There are a few important things to know about spectrograms:

1. In MATLAB, the function `specgram` will compute the spectrogram, as already explained in Lab C.3. Type `help specgram` to learn more about this function and its arguments.[8]

2. Spectrograms are calculated numerically and provide only an estimate of the time-varying frequency content of a signal. There are theoretical limits on how

---

[8] The *DSP First Toolbox* contains an equivalent function called `spectgr` that can be used in place of `specgram`.

well they can actually represent the frequency content of a signal. Lab C.11 will treat this problem when we use the spectrogram to extract the frequencies of piano notes.

C.4.2 Warm-up

The instructor verification sheet can be found on the CD-ROM and printed out as needed.

C.4.2.1 MATLAB Synthesis of Chirp Signals

1. The following MATLAB code will synthesize a chirp:

```
fsamp = 8000;
dt = 1/fsamp;
dur = 1.8;
tt = 0 : dt : dur;
psi = 2*pi*(100 + 200*tt + 500*tt.*tt);
xx = real( 7.7*exp(j*psi) );
sound( xx, fsamp );
```

Determine the range of frequencies (in Hz) that will be synthesized by this MATLAB script. Make a sketch by hand of the instantaneous frequency versus time. What are the minimum and maximum frequencies that will be heard? Listen to the signal to verify that it has the expected frequency content.

Instructor Verification (separate page)

2. Use the code provided in item 1 to help you write a MATLAB function that will synthesize a "chirp" signal according to the following comments:

```
function   xx = mychirp( f1, f2, dur, fsamp )
%MYCHIRP        generate a linear-FM chirp signal
%
%  usage:   xx = mychirp( f1, f2, dur, fsamp )
%
%          f1 = starting frequency
%          f2 = ending frequency
%         dur = total time duration
%       fsamp = sampling frequency   (OPTIONAL: default is 8000)
%
if( nargin < 4 )   %<-- Allow optional input argument
   fsamp = 8000;   %<-- Default sampling rate
end
```

Reminder: *When unsure about a command, use* `help`.

Generate a chirp sound to match the frequency range of the chirp in item 1. Listen to the chirp using the sound function. Also, compute the spectrogram of your chirp using the MATLAB function: specgram(xx,[],fsamp).

Instructor Verification (separate page)

C.4.3 Lab A: Chirps and Beats

C.4.3.1 Synthesize a Chirp   Use your MATLAB function mychirp to synthesize a "chirp" signal for your lab report. Use the following parameters:

1. A total time duration of 3 sec with a D-to-A conversion rate of $f_s = 8000$ Hz.
2. The instantaneous frequency starts at 15,000 Hz and ends at 300 Hz.

Listen to the signal. What comments can you make regarding the sound of the chirp (e.g., is it linear)? Does it chirp down, or chirp up, or both? Create a spectrogram of your chirp signal. Use the sampling theorem (from Chapter 4) to help explain what you hear and see.

C.4.3.2 Beat Notes   In Section 3.2 of Chapter 3, we analyzed the situation in which we had the sum of two sinusoidal signals of slightly different frequencies, i.e.,

$$x(t) = A \cos(2\pi(f_c - f_\Delta)t) + B \cos(2\pi(f_c + f_\Delta)t) \tag{C.4.5}$$

In this part, we will compute samples of such a signal and listen to the result.

1. Write an M-file called beat.m that implements (C.4.5) and has the following as its first lines:

```
function        [xx, tt] = beat(A, B, fc, delf, fsamp, dur)
%BEAT    compute samples of the sum of two cosine waves
%  usage:
%        [xx, tt] = beat(A, B, fc, delf, fsamp, dur)
%
%              A = amplitude of lower frequency cosine
%              B = amplitude of higher frequency cosine
%             fc = center frequency
%           delf = frequency difference
%          fsamp = sampling rate
%            dur = total time duration in seconds
%             xx = output vector of samples
%--OPTIONAL Output:
%             tt = time vector corresponding to xx
```

As a part of your report, include a copy of your M-files. You may want to call the sumcos function written in Lab C.2 to do the calculation. The function may also generate its own time vector. You may elect to not implement the second output vector tt, but it is quite convenient for plotting. To assist you in your experiments with beat notes, a new tool called beatcon has been created. This *user interface controller* actually calls your function beat.m. Therefore, before you invoke beatcon you should be sure that your M-file is free of errors. Once you have the function beat.m working properly, invoke the test tool by typing beatcon at the MATLAB prompt. A small control panel will appear on the screen with *buttons* and *sliders* that vary the different parameters for these exercises. Experiment with the beatcon control panel and use it to complete the remaining exercises in this section.

beatcon.m

2. Test the M-file written in item 1 with beatcon by using the values A=10, B=10, fc=1000, delf=10, fsamp=8000, and dur=1 sec. Plot the first 0.2 seconds of the resulting signal. Describe the waveform and explain its properties. Hand in a copy of your plot with measurements of the period of the "envelope" and period of the high-frequency signal underneath the envelope.

3. For this part, set delf to 10 Hz. Send the resulting signal to the D-to-A converter and listen to the sound (there is a *button* on beatcon that will do this for you automatically). Explain the nature of the sound based on the waveform plotted in item 2 and on the theory developed in Chapter 3.

4. Experiment with different values of the frequency difference f_Δ.

C.4.3.3   More on Spectrograms (Optional)

Beat notes provide an interesting way to investigate the time-frequency characteristics of spectrograms. Although some of the mathematical details are beyond the reach of this course, it is not difficult to understand that there is a fundamental trade-off between knowing which frequencies are present in a signal (or its spectrum) and knowing how those frequencies vary with time. As mentioned previously in Section C.4.1.4, a spectrogram estimates the frequency content over short sections of the signal. Long sections give excellent frequency resolution, but track sudden frequency changes poorly. Shorter sections have poor frequency resolution, but good tracking. This trade-off between the section length (in time) and frequency resolution is equivalent to Heisenburg's uncertainty principle in physics. More discussion of the spectrogram can be found in Chapter 9 and Lab C.11.

A beat-note signal may be viewed as a single frequency signal whose amplitude varies with time, *or* as the sum of two signals with different constant frequencies. Both views will be useful in evaluating the effect of window length when finding the spectrogram of a beat signal.

1. Create and plot a beat signal with

   (a) $f_\Delta = 32$ Hz

   (b) $T_{dur} = 0.26$ sec

(c) $f_s = 8000$ Hz, or $11{,}025$ Hz

(d) $f_0 = 2000$ Hz

2. Find the spectrogram using a window length of 2048 using the commands: `specgram(x,2048,fsamp); colormap(1-gray(256))`. Comment on what you see.

3. Find the spectrogram using a window length of 16 using the commands: `specgram(x,16,fsamp); colormap(1-gray(256))`. Comment on what you see.

C.4.4 Lab B: FM Synthesis of Instrument Sounds

Frequency modulation (FM) can be used to make interesting sounds that mimic musical instruments, such as bells, woodwinds, drums, etc. The goal in this lab is to implement one or two of these FM schemes and hear the results.

We have already seen that FM signals are of the form

$$x(t) = A \cos(\psi(t))$$

and that the instantaneous frequency (C.4.4) changes according to the oscillations of $\psi(t)$. If $\psi(t)$ is linear, $x(t)$ is a constant-frequency sinusoid whereas if $\psi(t)$ is quadratic, $x(t)$ is a chirp signal whose frequency changes linearly in time. FM music synthesis uses a more interesting $\psi(t)$, one that is sinusoidal. Since the derivative of a sinusoidal $\psi(t)$ is also sinusoidal, the instantaneous frequency of $x(t)$ will oscillate sinusoidally. This is useful for synthesizing instrument sounds because the proper choice of the modulating frequencies will produce a fundamental frequency and several overtones, as many instruments do.

The general equation for an FM sound synthesizer is:

$$x(t) = A(t) \cos\left(2\pi f_c t + I(t) \cos(2\pi f_m t + \phi_m) + \phi_c\right) \tag{C.4.6}$$

The signal's amplitude, $A(t)$, is a function of time so that the instrument sound can be made to fade out slowly or cut off quickly. Such a function is called an *envelope*. The parameter f_c is called the *carrier* frequency, which is a constant in the following expression for the instantaneous frequency:

$$
\begin{aligned}
f_i(t) &= \frac{1}{2\pi} \frac{d}{dt} \psi(t) \\
&= \frac{1}{2\pi} \frac{d}{dt} \left(2\pi f_c t + I(t) \cos(2\pi f_m t + \phi_m) + \phi_c\right) \\
&= f_c - I(t) f_m \sin(2\pi f_m t + \phi_m) + \frac{1}{2\pi} \frac{dI}{dt} \cos(2\pi f_m t + \phi_m) \tag{C.4.7}
\end{aligned}
$$

The constant f_c is the frequency that would be produced without any frequency modulation. The parameter f_m is called the *modulating* frequency. It expresses the rate of oscillation of $f_i(t)$. The parameters ϕ_m and ϕ_c are arbitrary phase constants, usually both set to $-\pi/2$ so that $x(0) = 0$.

The function $I(t)$ has a less obvious purpose than the other FM parameters in (C.4.6). It is technically called the *modulation index envelope*. To see what it does, examine the expression for the instantaneous frequency (C.4.7). The quantity $I(t) f_m$ multiplies a sinusoidal variation of the frequency. If $I(t)$ is constant or $\frac{dI}{dt}$ is relatively small, then $I(t) f_m$ gives the maximum amount by which the instantaneous frequency deviates from f_c. Beyond that, however, it is difficult to relate $I(t)$ to the sound made by $x(t)$ without some rather tedious mathematical analysis.

In our study of signals, we would like to characterize $x(t)$ as the sum of several constant-frequency sinusoids instead of a single signal whose frequency changes. In this regard, the following comments are relevant: When $I(t)$ is small (e.g., $I \approx 1$), low multiples of the carrier frequency (f_c) have high amplitudes. When $I(t)$ is large ($I > 4$), both low and high multiples of the carrier frequency have high amplitudes. The net result is that $I(t)$ can be used to vary the harmonic content of the instrument sound (called overtones). When $I(t)$ is small, mainly low frequencies will be produced. When $I(t)$ is large, higher harmonic frequencies can also be produced. Since $I(t)$ is a function of time, the harmonic content will change with time. For more details, see the paper by Chowning.[9]

C.4.4.1 Generating the Bell Envelopes
Now we take the general FM synthesis formula (C.4.6) and specialize it for the case of a bell sound. The amplitude envelope $A(t)$ and the modulation index envelope $I(t)$ for the bell are both decaying exponentials. That is, both have the following form:

$$y(t) = e^{-t/\tau} \tag{C.4.8}$$

where τ is a parameter that controls the decay rate of the exponential. Notice that $y(0) = 1$ and $y(\tau) = 1/e$, so τ is the time it takes a signal of the form (C.4.8) to decay to $1/e = 36.8\%$ of its initial value. For this reason, the parameter τ is called the *time constant*.

Use (C.4.8) to write a MATLAB function that will generate a decaying exponential to be used later in synthesizing a bell sound. The file header should look like this:

---

[9] John M. Chowning, "The Synthesis of Complex Audio Spectra by Means of Frequency Modulation," *Journal of the Audio Engineering Society*, vol. 21, no. 7, Sept. 1973, pp. 526–534.

```
function  yy = bellenv( tau, dur, fsamp )
%BELLENV produces envelope function for bell  sounds
%
%        usage: yy = bellenv(tau, dur, fsamp);
%
%              where tau = time constant
%                    dur = duration of the envelope
%                  fsamp = sampling frequency
%              returns:
%                    yy = decaying exponential envelope
%
%   note: produces exponential decay for positive tau
```

The function will be one or two lines of MATLAB code. The first line should define your time vector based on fsamp and dur, and the second generates the exponential (C.4.8).

The bell's amplitude envelope, $A(t)$, and modulation index envelope, $I(t)$, are identical, up to a scale factor.

$$A(t) = A_0 e^{-t/\tau} \qquad \text{and} \qquad I(t) = I_0 e^{-t/\tau}$$

Hence, the bellenv function can generate either envelope.

C.4.4.2 Parameters for the Bell   Now that we have the bell's amplitude and modulation index envelopes, we can create the actual sound signal for the bell by specifying all the parameters in the general FM synthesis formula (C.4.6). The frequencies f_c and f_m must be given numerical values. The ratio of carrier to modulating frequency is important in creating the sound of a specific instrument. For the bell, a good choice for this ratio is 1:2, e.g., $f_c = 110$ Hz and $f_m = 220$ Hz.

Now write a simple M-file bell.m that implements (C.4.6) to synthesize a bell sound. Your function should call bellenv.m to generate $A(t) = e^{-t/\tau}$ (assume $A_0 = 1$) and $I(t) = I_0 e^{-t/\tau}$.

```
function  xx = bell(ff, Io, tau, dur, fsamp)
%BELL      produce a bell sound
%
%     usage:   xx = bell(ff, Io, tau, dur, fsamp)
%
%     where:  ff = frequency vector (containing fc and fm)
%             Io = scale factor for modulation index
%            tau = decay parameter for A(t) and I(t)
%            dur = duration (in sec.) of the output signal
%          fsamp = sampling rate
```

C.4.4.3   The Bell Sound   Test your bell( ) function using the parameters of case 1 in the table. Play it with the sound( ) function at 11,025 Hz.[10] Does it sound like a bell? The value of $I_0 = 10$ for scaling the modulation index envelope is known to give a distinctive sound. Later on, you can experiment with other values to get a variety of bells.

CASE	f_c (Hz)	f_m (Hz)	I_0	τ (sec)	T_{dur} (sec)	f_s (Hz)
1	110	220	10	2	6	11,025
2	220	440	5	2	6	11,025
3	110	220	10	12	3	11,025
4	110	220	10	0.3	3	11,025
5	250	350	5	2	5	11,025
6	250	350	3	1	5	11,025

The frequency spectrum of the bell sound is complicated, but it does consist of spectral lines, which can be seen with a spectrogram. Among these frequencies, one spectral line will dominate what we hear. We can call this the *note frequency* of the bell. It is tempting to guess that the note frequency will be equal to f_c, but you will have to experiment to find the true answer. It might be f_m, or it might be something else—perhaps the fundamental frequency that is the greatest common divisor of f_c and f_m.

For each case in the table, do the following:

1. Listen to the sound by playing it with the sound( ) function.

2. Calculate the fundamental frequency of the "note" being played. Explain how you can verify, by listening, that you have the correct fundamental frequency.

3. Describe how you can hear the frequency content changing according to $I(t)$. Plot $f_i(t)$ versus t for comparison.

4. Display a spectrogram of the signal. Describe how the frequency content changes, and how that change is related to $I(t)$. Point out the "harmonic" structure of the spectrogram, and calculate the fundamental frequency, f_0.

5. Plot the entire signal and compare it to the envelope $A(t)$ generated by bellenv.

6. Plot about 100 to 200 samples from the middle of the signal and explain what you see, especially the frequency variation.

---

[10] A higher sampling rate of 11,025 Hz is used because the signal contains many harmonics, some of which might alias if a lower f_s were used. Experiment with lower values of f_s to see whether you can hear a difference, e.g., $f_s = 8000$ Hz.

If you are making a lab report, do the plots for two cases; choose one of the first four cases in the table and one of the last two. Write up an explanation for only the two that you choose.

C.4.4.4   Comments about the Bell
Cases 3 and 4 are extremes for choosing the decay rate τ. In case 3, the waveform does not decay very much over the course of 3 seconds and sounds a little like a sum of harmonically related sinusoids. With a "faster" decay rate, as in case 4, we get a percussion-like sound. Modifying the fundamental frequency f_0 (determined in item 2) should have a noticeable effect on the tone you hear. Try some different values for f_0 by changing f_c and f_m, but still in the ratio of 1:2. Describe what you hear.

Finally, experiment with different carrier-to-modulation frequency ratios. For example, in his paper, Chowning uses a fundamental frequency of $f_0 = 40$ Hz and a carrier-to-modulation frequency ratio of 5:7. Try this and a few other values. Which parameters sound best to you?

C.4.5 Woodwinds

As an alternative to the bell sounds, this section shows how different parameters in the same FM synthesis formula (C.4.6) will yield a clarinet sound, or other woodwinds.

C.4.5.1   Generating the Envelopes for Woodwinds
There is a function on the CD-ROM called woodwenv that produces the functions needed to create both the $A(t)$ and $I(t)$ envelopes for a clarinet sound. The file header looks like this:

woodwenv.m

```
function    [y1, y2] = woodwenv(att, sus, rel, fsamp)
%WOODWENV        produce normalized amplitude and modulation index
%                functions for woodwinds
%
%  usage: [y1, y2] = woodwenv(att, sus, rel, fsamp);
%
%     where  att = attack TIME
%            sus = sustain TIME
%            rel = release TIME
%          fsamp = sampling frequency (Hz)
%     returns:
%            y1 = (NORMALIZED) amplitude envelope (A0 = 1)
%            y2 = (NORMALIZED) modulation index envelope (I0 = 1)
%
%  NOTE: attack is exponential, sustain is constant,
%        release is exponential
```

The outputs from woodwenv are normalized so that the maximum value is 1 and the minimum is 0. Try the following statements to see what the function produces:

```
fsamp = 8000;
Ts = 1/fsamp;
tt = delta : Ts : 0.5;
[y1, y2] = woodwenv(0.1, 0.35, 0.05, fsamp);
subplot(2,1,1), plot(tt,y1), grid on
subplot(2,1,2), plot(tt,y2), grid on
```

C.4.5.2   Scaling the Clarinet Envelopes   Since the woodwind envelopes produced by woodwenv range from 0 to 1, some scaling is necessary to make them useful in the FM synthesis equation (C.4.6). In this section, we consider the general process of linear rescaling. If we start with a *normalized* signal $y_{norm}(t)$ and want to produce a new signal whose max is y_{max} and whose min is y_{min}, then we must map 1 to y_{max} and 0 to y_{min}. Consider the linear mapping

$$y_{new}(t) = \alpha\, y_{norm}(t) + \beta \qquad\qquad (C.4.9)$$

Determine the relationship between α and β and y_{max} and y_{min}, so that the max and the min of $y_{new}(t)$ are correct.

Test this idea in MATLAB by doing the following example (where $\alpha = 5$ and $\beta = 3$):

```
ynorm = 0.5 + 0.5*sin( pi*[0:0.01:1]);
subplot(2,1,1),  plot(ynorm)
alpha = 5;    beta = 3;
ynew = alpha*ynorm + beta;              %<------ Linear re-scaling
subplot(2,1,1),  plot(ynew)
max(ynorm),  min(ynorm)        %<--- ECHO the values
max(ynew),  min(ynew)
```

What happens if we make α negative?

Write a short one-line function that implements (C.4.9). Your function should have the following form: function y = scale(data, alpha, beta).

C.4.5.3   Clarinet Envelopes   For the clarinet sound, the amplitude $A(t)$ needs no scaling; the MATLAB function sound will automatically scale to the maximum range of the D-to-A converter. Thus, $A(t)$ equals the vector y1. From the plot of y1 shown in Fig. C.10, it should be obvious that this envelope will cause the sound to rise quickly to a certain volume, sustain that volume, and then quickly turn off.

The modulation index envelope, $I(t)$, however, does not equal y2. The range for $I(t)$ lies between 2 and 4, as in Fig. C.10. Furthermore, there is an inversion so that when y2 is 0, $I(t)$ should equal 4, and when y2 is 1, $I(t)$ should be 2. Using this information, solve for the appropriate α and β, then use scale to produce the modulation index envelope function (I) for a clarinet sound.

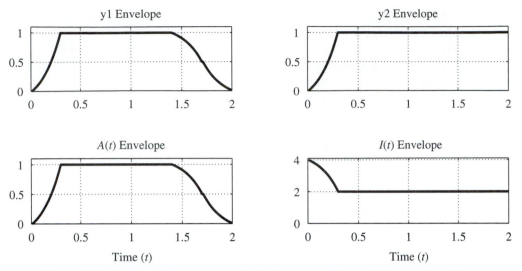

Figure C.10   Envelopes for the woodwinds. The functions $A(t)$ and $I(t)$ are produced by scaling y1 and y2, the outputs of woodwenv.

C.4.5.4   Parameters for the Clarinet

So far we have a general equation for FM signals, an amplitude envelope for the clarinet, and a modulation index envelope for the clarinet. To create the actual sound signal for the clarinet, we need to specify the additional parameters in (C.4.6). The ratio of carrier to modulating frequency is important in creating the sound of a specific instrument. For the clarinet, this ratio should be 2:3. The actual note frequency will be the greatest common divisor of the carrier and modulating frequencies. For example, when we choose $f_c = 600$ Hz and $f_m = 900$ Hz, the synthesized signal will have a fundamental frequency of $f_0 = 300$ Hz.

Write a simple M-file clarinet.m that implements the FM synthesis equation (C.4.6) to synthesize a clarinet note. Your function should generate the envelopes $A(t)$ and $I(t)$ using scale and woodwenv and then compute the waveform of the clarinet signal. The function header should look like this:

```
function  yy = clarinet(f0, Aenv, Ienv, dur, fsamp)
%CLARINET      produce a clarinet note signal
%
%       usage:  yy = clarinet(f0, Aenv, Ienv, dur, fsamp)
%
%       where:   f0 = note frequency
%              Aenv = the array holding the A(t) envelope
%              Ienv = the array holding the I(t) envelope
%               dur = the amount of time the signal lasts
%             fsamp = the sampling rate
```

C.4.5.5   Experiment with the Clarinet Sound   Using your `clarinet( )` function, create a 250-Hz clarinet note with $f_s = 8000$ or 11,025 Hz. Play it with the sound (`xnote, fsamp`) function. Does it sound like a clarinet? How can you verify that its fundamental frequency is at 250 Hz?

Explain how the modulation index $I(t)$ will affect the frequency content versus time of the clarinet sound. Describe how you can hear the frequency content changing according to $I(t)$. Plot the instantaneous frequency $f_i(t)$ versus t for comparison.

Plot the entire signal and compare it to the amplitude envelope function `y1` in Fig. C.10. Plot about 100 to 200 samples from the middle of the signal and explain what you see.

Finally, synthesize other note frequencies. For example, make the C-major scale (defined in Lab C.3) consisting of seven consecutive notes beginning with middle-C.

C.5 LABORATORY: FIR FILTERING OF SINUSOIDAL WAVEFORMS

The goals of this lab are to learn how to implement FIR filters in MATLAB, and to study the response of FIR filters to inputs such as complex exponentials. In addition, we will use FIR filters to study properties such as linearity and time invariance.

C.5.1 Overview of Filtering

For this lab, we will define an FIR filter as a discrete-time system that converts an input signal $x[n]$ into an output signal $y[n]$ by means of the weighted summation

$$y[n] = \sum_{k=0}^{M} b_k\, x[n-k] \tag{C.5.1}$$

Equation (C.5.1) gives a rule for computing the nth value of the output sequence from certain values of the input sequence. The filter coefficients $\{b_k\}$ are constants that define the filter's behavior. As an example, consider the system for which the output values are given by

$$y[n] = \tfrac{1}{3}x[n] + \tfrac{1}{3}x[n-1] + \tfrac{1}{3}x[n-2] \tag{C.5.2}$$

$$= \tfrac{1}{3}\{x[n] + x[n-1] + x[n-2]\}$$

This equation states that the nth value of the output sequence is the average of the nth value of the input sequence $x[n]$ and the two preceding values, $x[n-1]$ and $x[n-2]$. For this example, the b_ks are $b_0 = \tfrac{1}{3}$, $b_1 = \tfrac{1}{3}$, and $b_2 = \tfrac{1}{3}$.

firfilt.m

MATLAB has a built-in function for implementing the operation in (C.5.1), namely, the function `filter( )`, but we have also supplied another M-file called `firfilt( )` for the special case of FIR filtering. The `filter` function implements a wider class of filters than just the FIR case. Technically speaking, the `firfilt` function implements an operation called *convolution*, although we will not be concerned with the meaning of that terminology right now. The following MATLAB statements implement the 3-point averaging system of (C.5.2):

```
nn = 0:99;               %<--Time indices
xx = cos( 0.08*pi*nn );  %<--Input signal
bb = [1/3 1/3 1/3];      %<--Filter coefficients
yy = firfilt(bb, xx);    %<--Compute the output
```

In this case, the input signal xx is a vector containing samples of a cosine function. In general, the vector bb contains the filter coefficients $\{b_k\}$ needed in (C.5.1). These are loaded into the bb vector as follows:

```
bb = [b0, b1, b2, ... , bM].
```

In MATLAB, all sequences have finite length because they are stored in vectors. If the input signal has, for example, L samples, we would normally store only the L nonzero samples, and would assume that $x[n] = 0$ for n outside the interval of L samples; i.e., we do not have to store the zero samples unless it suits our purposes. If we process a finite-length signal through (C.5.1), then the output sequence $y[n]$ will be longer than $x[n]$ by M samples. Whenever firfilt( ) implements (C.5.1), we will find that

```
length(yy) = length(xx)+length(bb)-1
```

In the experiments of this lab, you will use firfilt( ) to implement FIR filters and begin to understand how the filter coefficients define a digital filtering algorithm. In addition, this lab will introduce examples to show how a filter reacts to different frequency components in the input.

C.5.1.1   Frequency Response of FIR Filters   The output or *response* of a filter for a complex sinusoid input, $e^{j\hat{\omega}n}$, depends on the frequency, $\hat{\omega}$. Often a filter is described solely by how it affects different frequencies. This is called the *frequency response*.

For example, the frequency response of the 2-point averaging filter

$$y[n] = \tfrac{1}{2}x[n] + \tfrac{1}{2}x[n-1]$$

can be found by using a general complex exponential as an input and observing the output or response.

$$x[n] = Ae^{j\hat{\omega}n+\phi} \tag{C.5.3}$$

$$y[n] = \tfrac{1}{2}Ae^{(j\hat{\omega}n+\phi)} + \tfrac{1}{2}Ae^{(j\hat{\omega}(n-1)+\phi)} \tag{C.5.4}$$

$$= Ae^{(j\hat{\omega}n+\phi)}\tfrac{1}{2}\left\{1 + e^{-j\hat{\omega}}\right\} \tag{C.5.5}$$

In (C.5.5), there are two terms, the original input, and a term that is a function of $\hat{\omega}$. This second term is the frequency response and it is commonly denoted by $H(e^{j\hat{\omega}})$.[11]

$$H(e^{j\hat{\omega}}) = \tfrac{1}{2}\left\{1 + e^{-j\hat{\omega}}\right\} \tag{C.5.6}$$

Once the frequency response, $H(e^{j\hat{\omega}})$, has been determined as a function of $\hat{\omega}$, the effect of the filter on any complex exponential may be determined by evaluating $H(e^{j\hat{\omega}})$ at the corresponding frequency. The result will be a complex number whose phase describes the phase shift of the complex sinusoid and whose magnitude describes the gain applied to the complex sinusoid.

---

[11] The notation $H(e^{j\hat{\omega}})$ is used in place of $\mathcal{H}(\hat{\omega})$ for the frequency response because we will eventually connect this notation with the z-transform.

The frequency response of a general FIR linear time-invariant system is

$$H(e^{j\hat{\omega}}) = \sum_{k=0}^{M} b_k e^{-j\hat{\omega}k} \tag{C.5.7}$$

MATLAB has a built-in function for computing the frequency response of a discrete-time LTI system. It is called freqz( ). The following MATLAB statements show how to use freqz to compute and plot the magnitude (absolute value) of the frequency response of a 2-point averaging system as a function of $\hat{\omega}$ in the range $-\pi \leq \hat{\omega} \leq \pi$:

```
bb = [1, -1];    %<-- Filter Coefficients
ww = -pi:(pi/100):pi;
H = freqz(bb, 1, ww);
plot(ww, abs(H))
```

We will always use capital H for the frequency response. For FIR filters of the form of (C.5.1), the second argument of freqz must always be set equal to 1.

C.5.2 Warm-up

The instructor verification sheet can be found on the CD-ROM and printed out as needed.

C.5.2.1 Frequency Response of the 3-Point Averager
In Chapter 6, we examined filters that average input samples over a certain interval. These filters are called "running-average" filters, or "averagers," and they have the following form:

$$y[n] = \frac{1}{L+1} \sum_{k=0}^{L-1} x[n-k] \tag{C.5.8}$$

1. Show that the frequency response for the three-point running average operator is given by:

$$H(e^{j\hat{\omega}}) = \frac{2\cos\hat{\omega} + 1}{3} e^{-j\hat{\omega}} \tag{C.5.9}$$

2. Compute (C.5.9) directly in MATLAB. Use a vector that includes 400 samples between $-\pi$ and π for $\hat{\omega}$. Since the frequency response is a complex-valued

quantity, use abs( ) and angle( ) to extract the magnitude and phase of the frequency response for plotting. Plotting the real and imaginary parts of $H(e^{j\hat{\omega}})$ is not very informative.

3. The following MATLAB statements will compute $H(e^{j\hat{\omega}})$ numerically and plot its magnitude and phase versus $\hat{\omega}$.

```
bb = 1/3*ones(1,3);
ww = -pi:(pi/200):pi;
H = freqz( bb, 1, ww );
subplot(2,1,1)
plot( ww, abs(H) )   %<--- Magnitude
subplot(2,1,2)
plot( ww, angle(H) )   %<--- Phase
xlabel('NORMALIZED FREQUENCY')
```

The function freqz evaluates the frequency response for all frequencies in the vector ww. It uses the summation in (C.5.7), not the formula in (C.5.9). The filter coefficients are defined in the assignment to vector bb. How do your results compare with item 2?

Instructor Verification (separate page)

C.5.3 Lab: FIR Filters

In the following sections, we will study how a filter affects sinusoidal inputs, and begin to understand the performance of the filter as a function of the input frequency. You will see that:

1. Filters of the form of (C.5.1) can modify the amplitude and phase of a cosine wave, but they do not modify the frequency.

2. For a sum of cosine waves, the system modifies each component independently.

3. Filters can completely remove one or more components of a sum of cosine waves.

C.5.3.1 Filtering Cosine Waves   We will be interested in filtering discrete-time sinusoids of the form

$$x[n] = A\cos(\hat{\omega}n + \phi) \qquad \text{for } n = 0, 1, 2, \ldots, L-1 \qquad (C.5.10)$$

The *discrete-time frequency* for a discrete-time cosine wave, $\hat{\omega}$, always satisfies $0 \leq \hat{\omega} \leq \pi$. If the discrete-time sinusoid is produced by sampling a continuous-time cosine, the discrete-time frequency is $\hat{\omega} = \omega T_s = 2\pi f/f_s$, as discussed in Chapter 4 on *sampling*.

C.5.3.2 First-Difference Filter   Generate 50 samples of a discrete-time cosine wave with $A = 7$, $\phi = \pi/3$, and $\hat{\omega} = 0.125\pi$. Store this signal in the vector xx so it can also be used in succeeding parts. Now use `firfilt( )` to implement the following filter with the signal xx as input:

$$y[n] = 5x[n] - 5x[n-1] \tag{C.5.11}$$

This is called a *first-difference* filter, but with a gain of 5. In MATLAB you must define the vector bb needed in `firfilt`.

1. Note that $y[n]$ and $x[n]$ are not the same length. What is the length of the filtered signal, and why is it that length? (If you need a hint, refer to Section C.5.1.)

2. Plot the first 50 samples of both waveforms $x[n]$ and $y[n]$ on the same figure, using `subplot`. Use the `stem` function to make a discrete-time signal plot, but label the x-axis to span the range $0 \le n \le 49$.

3. Verify the amplitude and phase of $x[n]$ directly from its plot in the time domain.

4. From the plot, observe that, with the exception of the first sample $y[0]$, the sequence $y[n]$ seems to be a scaled and shifted cosine wave of the *same* frequency as the input. Explain why the first sample is different from the others.

5. Determine the frequency, amplitude, and phase of $y[n]$ directly from the plot. Ignore the first output point, $y[0]$.

6. Characterize the filter performance at the input frequency by computing the relative amplitude and phase, i.e., the ratio of output amplitude to input amplitude and the difference of output and input phases.

7. In order to compare your measured results to the theory developed in Chapter 6 for this system, derive the mathematical expression for the output when the input signal is a complex exponential $x[n] = e^{j\hat{\omega}n}$. From this formula, determine how much the amplitude and phase should change for $x[n]$, which has a frequency of $\hat{\omega} = 0.125\pi$.

C.5.3.3   Linearity of the Filter

1. Now multiply the vector xx from Section C.5.3.2 by 2 to get xa = 2*xx. Generate the signal ya by filtering xa with the first difference filter given by (C.5.11). Repeat the relative amplitude and phase measurements described in Section C.5.3.2.

2. Now generate a new input vector xb corresponding to the discrete-time signal

$$x_b[n] = 8\cos(0.25\pi n)$$

and then filter it with the first-difference operator to get $y_b[n]$. Then repeat the relative amplitude and phase measurements as before. In this case, the measurement of phase might be a bit tricky because there are only a few samples per period. Record how the amplitude, phase, and frequency of the output yb change compared to the input.

3. Now form another input signal xc that is the sum of xa and xb. Run xc through the filter to get yc and then plot yc. Compare yc to a plot of ya + yb. Are they equal? Explain any differences that you observe.

C.5.3.4   Time Invariance of the Filter   Now time-shift the input vector xx by 3 time units to get the sequence

$$x_s[n] = 7\cos(0.125\pi(n-3) + \pi/3) \qquad \text{for } n = 0, 1, 2, 3, \dots$$

and then filter $x_s[n]$ using the first-difference operator to get $y_s[n]$. Compare ys to yy, the output when the input is xx. Find a shift of yy (in number of samples) so that it lines up perfectly with ys.

C.5.3.5   Cascading Two Systems   More complicated systems are often made up from simple building blocks. In this system, a nonlinear system (squaring) is cascaded with an FIR filter.

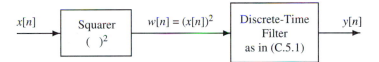

1. First, assume that this system is described by the two equations

$$w[n] = (x[n])^2 \qquad \text{(Squarer)}$$

$$y[n] = w[n] - w[n-1] \qquad \text{(First difference)}$$

Implement this system using MATLAB. Use as input the same vector xx as in Section C.5.3.2. In MATLAB, the elements of a vector xx can be squared by either the statement xx.*xx or xx.^2.

2. Plot all three waveforms $x[n]$, $w[n]$, and $y[n]$ on the same figure with subplot.

3. Make a sketch[12] of the spectrum of the three signals $\{x[n], w[n], y[n]\}$. Recall that the "squarer" is nonlinear and that it is therefore possible for the frequency spectrum of $w[n]$ to contain frequency components not present in $x[n]$.

4. Observe the time–domain output, $w[n]$, of the "squarer." Can you "see" the additional frequencies introduced by the squaring operation?

5. Use the linearity results to explain what happens as the signal $w[n]$ then passes through the first-difference filter. *Hint:* Track each frequency component through separately.

---

[12] This should, as the term sketch implies, be done by hand analysis. *Do not* use specgram or spectgr.

6. Now replace the first-difference filter in the figure with the second-order FIR filter:

$$y_2[n] = w[n] - 2\cos(0.25\pi)w[n-1] + w[n-2] \qquad (C.5.12)$$

Implement the squaring and filtering to produce a new output $y_2[n]$. Determine which frequencies are present in the output signal. Explain how this new filter is able to remove a frequency component by calculating $y_2[n]$ when $w[n] = e^{j0.25\pi n}$ in (C.5.12), and by plotting the time-frequency response of the filter as defined by (C.5.12). In addition, sketch the spectrum of $y_2[n]$.

C.6   LABORATORY: FILTERING SAMPLED WAVEFORMS

In the experiments of this lab, we will use `firfilt( )` to implement filters and begin to understand how the filter's frequency response relates to the action of the filter for smoothing or sharpening. We will begin to see how a filter reacts to different frequency components in the input. In addition, you will verify that the cascade connections of LTI systems can be reordered without changing the overall frequency response.

C.6.1   Overview of Linear Filters

An FIR (finite impulse response) system is described by the difference equation

$$y[n] = \sum_{k=0}^{M} b_k x[n - k] \qquad \text{(C.6.1)}$$

Equation (C.6.1) gives a formula for computing the nth value of the output sequence from values of the input sequence. Recall from Lab C.5 that (C.6.1) is implemented by the following MATLAB statement:

$$yy = firfilt(bb, xx);$$

firfilt.m

where it is assumed that xx is a vector of input samples and the vector bb contains the b_k coefficients of (C.6.1) stored in the following way: [b0, b1, b2, ... , bM]. The experiments of this lab show how to use `firfilt( )` to implement filters on several interesting signals.

The frequency response of a general FIR linear time-invariant system is[13]

$$H(e^{j\hat{\omega}}) = \sum_{k=0}^{M} b_k e^{-j\hat{\omega}k} \qquad \text{(C.6.2)}$$

Recall that you can use the `freqz( )` function in MATLAB to compute the frequency response of a discrete-time LTI system. The following MATLAB statements

---

[13] The notation $H(e^{j\hat{\omega}})$ is used in place of $\mathcal{H}(\hat{\omega})$ for the frequency response because we will eventually connect this notation with the z-transform.

compute and plot the magnitude (absolute value) and phase of the frequency response of a first-difference system as a function of $\hat{\omega}$ in the range $-\pi \le \hat{\omega} \le \pi$:

```
bb = [1, -1];    %<--- Filter Coefficients
ww = -pi:(pi/100):pi;
H = freqz(bb, 1, ww);
subplot(2,1,1);
plot(ww, abs(H))
subplot(2,1,2);
plot(ww, angle(H))
```

For FIR filters of the form (C.6.1), the second argument of `freqz` must always be set equal to 1.

C.6.2 Warm-up

Instructor Verification Sheet

The instructor verification sheet can be found on the CD-ROM and printed out as needed.

To begin, start MATLAB and then load the data for this lab as follows:

$$\text{load lab6dat}$$

lab6dat

This loads a data file `lab6dat.mat` containing several filters and signals. The variables in this data file are:

`x1`: a stair-step signal such as one might find in one sampled scan line from a TV test-pattern image.

`xtv`: an actual scan line from a digital image.

`x2`: a speech waveform sampled at 8000 samples/sec.

`h1`: the coefficients for an FIR discrete-time filter of the form of (C.6.1).

`h2`: a second FIR filter.

You will use these in the following experiments.

C.6.2.1 Properties of Discrete-Time Filters

1. The frequency responses of discrete-time filters are *always* periodic with period equal to 2π. Explain why this is so by stating a definition of the frequency response and then considering two input sinusoids whose frequencies are $\hat{\omega}$ and $\hat{\omega} + 2\pi$.

$$x_1[n] = e^{j\hat{\omega}n} \qquad \text{versus} \qquad x_2[n] = e^{j(\hat{\omega}+2\pi)n}$$

Prove that the outputs from (C.6.1) will be identical.

Instructor Verification (separate page)

2. When several systems are cascaded (connected so that the output of one is the input to the next one), it is possible to calculate the overall frequency response by multiplying the individual frequency responses. Since the frequency response for a general FIR linear time-invariant system given by (C.6.2) can be considered a polynomial in the variable $e^{j\hat{\omega}}$, multiplying two frequency responses amounts to nothing more than multiplying two polynomials.

   As a quick review, multiply the following two polynomials by hand:

$$x + 0.5x^2 - 2x^3 \quad \text{and} \quad 1 + x - 0.25x^3 \tag{C.6.3}$$

3. Now use the MATLAB command `firfilt` to convolve the two sequences whose elements are the coefficients of the polynomials in item 2. Remember to use a coefficient of 0 for those powers of x that are missing in the polynomial. How do your results compare with the coefficients of the multiplied polynomial of the previous step? Read the `help` on `firfilt` that mentions that convolving two vectors of polynomial coefficients is equivalent to multiplying the polynomials.

4. Use polynomial multiplication to solve for the overall frequency response of a cascade of a 3-point averager and a 5-point averager. *Hint*: Use polynomial multiplication on the coefficients, then evaluate the result using `freqz( )`.

Instructor Verification (separate page)

C.6.3 Laboratory: Sampling and Filters

The experiments in this lab will demonstrate several important facts about LTI filtering of sampled signals. Keep in mind that:

- Filters of the form of (C.6.1) can modify the frequency spectrum of any signal (not just complex exponentials) in interesting ways.
- The order of cascaded LTI systems does not affect the overall response.
- Lowpass filters "smooth" a signal and highpass filters "roughen" a signal, so listening tests on filtered speech give sounds that are either "muffled" or "crisper."

C.6.3.1   Filtering a Stair-Step Signal   In this experiment we are going to investigate the two systems shown in Figs. C.11 and C.12. In these two systems, the system called "first difference" is defined by the difference equation:

$$y[n] = x[n] - x[n-1] \tag{C.6.4}$$

and the system called "5-point averager" is defined by the equation:

$$y[n] = \tfrac{1}{5} \sum_{k=0}^{4} x[n-k] \tag{C.6.5}$$

The first test signal used will be a stair-step signal in which the signal is constant for different intervals, but the constant value of the signal in each flat region is different.

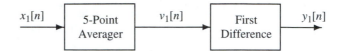

Figure C.11 First cascade system; averaging operator followed by first difference.

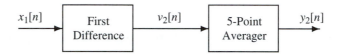

Figure C.12 Second cascade system; differencing operator followed by averaging.

C.6.3.2 Implementation of Five-Point Averager Use the MATLAB function fir-filt( ), which can be found in the *DSP First Toolbox*, to implement the 5-point averager, i.e., compute v1.

1. Plot x1 and v1 in the same figure window using a two-panel subplot. The signals have different lengths, but the plot should start at the same index, i.e., $n = 0$. Give a qualitative description of how the 5-point averager system changed the input signal. Estimate the time shift between the input and output signals. Express this delay in number of samples.

2. Use freqz( ) to compute the frequency response of the 5-Point Averager and plot its magnitude as a function of radian frequency for $-\pi \leq \hat{\omega} \leq \pi$. From the shape of the frequency–response curve, determine which frequency region is "passed" by the filter? Which frequencies are rejected? Relate the frequency response to the qualitative description of the time–domain response.

3. Plot the phase response versus radian frequency. Measure the slope of the phase response and compare this slope to the time shift between the input and output signals.

C.6.3.3 Implementation of First-Difference System

1. Plot x1 and v2 in the same figure window using a two-panel subplot.

2. Describe qualitatively how the first-difference system changed the input signal. Where are the peaks of v2? What is the nature of the input x1 at the places where the peaks of v2 occur?

3. Use freqz( ) to compute the frequency response of the first difference and plot its magnitude as a function of frequency for $-\pi \leq \hat{\omega} \leq \pi$. Is the shape of

the frequency-response curve consistent with your interpretation of what the system did to the input signal? Is the first difference a highpass or a lowpass filter?

C.6.3.4 Implementation of First Cascade (Fig. C.11)

1. Use the MATLAB function `firfilt( )` to implement the overall system of Fig. C.11 by first computing the output of the 5-point averager, v1, and then using v1 as input to the first-difference system in Fig. C.11 to compute the output y1.

2. Use `freqz( )` to compute the frequency response of the cascaded system and plot its magnitude and phase as a function of frequency for $-\pi \leq \hat{\omega} \leq \pi$.

C.6.3.5 Implementation of Second Cascade (Fig. C.12)

1. Use the MATLAB function `firfilt( )` to implement the overall system of Fig. C.12 by first computing the output of the first difference, v2, and then using v2 as input to the 5-point averager system in Fig. C.12 to compute the output y2.

2. Use `freqz( )` to compute the frequency response of the cascaded system and plot its magnitude and phase as a function of frequency for $-\pi \leq \hat{\omega} \leq \pi$. Compare to the overall frequency response computed for Fig. C.11.

C.6.3.6 Comparison of Systems of Figs. C.11 and C.12   Execute the MATLAB statement `sum( (y1-y2).*(y1-y2) )`. Write the mathematical expression that is being evaluated by MATLAB. Discuss how this calculation measures the error between the two alternative implementations and then discuss the implications of the result.

lab6dat

C.6.3.7 Filtering the Speech Waveform   A sampled-speech waveform is stored in the variable x2 in the file `lab6dat.mat`. Two sets of filter coefficients are stored in h1 and h2 (i.e., these are the "b_ks" for two different filters). Use `length` to find out how many filter coefficients are contained in h1 and h2. In this experiment, we will test these filters on the speech signal.

1. Filter the speech signal with filter h1 using the statements

   ```
   y1 = firfilt(h1, x2);
   inout(x2, y1, 3000, 1000, 3)
   ```

inout.m

The M-file `inout( )` will plot two very long signals together on the same plot. It formats the plot so that the input signal occupies the first, third, and fifth lines, etc, while the output is on the second, fourth, and sixth lines, etc. Type `help inout` to learn more.

   Compare the input and output signals. Is the output "rougher" or "smoother" than the input signal?

2. Use `freqz( )` to plot the frequency response of the system defined by the coefficients h1 as a function of frequency for $-\pi \leq \hat{\omega} \leq \pi$. Why is h1 called a "lowpass filter"?

3. Since the vector of filter coefficients is rather long, a `stem` plot of h1 can be informative to show the nature of the filter. Use the `stem` plot to find a point of symmetry in the coefficients.

4. Filter the speech signal with filter h2 and plot the input and output using the statements

```
y2 = firfilt(h2,x2);
inout(x2, y2, 3000, 1000, 3)
```

Compare the input and output, and state whether the output is "rougher" or "smoother" than the input signal.

5. Use `freqz( )` to plot the frequency response of the system defined by the coefficients h2 as a function of frequency for $-\pi \leq \hat{\omega} \leq \pi$. Why is h2 called a "highpass filter"?

6. Make a `stem` plot of h2 and look for a symmetry in the coefficients.

7. Make an "A-B-C" listening comparison by executing the following statements:[14]

```
sound([x2; y1; y2], 8000)
```

Comment on your perception of the filtered outputs versus the original.

8. What do you expect to hear when you execute the following statement? Why?

```
sound([x2; (y1+y2)], 8000)
```

Was your expectation confirmed? If you added the two frequency responses together

$$H_1(e^{j\hat{\omega}}) + H_2(e^{j\hat{\omega}})$$

what would you expect the answer to be?

---

[14] In MATLAB version 5, use `soundsc( )`.

C.7 LABORATORY: EVERYDAY SINUSOIDAL SIGNALS

This lab introduces two practical applications where sinusoidal signals are used to transmit information: a touch-tone dialer and amplitude modulation (AM) for radio. In both cases, FIR filters can be used to extract the information encoded in the waveforms.

C.7.1 Background

This lab has two parts. Part A investigates the generation and detection of the signals used to dial the telephone. Part B is concerned with modulation and demodulation of AM (amplitude modulation) waveforms such as those used in AM radio.

C.7.1.1 Background A: Telephone Touch Tone[15] Dialing   Telephone touch pads generate *dual tone multifrequency* (DTMF) signals to dial a telephone. When any key is pressed, the tones of the corresponding column and row (in Fig. C.13) are generated, hence dual tone. As an example, pressing the **5** button generates the tones 770 Hz and 1336 Hz summed together.

Frequencies	1209 Hz	1336 Hz	1477 Hz
697 Hz	1	2	3
770 Hz	4	5	6
852 Hz	7	8	9
941 Hz	*	0	#

Figure C.13 DTMF encoding table for touch tone dialing. When any key is pressed, the tones of the corresponding column and row are generated.

The frequencies in Fig. C.13 were chosen to avoid harmonics. No frequency is a multiple of another, the difference between any two frequencies does not equal any of the frequencies, and the sum of any two frequencies does not equal any of the frequencies.[16] This makes it easier to detect exactly which tones are present in the dial signal in the presence of line distortions.

---

[15] Touch Tone is a registered trademark of AT&T.

[16] More information can be found at: `http://www.shout.net/~wildixon`.

C.7.1.2   DTMF Decoding   There are several steps to decoding a DTMF signal:

1. Divide the signal into shorter time segments representing individual key presses.
2. Determine which two frequency components are present in each time segment.
3. Determine which button was pressed, **0–9**, *, or **#**.

It is possible to decode DTMF signals using a simple FIR filter bank. The filter bank in Fig. C.14 consists of filters that each pass only one of the DTMF frequencies and whose inputs are the same DTMF signal.

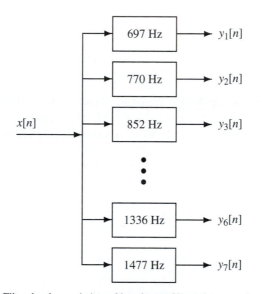

Figure C.14   Filter bank consisting of band-pass filters that pass frequencies corresponding to the seven DTMF component frequencies listed in Fig. C.13.

When the input to the filter bank is a DTMF signal, the outputs of two filters should be larger than the rest. The two corresponding frequencies must be detected in order to determine the DTMF code. A good measure of the output levels is the average power at the filter outputs. This is calculated by squaring the filter outputs and averaging over a short time interval. More discussion of the detection problem can be found in Section C.7.4.

C.7.1.3   Background B: Amplitude Modulation (AM)   Amplitude modulation is often used to transmit a signal with low-frequency content using a high-frequency transmission channel. A common example is AM radio. In AM radio, a relatively low-frequency signal such as a speech signal (which has frequencies between 50 Hz and 4 kHz) is transmitted by radio waves at frequencies around 1 MHz.

Amplitude modulation is performed by multiplying a high-frequency signal (called the *carrier*) by a low-frequency *message* signal $m(t)$:

$$x(t) = (1 + m(t)) \cos(2\pi f_c t + \phi) \qquad \text{(C.7.1)}$$

where the carrier signal corresponds to the $\cos(2\pi f_c t + \phi)$ term. Although the message signal $m(t)$ may be very complicated, a good understanding of AM can be obtained by analyzing AM signals of the form:

$$x(t) = (1 + A \cos(2\pi f_m t)) \cos(2\pi f_c t) \qquad \text{(C.7.2)}$$

i.e., the message signal is a cosine, $m(t) = A \cos(2\pi f_m t)$. A straightforward expansion of (C.7.2) shows that:

$$x(t) = \cos(2\pi f_c t) + \frac{A}{2} \cos(2\pi (f_c + f_m)t)) + \frac{A}{2} \cos(2\pi (f_c - f_m)t) \qquad \text{(C.7.3)}$$

In communications jargon, the signal components at $f = f_c \pm f_m$ are called the *sidebands* and the signal component with a frequency of f_c is called the carrier. When $f_c \gg f_m$ the spectrum of the AM signal looks like that shown in Fig. C.15. Note that an AM signal in (C.7.3) is very similar to the beat signals studied earlier in Lab C.4 and Chapter 3, except for the addition of the carrier term.

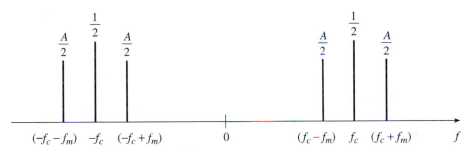

Figure C.15   Spectrum of an amplitude-modulated (AM) tone.

More complicated message signals may also be analyzed. If $m(t)$ in (C.7.2) is made up of multiple sinusoidal components, those components are each shifted in frequency, as was the single sinusoid in (C.7.3).

$$x(t) = \left(1 + \sum_k A_k \cos(2\pi f_k t)\right) \cos(2\pi f_c t) \qquad \text{(C.7.4)}$$

$$= \cos(2\pi f_c t) + \sum_k \frac{A_k}{2} \cos(2\pi (f_c - f_k)t) + \sum_k \frac{A_k}{2} \cos(2\pi (f_c + f_k)t)$$

Thus we would have many spectral lines in the sidebands.

C.7.1.4   AM Demodulation   *Demodulation* is the process of recovering the message waveform from a modulated signal such as AM. There are numerous methods of demodulating a signal, but only two will be discussed here. This lab focuses on an LTI filtering approach, but a more common approach is presented first for comparison.

C.7.1.5   Envelope Detection (Peak Tracking)   Inexpensive AM radios use a capacitor, resistor, and diode to perform the AM demodulation (see Fig. C.16). The idea is to get a waveform that approximately follows the peaks of the AM waveform. During each positive cycle of the AM signal, the capacitor is charged through the diode. Then, during the negative part of the cycle, the resistor discharges the capacitor slowly so that the demodulated waveform can also follow the envelope of the modulated waveform as the peaks decrease in amplitude. The calling format of a MATLAB function to perform this type of demodulation is shown in Fig. C.17.

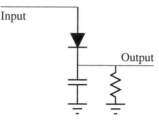

Figure C.16   Simple capacitor, resistor, diode-type AM demodulator, as used in early AM radios.

A close-up of the output waveform from the simple demodulator of Fig. C.16 is shown in Fig. C.18. Also shown are the message waveform, $m(t) = 2\sin(2\pi(25)t)$, and the carrier at 371 Hz, which make up the AM signal:

$$x(t) = (1 + 0.4m(t))\cos(2\pi(371)t)$$

The demodulated signal is not perfect, but it does approximate the sinusoidal shape of the message signal once you subtract the DC level of one. The jagged appearance is due to the exponential discharge of the RC circuit.

```
function dd = amdemod(xx,fc,fs,tau)
%
% where
%       xx = the input AM waveform to be demodulated
%       fc = carrier frequency
%       fs = sampling frequency
%      tau = time constant of the RC circuit normalized by fs
%            (OPTIONAL: default value is tau=0.97)
%       dd = demodulated message waveform
```

Figure C.17   Arguments for the MATLAB function amdemod, which simulates an AM demodulator based on a simple capacitor, resistor, diode-type circuit in Fig. C.16.

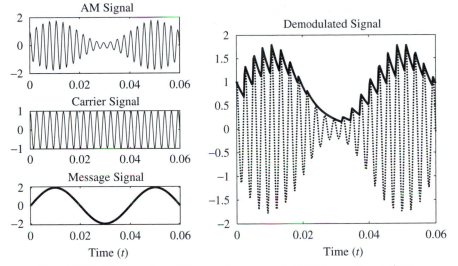

Figure C.18   Close-up view of the waveform generated by the capacitor, resistor, diode-type AM demodulator with $\tau = 0.96$. The dark, jagged line is the demodulator output, and the dotted line is a close-up of the cycles of the AM waveform.

C.7.1.6   LTI filter-based demodulation   It is possible to recover the message signal by *modulating the modulated signal* and then filtering. This is the basic principle used in commercial radios nowadays. Given an AM signal of the form (C.7.2), we can isolate the message signal by multiplying $x(t)$ by $\cos(2\pi f_c t)$.

$$x(t)\cos(2\pi f_c t) = (1 + m(t))\cos(2\pi f_c t)\ \cos(2\pi f_c t) \qquad (C.7.5)$$

$$= (1 + m(t))\left(\tfrac{1}{2} + \tfrac{1}{2}\cos(2\pi(2f_c)t)\right) \qquad (C.7.6)$$

$$= \tfrac{1}{2}m(t) + \tfrac{1}{2} + \tfrac{1}{2}(1 + m(t))\cos(2\pi(2f_c)t) \qquad (C.7.7)$$

Notice that the message signal is now available outside of the product term. There are still two terms in (C.7.7) that must be eliminated. The $\tfrac{1}{2}$ is a DC offset that can be subtracted out or ignored; the other term is a very-high-frequency term (at $f = 2f_c$) that we will eliminate by filtering. The scale factor of $\tfrac{1}{2}$ multiplying $m(t)$ can be compensated for by doubling the final output.

C.7.1.7   Notch Filters for Demodulation   A notch filter will be used as part of the demodulation process. Notch filters are filters that completely eliminate some frequency other than $\hat{\omega} = 0$ or $\hat{\omega} = \pi$. It is possible to make a notch filter with as few as three coefficients. If $\hat{\omega}_{not}$ is the desired notch frequency, then the following length-3 FIR filter

$$y[n] = x[n] - 2\cos(\hat{\omega}_{not})x[n-1] + x[n-2] \qquad (C.7.8)$$

will have a zero at $\hat{\omega} = \hat{\omega}_{not}$ in its frequency response. For example, a filter designed to completely eliminate signals of the form $Ae^{j0.5\pi n}$ would have coefficients

$$b_0 = 1, \quad b_1 = 2\cos(0.5\pi) = 0, \quad b_2 = 1$$

However, our specifications will be given in terms of continuous-time frequency, e.g., we will eliminate the spectral component at f_{not}. We must convert to discrete-time frequency by using the frequency scaling that is due to sampling:

$$\hat{\omega}_{not} = 2\pi \frac{f_{not}}{f_s}$$

where f_s is the sampling frequency.

C.7.2 Warm-up A: DTMF Synthesis

Instructor
Verification
Sheet

The instructor verification sheet can be found on the CD-ROM and printed out as needed

C.7.2.1   DTMF Dial Function   Write a function, dtmfdial, to implement a DTMF dialer defined in Fig. C.13. A skeleton of dtmfdial.m, including the help comments, is given in Fig. C.19; you must complete the code so that it implements the following:

1. The input to the function is a vector of numbers that may range between 1 and 12, with 1 to 10 corresponding to the digits (10 corresponds to **0**), 11 being the * key, and 12 being the # key.

2. The output should be a vector containing the DTMF tones, sampled at 8 kHz. The duration of the tones should be about 0.5 sec, and a silence, about 0.1 sec long, should separate each tone pair.

```
function tones = dtmfdial(nums)
%DTMFDIAL   Create a vector of tones which will dial
%              a DTMF (Touch Tone) telephone system.
%
% usage:   tones = dtmfdial(nums)
%       nums  = vector of numbers ranging from 1 to 12
%       tones = vector containing the corresponding tones.
%
if (nargin < 1)
     error('DTMFDIAL requires one input');
end
fs = 8000;   %<-- This MUST be 8000, so dtmfdeco( ) will work.
   .
   .
   .
```

Figure C.19   Skeleton of dtmfdial.m. A DTMF phone dialer.

Your function should create the appropriate tone sequence to dial an arbitrary phone number. When played through a telephone handset, the output of your function will be able to dial the phone. You may use specgram or the function dtmfchck (in the *DSP First Toolbox*) to check your work.[17]

Instructor Verification (separate page)

C.7.3 Warm-up B: Tone Amplitude Modulation

1. Derive (C.7.3) from (C.7.2). *Hint*: Express the cosine terms as sums of complex exponentials using Euler's identity.
2. Create a test AM signal with the following characteristics:
   (a) The carrier tone, cc, must have a frequency of 1200 Hz, a duration of 1 sec, a sample rate of 8000 Hz, and a phase of 0.[18]
   (b) The message signal, mm, should be a 100-Hz tone with an amplitude of 0.8.
3. Plot the first 200 points of the modulated signal and the message signal on the same plot. Use different colors or line types for the two signals (see help plot).
4. Compare the spectra of the message, the carrier, and the modulated signals using the following command for each:

showspec.m

$$\text{showspec}(\_, \ 8000);$$

Instructor Verification (separate page)

C.7.4 Laboratory A: DTMF Decoding

A DTMF decoding system needs a bandpass filter to isolate individual frequency components, and a detector to determine whether or not a given component is present. The detector must "score" each possibility and determine which frequencies most likely are present. In a practical system where noise and interference are present, this scoring process is a crucial part of the system design, but we will work only with noise-free signals to understand the basic functionality in the decoding system.

C.7.4.1   Filter Design   The filters that will be used in the filter bank (Fig. C.14) are a simple type constructed with sinusoidal impulse responses. In Section 7.7, a simple

---

[17] In MATLAB, the demo called **phone** also shows the waveforms and spectra generated in a DTMF system.

[18] Although we will not cover it in this lab, the phase of the carrier is important when using the demodulation described above; specifically, the demodulation tone must have the same phase (and frequency) as the carrier. There are some sophisticated ways of ensuring that the demodulation remains in phase with the carrier tone, but we will not cover them here. Instead, we will use a cosine function with zero phase for both the carrier and the demodulation tone.

bandpass filter design method was presented in which the impulse response of the filter is simply a finite-length cosine of the form:

$$h[n] = \frac{2}{L} \cos\left(\frac{2\pi f_b n}{f_s}\right), \qquad 0 \le n < L$$

where L is the filter length, and f_s is the sample frequency. The parameter f_b defines the frequency location of the passband, e.g., we pick $f_b = 697$ if we want to isolate the 697-Hz component. The bandwidth of the bandpass filter is controlled by L; the larger the value of L, the narrower the bandwidth.

1. Generate a bandpass filter, h770, for the 770-Hz component with $L = 64$ and $f_s = 8000$. Plot the filter coefficients in the first panel of a two-panel subplot using the stem( ) function.

2. Generate a bandpass filter, h1336, for the 1336-Hz component with $L = 64$ and $f_s = 8000$. Plot the filter coefficients in the second panel of a two-panel subplot using the stem( ) function.

3. Use the following commands to plot the frequency response (magnitude) of h770

```
fs = 8000;
ww = 0:(pi/256):pi;   %<--- only need positive freqs
ff = ww/(2*pi)*fs;
H = freqz(h770,1,ww);
plot(ff,abs(H)); grid on;
```

4. Indicate the locations of each of the DTMF frequencies (697, 770, 852, 941, 1209, 1336, and 1477 Hz) on the plot from item 3. *Hint*: Use the hold and stem( ) commands.

5. Comment on the selectivity of the bandpass filter h770, i.e., use the frequency response to explain how the filter passes one component while rejecting the others. Is the filter's passband narrow enough?

6. Plot the magnitude response of the h1336 filter and compare its passband to that of the h770 filter.

C.7.4.2 A Scoring Function   The final objective is decoding, a process that requires a binary decision on the presence or absence of the individual tones. In order to make the signal detection an automated process, we need a *score* function that rates the different possibilities.

1. Complete the dtmfscor function based on the skeleton given in Fig. C.20. Assume that the input signal xx to the dtmfscor function is actually a short segment from the DTMF signal. The task of breaking up the signal so that each segment corresponds to one key will be done by another function prior to calling dtmfscor.

The implementation of the FIR bandpass filter is done with the conv function, but we could also use firfilt. The running time of the convolution function is proportional to the filter length L. Therefore, the filter length L must satisfy two competing constraints: L should be large so that the bandwidth of the BPF is narrow enough to isolate individual frequencies, but making it too large will cause the program to run slowly.

```
function  ss = dtmfscor(xx, freq, L, fs)
%DTMFSCOR
%           ss = dtmfscor(xx, freq, L, [fs])
%      returns 1 (TRUE) if freq is present in xx
%              0 (FALSE) if freq is not present in xx.
%      xx = input DTMF signal
%    freq = test frequency
%       L = length of FIR bandpass filter
%      fs = sampling freq (DEFAULT is 8000)
%
%  The signal detection is done by filtering xx with a length-L
%  BPF, hh, squaring the output, and comparing with an arbitrary
%  set point based on the average power of xx.
%
if (nargin < 4), fs = 8000; end;

hh =        %<========= define the bandpass filter coeffs here
ss = (mean(conv(xx,hh).^2) > mean(xx.^2)/5);
```

Figure C.20   Skeleton of the dtmfscor.m function.

2. Explain the last line in dtmfscor.m:

$$ss = (mean(conv(xx,hh).^2) > mean(xx.^2)/5);$$

C.7.4.3 DTMF Decode Function

The DTMF decode function, dtmfdeco will use dtmfscor to determine which key was pressed based on an input DTMF signal. The skeleton of this function in Fig. C.21 includes the help comments and the table of tone pairs from Fig. C.13. You must add the logic to decide which key is present.

Assume that the input signal xx to the dtmfscor function is actually a short segment from the DTMF signal. The task of breaking up the signal so that each segment corresponds to one key has already been done by the function that calls dtmfdeco.

There are several ways to write the dtmfdeco function, but you should avoid excessive use of "if" statements to test all 12 cases. *Hint:* Use MATLAB's vector logicals (see help relop) to implement the tests in a few statements.

```
function key = dtmfdeco(xx,fs)
  %DTMFDECO    key = dtmfdeco(xx,[fs])
  %  returns the key number corresponding to the DTMF waveform, xx.
  %      fs = sampling freq (DEFAULT = 8000 Hz if not specified.
  %
  if (nargin < 3), fs = 8000; end;
  tone_pairs = ...
  [ 697  697  697  770  770  770  852  852  852  941  941  941;
    1209 1336 1477 1209 1336 1477 1209 1336 1477 1336 1209 1477 ];
  .
  .
```

Figure C.21   Skeleton of dtmfdeco.m.

C.7.4.4   Telephone Numbers   A function, dtmfmain, supplied with the *DSP First* CD-ROM will run the entire DTMF system consisting of the three M-files you have written: dtmfdial.m, dtmfscor.m, and dtmfdecod.m.   If you are presenting this project in a lab report, the dtmfmain function can be used to demonstrate a working version of your programs.

dtmfmain.m

An example of using dtmfmain is shown here:

```
>> dtmfmain( dtmfdial([1:12]) )
ans =
    1    2    3    4    5    6    7    8    9    10   11   12
```

For this function to work correctly, all three M-files must be on the MATLAB path. It is also essential to have short pauses in between the tone pairs so that dtmfmain can parse out the individual signal segments.

C.7.5 Laboratory B: AM Waveform Detection

As discussed in Section C.7.1.3, there are several ways to perform AM waveform detection, or *demodulation*. Inexpensive AM radios use a method of peak tracking to recover the envelope of the modulated waveform. Better detectors use a combination of additional modulation and filtering.[19] This is the method that we will use in this lab.

  1. Demodulate the AM test signal created in the warmup, Section C.7.1.3, using the filter-based method as follows:

---

[19] You may wish to try the MATLAB demod function on the signal used in the Warm-up section of the lab. This function is part of the MATLAB Signal Processing Toolbox.

(a) Multiply the AM test signal by the carrier as described in (C.7.7) and look at the spectrum using the showspec command. Identify each of the spectral peaks with terms in (C.7.7).

(b) Create a three-term notch filter designed to eliminate the high-frequency component at 2400 Hz. Plot the frequency response of this notch filter to verify that you have done this correctly.

(c) Filter the product signal created in step (a) with your notch filter. The result should look similar to the original message signal.

amdemod.m

2. Demodulate the AM test signal using the amdemod function, which implements the peak following circuit of Figs. C.16 and C.17. Experiment with the decay rate (the fourth input parameter to amdemod) to find the best setting. Usually values of τ near 0.9 will give good results, but look at the output for $\tau = 1.2$ and $\tau = 0.4$.

3. Create a three-panel subplot with containing plots of the *middle 200 points* from the following signals:

   (a) The original message signal

   (b) The demodulated signal found in item 1

   (c) The demodulated signal found in item 2

   Compare the quality of the two demodulated signals and judge how well they follow the original message signal in the time domain.

4. Compare the spectra of the message and the two demodulated signals using the following command for each:

$$\text{showspec}(\_, 8000);$$

5. Explain how the filter method could be modified to produce a better-quality signal. *Hint*: Look at the spectrum of the demodulated signal to see whether the high-frequency component at 2400 Hz is completely gone. If not, then filter the signal again with the same notch filter used in the demodulation of item 1. Is this equivalent to using a higher-order notch filter with more coefficients?

C.7.6 Optional: Amplitude Modulation with Speech

Load the speech signal and other MATLAB data with the command

lab7dat

$$\text{load lab7dat}$$

This MATLAB data file contains three variables:

cc: a carrier signal at 24 kHz with a sampling rate of 96 kHz.

mm: a speech signal at the same sampling rate as the carrier signal.

ss: the speech signal at its original sampling rate of 8 kHz.

1. Create a modulated signal, aa, from cc and mm with the command:

$$\text{aa} = (1 + \text{mm}) . * \text{cc};$$

2. In a 3-panel subplot, show the spectra of cc, mm, and aa. Comment briefly on the relationship between them.

3. Demodulate the speech waveform using the function amdemod; call the demodulated waveform dd. The sampling rate of the demodulated waveform is very high, much higher than it needs to be. Since the frequency content of dd is now almost entirely below 4 kHz, we may sample it at 8 kHz for playback. This is done by taking only every twelfth sample from dd, and throwing away the rest:

$$\text{ds} = \text{dd}(1{:}12{:}\text{length}(\text{dd}));$$

4. Listen to ds and ss, and comment on what you hear.

C.8   LABORATORY: FILTERING AND EDGE DETECTION OF IMAGES

In this lab we introduce *digital images* as a signal type for the application of filtering. The lab has a project on lowpass and highpass filtering of images with an investigation of the frequency content of an image. In addition, the problem of edge detection is introduced, where a combination of linear and nonlinear filters can be used to solve the problem.

C.8.1 Overview

C.8.1.1   Digital Images   An image can be represented as a function $x(t_1, t_2)$ of two continuous variables representing the horizontal (t_2) and vertical (t_1) coordinates of a point in space. For monochrome images, the signal $x(t_1, t_2)$ would be a scalar function of the two spatial variables, but for color images the function would be a vector function of the two variables.[20] Moving images (TV) would add a time variable to the two spatial variables. Monochrome images are displayed using black and white and shades of gray, so they are called *gray-scale* images. In this lab we will consider only sampled gray-scale still images.

A sampled gray-scale still image would be represented as a two-dimensional array of numbers of the form

$$x[m, n] = x(mT_1, nT_2) \qquad 1 \leq m \leq M, \text{ and } 1 \leq n \leq N$$

Typical values of M and N are on the order of 256 or 512; e.g., a 512×512 image has nearly the same resolution as a standard TV image. In MATLAB we can represent an image as a matrix consisting of M rows and N columns. The matrix entry at (m, n) is the sample value $x[m, n]$, called a *pixel* (short for picture element). In MATLAB, the pixel at $(1, 1)$ is in the upper left corner of the image.

An important property of light images such as photographs and TV pictures is that their pixel values are always non-negative and finite in magnitude, i.e.,

$$0 \leq x[m, n] \leq X_{max}$$

This is because light images are formed by measuring the intensity of reflected or emitted light, which must always be a positive finite quantity. When stored in a computer or displayed on a monitor, the values of $x[m, n]$ are usually scaled so that an 8-bit integer representation can be used. With 8-bit integers, the maximum value can be $X_{max} = 2^8 - 1 = 255$, and there are 256 gray levels for the display.

C.8.1.2   Displaying Images   As you will discover, the correct display of an image on a gray-scale monitor can be tricky, especially after some processing has been

---

[20] For example, an RGB color system needs a three-element vector containing values for red, green, and blue at each spatial location.

performed on the image. We have provided the function show_img.m in the *DSP First Toolbox* to handle most of these problems, but it will be helpful if the following points are noted:

1. All image values must be non-negative for the purposes of display. Filtering may introduce negative values, especially if differencing is used (e.g., a highpass filter).

2. The default format for most gray-scale displays is 8 bits, so the pixel values $x[m, n]$ in the image are integers in the range $0 \leq x[m, n] \leq 255 = 2^8 - 1$.

3. The MATLAB functions max and min can be used to find the largest and smallest values in the image.

4. The functions round, fix, and floor can be used to quantize pixel values to integers.

5. The actual display on the monitor is created with the show_img function. The appearance of the image can be altered by running the pixel values through a "color map." In our case, all three primary colors (red, green and blue, or RGB) are used equally, so we get a "gray map." In MATLAB the gray color map is created with

$$\text{colormap(gray(256))}$$

which gives a 256×3 matrix where all 3 columns are equal. The colormap gray(256) creates a linear mapping, so that each input pixel amplitude is rendered with a screen intensity proportional to its value (assuming that the monitor is calibrated). For our experiments, nonlinear color mappings would introduce an extra level of complication, so we will not use them.

6. When the image values lie outside the range [0, 255], or when the image is scaled so that it occupies only a small portion of the range [0, 255], the display may have poor quality. We can analyze this condition by using the MATLAB function hist to plot how often each pixel value occurs (called a histogram). This will indicate whether some values at the extremes are seldom used and, therefore, can be "thrown away."

Based on the histogram, we can adjust the pixel values prior to display. This can be done in two different ways:

7. *Clipping the image:* When some of the pixel values lie outside the range [0, 255], but the scaling needs to be preserved, the pixel values should be clipped. All negative values will be set to zero, and anything above 255 will be set equal to 255. The function clip is provided to do this job.

8. *Automatically rescaling the image:* This requires a linear mapping of the pixel values:[21]

$$x_s[m, n] = \alpha x[m, n] + \beta$$

---

[21] The MATLAB function show_img will perform this scaling while making the image display.

The scaling constants α and β can be derived from the min and max values of the image, so that all pixel values are recomputed using:

$$x_s[m, n] = \left\lfloor 256 \, \frac{x[m, n] - x_{\min}}{x_{\max} - x_{\min}} \right\rfloor$$

where $\lfloor x \rfloor$ is the floor function, i.e., the greatest integer less than or equal to x.

C.8.1.3   Image Filtering   It is possible to filter image signals, just as it is possible to filter one-dimensional signals. One method of filtering two-dimensional signals (images) is to filter each row with a one-dimensional filter and then to filter each of the resulting columns with a one-dimensional filter. This is the approach taken in this lab.

C.8.2 Warm-up: Display of Images

IMAGE DATA
FILES

The images needed for this lab are available as *.mat files. Any file with the extension .mat is in MATLAB format, and can be loaded by using the load command. To find some of these files, look for *.mat in the *DSP First Toolbox* or in the MATLAB directory called toolbox/matlab/demos. Some of the image files are named lenna.mat, echart.mat and zone.mat, but there are others within MATLAB's demos. The default size is usually 256×256, but alternative versions are sometimes available at 512×512 under names such as lenna_512.mat. After loading, use the MATLAB function whos to determine the name of the variable that holds the image and its size.

Although MATLAB has several functions for displaying images on the CRT of the computer, we have written a special function show_img( ) for this lab. It is the visual equivalent of sound( ), which we used when listening to speech and tones; i.e., show_img( ) is the D-to-C converter for images. This function handles the scaling of the image values and allows you to open up multiple image display windows. Here is the help on show_img:

show_img.m

```
function [ph] = show_img(img, figno, scaled, map)
%SHOW_IMG     display an image with possible scaling
% usage:  ph = show_img(img, figno, scaled, map)
%     img = input image
%     figno = figure number to use for the plot
%               if 0, re-use the same figure
%               if omitted a new figure will be opened
% optional args:
%     scaled = 1 (TRUE) to do auto-scale (DEFAULT)
%               not equal to 1 (FALSE) to inhibit scaling
%     map = user-specified color map
%      ph = figure handle returned to caller
%----
```

Notice that, unless the input parameter figno is specified, a new figure will be opened. This is extremely important if you want to display several images together using subplot. Using subplot is the preferred method of *printing* multiple images (e.g., for comparisons in lab reports) because one page can easily hold four images.

C.8.2.1 Display Test In order to probe your understanding of image display, generate a simple test image in which all of the rows are identical.

1. Create a test image that is 256×256 by making each horizontal line a discrete-time sinusoid with a period of 50 samples. Note that the row index is the vertical axis and the column index is the horizontal axis.

2. Extract one row from the image and make a plot to verify that the horizontal line is a sinusoid.

3. Use show_img to display the test image from item 1, and then explain the gray-scale pattern that you observe. If the sinusoid was generated with values between +1 and -1, explain the scaling used to make the 8-bit image suitable for screen display. In other words, what is the (integer) gray level for displaying zero? for -0.3?

Instructor Verification (separate page)

4. Now load and display the lenna image from lenna.mat. The command load lenna will put the sampled image into the array xx.

5. Make a plot of the 200th row of the lenna image. Observe the maximum and minimum values.

6. Consider a way in which you might display the negative of the lenna image by rescaling the pixels. In other words, derive a linear mapping to remap the pixel values so that white and black are interchanged.

Instructor Verification (separate page)

C.8.3 Laboratory: Filtering Images

In Labs C.5 and C.6, you experimented with one-dimensional filters, such as running averagers and first-difference filters, applied to one-dimensional signals. These filters can be applied to images if we regard each row (or column) of the image as a one-dimensional signal.

C.8.3.1 One-Dimensional Filtering For example, the 50th row of an image is the N-point sequence xx[50,n] for $1 \le n \le N$. We can filter this sequence with a one-dimensional filter using the conv operator, but how can we filter the entire image? Answering that question is the objective of this lab.

1. Load in the image echart.mat with the load command. Extract the 33rd row from the bottom of the image using the statement

$$\text{x1 = echart(256-33,:);}$$

Filter this one-dimensional signal with a 7-point averager, and plot both the input and the output in the same figure using a two-panel subplot. Observe whether or not the filtered waveform is "smoother" or "rougher" than the input. Explain.

2. Now extract the 99th column of the image, and filter it with a first differ-ence. Plot the input and output together for comparison. Once again, ob-serve whether or not the filtered waveform is "smoother" or "rougher" than the input. Explain.

C.8.3.2   Blurring an Image   From items 1 and 2 in section C.8.3.1, perhaps you can see how you could use a for loop to write an M-file that would filter all the rows. This would create a new image made up of the filtered rows:

$$y_1[m, n] = \frac{1}{7} \sum_{\ell=0}^{6} x[m, n - \ell] \qquad \text{for } 1 \leq m \leq M$$

However, this image $y_1[m, n]$ would be filtered only in the horizontal direction. Filtering the columns would require another for loop, and finally you would have the completely filtered image:

$$y_2[m, n] = \frac{1}{7} \sum_{k=0}^{6} y_1[m - k, n] \qquad \text{for } 1 \leq n \leq N$$

In this case, the image $y_2[m, n]$ has been filtered in both directions by a 7-point averager.

These filtering operations involve a lot of conv calculations, so the process can be slow. Fortunately, MATLAB has a built-in function called conv2( ) that will do all of this work with a single call. It performs a more general filtering operation than row or column filtering, but since it can do these simple one-dimensional operations it will be very helpful in this lab.

1. To filter the image in the horizontal direction using a 7-point averager, we form a *row* vector of filter coefficients and use the following statement:

```
bh = ones(1,7)/7;
y1 = conv2(xx, bh);
```

In other words, the filter coefficients bh for the 7-point averager *stored in a row vector* will cause conv2( ) to filter all rows in the *horizontal* direction. Display the input image xx and the output image y1 on the screen side-by-side. Compare the two images and give a qualitative description of what you see. Extract row 33 (from the bottom again) from the output image (y1(256−33, :)) and compare it to the output obtained in item 1 of Section C.8.3.1.

2. Now filter the image y1 in the vertical direction with a 7-point running averager to produce the image y2. This is done by calling `conv2` with a *column* vector of filter coefficients. Plot all three of the images xx, y1, and y2 on the screen at the same time. Now describe what you see. Where or when have you seen this effect before?

3. What do you think will happen if you repeat items 1 and 2 for a 21-point moving averager? Try it, and compare the results of the 21-point and 7-point averaging filters. Which one causes a more severe degradation of the original image?

C.8.3.3   More Image Filters

1. Load the `lenna` image into MATLAB, and display the image using the `show_img` command.

2. Filter the `lenna` image with each of the following filters. Remember to filter both the rows and columns, unless otherwise directed.

   (a) `a1 = [1 1 1]/3;`

   (b) `a2 = ones(1,7)/7;`

   (c) `a3 = [1 -1];` For this filter, filter only the rows. Note that this filter will yield negative values. Before displaying the resulting image, it must be scaled to fit back into the allowable range of [0, 255]. This will be done automatically by the `show_img( )` command.

   (d) `a4 = [-1 3 -1];` For this filter, filter only the rows.

   (e) `a5 = [-1 1 1 1 -1];`

   Comment on the effect of each filter, paying special attention to regions of the image with lots of detail such as the feathers. Answer the following questions by showing representative images that exhibit the characteristics of highpass or lowpass filtering.

   - Which filters appear to be lowpass filters?
   - What is the general effect of a lowpass filter on an image?
   - Which ones are highpass filters?

C.8.3.4   Frequency Content of an Image

Filters can be used to investigate the frequency content of an image. From what we have seen so far, lowpass filtering of an image causes blurring. In this section, we will prove that highpass (or bandpass) filtering will "sharpen" an image. In Section C.8.3.5, we will examine a system for edge detection that also uses highpass filters. For this exercise, you should use `baboon.mat` as the test image.

1. To make this demonstration, we need some filters that give a smooth frequency response. For this purpose we will use "Gaussian-shaped" functions for the

FIR filter coefficients $\{b_k\}$. First of all, we need coefficients for a lowpass filter:

$$b_k = \begin{cases} 1.1 & \text{for } k = 10 \\ e^{-0.075(k-10)^2} & \text{for } k = 0, 1, 2, \ldots 20 \text{ and } k \neq 10 \\ 0 & \text{elsewhere} \end{cases}$$

For this filter, plot the impulse response as a stem plot, and then plot the magnitude of the frequency response. Put them on the same page with a two-panel subplot. *Hint*: Remember that MATLAB starts indexing at one; the formula for b_k will have to be adjusted by one.

2. Produce a bandpass filter by the following definition:

$$\tilde{b}_k = \begin{cases} \cos(0.4\pi(k - 10))e^{-0.13(k-10)^2} & \text{for } k = 0, 1, 2, \ldots 20 \\ 0 & \text{elsewhere} \end{cases}$$

In this case, the filter coefficients $\{\tilde{b}_k\}$ are a Gaussian multiplied by a cosine. Once again, plot the impulse response as a stem plot, and also plot the magnitude of the frequency response. Put them on the same page with a two-panel subplot. Explain why this filter is called a bandpass filter (BPF).

3. Load the test image from baboon.mat. Filter this image along both its rows and columns with the LPF from item 1, and save the result as $y[m, n]$ for item 5. Display the image to see the effect of the filter. Determine whether the frequency content of $y[m, n]$ is mostly low-frequency or high-frequency.

4. Filter the test image along both its rows and columns with the BPF $\{\tilde{b}_k\}$, and save the result as $v[m, n]$ for item 5. Once again, determine whether the frequency content of $v[m, n]$ is mostly low-frequency or high-frequency.
   Note: If you try to display this image, you will experience problems because it has negative values that must be rescaled to get a positive image for the CRT. In addition, the resulting display looks mostly gray because its contrast has been compressed significantly owing to a small number of very large or very small values. Use hist to plot the distribution of pixel values, and then clip the image to eliminate a few of the very large and very small values.

5. The blurred image in $y[m, n]$ from item 3 can be improved by adding in some of the band-pass filtered image $v[m, n]$ from item 4. The system in Fig. C.22 implements the following linear combination:

$$(1 - \alpha)y[m, n] + \alpha v[m, n]$$

where α is a fraction, i.e., $0 \leq \alpha \leq 1$. In effect, α controls how much of the final result will come from the bandpass filtered image. Try to find the best value of α so that the combination looks as close as possible to the original. Explain why the result looks sharper.

For display, you can put four images together on one page with a four-panel subplot, i.e., `subplot(2,2,*)`. When using `show_img` in subplots, set the second argument to 0 to stay in the same figure window (see `help show_img`).

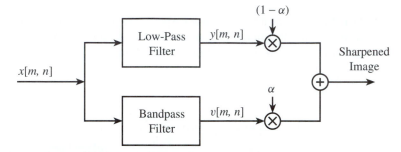

Figure C.22   Test set-up for varying low-frequency and high-frequency content of an image. The parameter α should be between 0 and 1.

C.8.3.5   The Method of Synthetic Highs   "Deblurring" is a rather difficult process to carry out, but the experiments of Section C.8.3.4 lead to the conclusion that:

> *We need to add some high-frequency information to sharpen an image.*

Therefore, we can use this idea to design a practical "deblurring" system. We cannot directly use the method of Section C.8.3.4 because the image $v[m, n]$ came from bandpass filtering the original. In a realistic "deblurring" problem, only the blurred image would be available as input to the deblurring system. Somehow, we must generate some high-frequency content from the blurred image. One procedure for doing this is the method of "synthetic highs" shown in Fig. C.23.

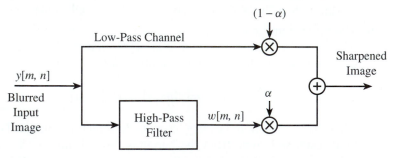

Figure C.23   Method of synthetic highs. High-frequency content of the image is generated by using a second-difference filter.

1. Construct a highpass filter as the second-difference filter:

$$H(z) = (1 - z^{-1})^2$$

   Plot the frequency response (magnitude) of this filter.

2. Compute a synthetic-high image by passing the blurred image $y[m, n]$ through the second difference to obtain $w[m, n]$. Be sure to process both the rows and the columns.

3. Explain the frequency content of $w[m, n]$ by considering that it was constructed by passing the original through the cascade of the blurring filter $\{b_k\}$ and then the second difference. Plot the frequency response of that cascade, and compare to the frequency response of the bandpass filter $\{\tilde{b}_k\}$ from Section C.8.3.4.

4. *Sharpening with synthetic highs:* Now use the same linear combination as before:

$$(1 - \alpha)y[m, n] + \alpha w[m, n]$$

   where α is a fraction, i.e., $0 \leq \alpha \leq 1$. The value of α controls how much high-frequency content to add for the purpose of sharpening.

5. Try to find the best choice of α to get a sharpened result. Comment on how well your processing works, realizing that it is not a perfect process. Show an example of the sharpened image versus the blurred image.

C.8.3.6   Nonlinear Filters   As you have seen in a previous sections of this lab, linear FIR filters are useful in reconstructing images. They can also blur or sharpen images through lowpass or highpass filters. However, other types of filters also are useful in image processing. Nonlinear filters are useful because the frequency content of the output image can be changed dramatically from that of the input image, yet the two are still mathematically and visually related. For example, when an artist touches up a photograph, the output image is "visually related" to the input. A common problem in image processing is to find a sequence of linear and nonlinear filters that corresponds to a certain visual relation. The problem posed in this section is: How do we get a computer to trace the outlines of objects in an image? A good solution turns out to require a highpass linear filter followed by a nonlinear threshold system.

C.8.3.7   Edges in an Image   The outline of an object is usually marked by a sudden change in color or gray level. The outline of a black object against a white background would be the set of points where neighbors have a different color or shade of gray. A black point in the image that has a white point to its right would thus be on the edge of the object. Most images do not have solid black and white regions, though.

Since the gray-scale variations within each region are usually gradual, but the edges are supposed to be rapid variations in color, we suspect that a simple highpass filter could detect edges. Try the first-difference filter on an image, such as echart, which has only black and white regions. Filter both the rows and the columns. Display the result as a gray-scale image. How well does the filter work on vertical and horizontal edges? What about edges at an angle? Explain.

C.8.3.8   The Slope–Threshold Function   If we accept the assumption that color (or gray-scale) should change rapidly at an edge, all we have to do is mark the points whose neighbors have greatly different values. For a single pixel in the image, this means comparing its value to the values of its neighbors. If the absolute value of the difference between one point and the next in any direction is greater than some threshold value, the pixel is probably on an edge. Figure C.24 (right) shows a binary (black-and-white) image where the pixels greater than the threshold are black.

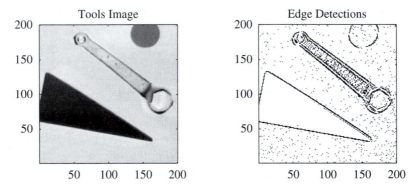

Figure C.24   Binary image (right) of edge detections on the 200×200 "Tools" gray-scale image (left).

This threshold test must be applied to every point in an image, but in MATLAB, it is much easier to apply a vector test to an entire matrix. So, we need to create matrices full of all the differences between pixels and their nearest neighbors, and then compare each entire matrix to the threshold value.

The easiest way to do this is with the first-difference filter. Applying the horizontal first-difference filter to the image produces a matrix of the differences between each point and the one to its left. However, the filtering operation inserts an extra column (see help filter2). Which column is extra? Can you remove it? You should be able to produce a matrix yh of exactly the same size as the original image where each point in the matrix is exactly the difference between the corresponding point and the one to its left in the original image. In addition, we need to compute the vertical first difference (yv) for the edge detector.

Now you should have the horizontal (yh) and vertical (yv) difference matrices trimmed so that they line up with the original image, i.e., xx(1,1)-xx(1,2) equals yh(1,1) and so on. The final step is to compare their absolute values to a threshold value. In MATLAB, logical operations such as a>b, a<b, etc., produce result values equal to 1 for true and 0 for false. The logical operation a|b means "a or b," and it will be useful here. The statement

$$yy = (abs(yh)>thr)|(abs(yv)>thr);$$

where thr is the threshold value, should produce a matrix yy filled with zeros where points in the original image do not lie on edges and ones where they do. For an 8-bit screen display using show_img, you should scale the output so that the minimum and maximum values are at least 200 apart. For printing, you should invert the colormap so that most of the output is white.

If the threshold is zero, any point that is not surrounded by points of the same gray-level is marked. This is too sensitive for an image with a good variety of color, i.e., a photograph. If the threshold value is too high, the edge of a gray object on a black background may not be marked. The threshold value should be carefully chosen for this operation. For the 8-bit images lenna or echart, a good value might be between 8 and 18. Experiment with the threshold value to get a good-looking result; it should be possible to make the output look like a line drawing that outlines the head, hat, and feathers in lenna.mat.

C.8.3.9   What's Nonlinear about Edge Detection?   Although the horizontal and vertical first-difference filters that we used are LTI systems, the logical operation is a nonlinear system. The threshold comparison converts each line of the image into a type of "square wave" function. Plot row 94 of the output image yy along with the corresponding row of the input image xx. Is edge detection sensitive to low-frequency content or high-frequency content, or neither? Explain.

C.9 LABORATORY: SAMPLING AND ZOOMING OF IMAGES

In this lab we study the application of FIR filtering to the image zooming problem, where lowpass filters are used to do the interpolation needed for high-quality zooming. The zooming problem is basically the same as the D-to-A reconstruction problem treated in Chapter 4.

C.9.1 Overview

If you have not already done Lab C.8, please read Section C.8.1.1 of that lab for important information on the display and printing of images.

C.9.2 Warm-up: Linear Interpolation

In Section C.9.3.2, we will be interested in image zooming, which is an interpolation problem. Before processing images, however, we consider the following one-dimensional interpolation problem.

1. Generate a sequence of data samples with two zeros between each non-zero sample:

   ```
   xss = zeros(1,19);
   samp = [1 3 -2 4 2 -1 -3];
   xss(1:3:19) = samp;
   ```

2. Now process this sequence through an FIR filter with "triangular" coefficients.

   ```
   coeffs = [1/3, 2/3, 1, 2/3, 1/3];
   output = firfilt(xss,coeffs);
   ```

   What observations can you make regarding the similarities between the output sequence and the nonzero samples of the original sequence (xss)? Do you notice a relative shift between these two sequences? Measure the length of this time shift in samples. Explain why the output of the triangle FIR filter is a *linearly interpolated* version of the input sequence.

3. Write the difference equation for the "triangular" filter defined by coeffs in item 2. Now calculate, *by hand*, the output of the filter to the input sequence

$$x[n] = \delta[n] + 0.5\delta[n-3] \qquad (C.9.1)$$

$$= \begin{cases} 1, & \text{when } n = 0 \\ 0.5, & \text{when } n = 3 \\ 0, & \text{elsewhere} \end{cases} \qquad (C.9.2)$$

Do not use MATLAB to complete this exercise. The purpose is to understand exactly how the linearly interpolated output is generated when you implement the difference equation. Pay close attention to how the filter coefficients align with the original data samples.

Instructor Verification (separate page)

In summary, we notice that linear interpolation involves two steps. The first step is zero filling, where the number of zeros inserted between each sample determines the number of interpolated values. The second step is FIR filtering with the appropriate triangle coefficients.

C.9.3 Laboratory: Sampling of Images

The images that are stored on the computer have to be sampled images because they are stored in an $M \times N$ array. The sampling rate in the two spatial dimensions was chosen at the time the image was digitized (in units of samples per inch if the original was a photograph). However, we can simulate a *lower* sampling rate by simply throwing away samples in a regular manner. If every other sample is removed, the sampling rate will be halved. Sometimes this is called *subsampling* or *downsampling*. So, for example, if we have a vector x1 representing a signal such as a row of an image, we can reduce the sampling rate by a factor of 4 by simply taking every fourth sample. In MATLAB, this is easy with the colon operator, i.e., xs = x1(1:4:length(x1)). The vector xs is one fourth the length of x1.

An alternative sampling strategy is to take every fourth sample, but also place zeros in between. The M-file below called imsample( ) will perform this type of sampling on an image, e.g., xs = imsample(xx,4).

1. Explain how the function imsample.m works:

```
function yy = imsample(xx, P)
%IMSAMPLE    Function for sub-sampling an image
% usage:    yy = imsample(xx, P)
%           xx = input image to be sampled
%            P = sub-sampling period (an integer like 2,3,etc)
%           yy = output image
%
[M,N] = size(xx);
S = zeros(M,N);
S(1:P:M,1:P:N) = ones(length(1:P:M), length(1:P:N));
yy = xx .* S;
```

2. Execute the statement xs = imsample(xx,4); and use show_img( ) to plot the images xs and xx. From the plot, you should see that imsample( ) throws away samples by setting them to zero. The samples that it keeps remain in their original spatial locations. With the zero samples included, the "sampled image" has the same spatial dimensions as the original when displayed on the screen.

3. *Downsampling* throws away samples, so it will shrink the size of the image. This is what is done by the following scheme:

$$xp = xx(1:p:M,1:p:N);$$

One potential problem with downsampling is that aliasing may occur. This can be illustrated in a dramatic fashion with the zone image, or the lenna image.

Perform downsampling by a factor of 2 on the zone.mat image. Compare the smaller downsampled image to the original.[22] The visual differences are due to aliasing. This may be suprising since no new pixel values are being created by the downsampling process. Describe in words the visual differences that you observe. Can you relate the changes to high-frequency content in the original image?

Downsample the lenna image by a factor of 2. Notice that this image seems to be relatively unaffected by the downsampling by 2 process, what can you say about the frequency content of the lenna image as opposed to the zone image?

C.9.3.1   Reconstruction of Images   When an image has been sampled, we can fill in the missing samples by interpolation. For images, this would be analogous to the examples shown in Chapter 4 for sine-wave interpolation, which is part of the reconstruction process in a D-to-A converter. We could use a "square pulse" or a "triangular pulse" or other pulse shapes.

RECON-
STRUCT
MOVIES

For the following reconstruction experiments, use either the tools image or the lenna image. It would be wise to put the original image and all the filtered reconstructions on the screen in different figure windows for easy comparison.

1. The simplest interpolation would be the square pulse which produces a "zero-order hold." We could fill in the gaps between samples in each row with the following statement:

```
xs = imsample(xx,4);
bs = ones(1,4);          %--- Length-4 square pulse
yhold = conv2(xs,bs);    %--- 2-D FIR filtering (horizontally)
```

Try this and display the image yhold.

2. Now filter the columns of yhold to fill in the missing points in each column and compare the result to the original image xx.

---

[22] One difficulty with showing aliasing is that we must display the pixels of the image exactly. This almost never happens because most monitors and printers will perform some sort of interpolation to adjust the size of the image to match the resolution of the device. In MATLAB, we can override these size changes by using the function trusize, which is part of the *DSP First Toolbox*.

3. *Linear interpolation:* Use what you learned about linear interpolation in the warm-up section to determine an FIR filter that will perform linear interpolation on the rows and columns of the subsampled (by 4) signal. This filter must have coefficients $\{b_k\}$ that follow a triangle shape, and the filter order must be $M = 6$.

4. Carry out the linear interpolation operations using MATLAB's conv2 function. Call the interpolated output ylin. Compare ylin to the original image xx and to the square pulse interpolated image from item 2. Comment on the visual appearance of the two "reconstructed" images.

5. Compute the frequency response of your linear interpolator used in item 4. Only the one-dimensional FIR filter that acts on the rows must be analyzed. Plot its magnitude (frequency) response for $-\pi \leq \hat{\omega} \leq \pi$.

6. *Smoothing with lowpass filtering:* At this point, you should be thinking that interpolation is very similar to lowpass filtering. To test this hypothesis, create a lowpass filtered version xs_filt of the sampled image by filtering the rows and columns with a 23-point FIR filter, whose coefficients are given by a modified "sinc" formula:

$$b_k = \frac{\sin(\pi(k - 11)/4)}{\pi(k - 11)/4} w_k \qquad k = 0, 1, 2, \ldots, 22$$

where w_k is given by:

$$w_k = 0.54 - 0.46 \cos\left(\frac{2\pi k}{22}\right) \qquad k = 0, 1, 2, \ldots, 22$$

Remember that MATLAB will have problems evaluating b_{11} because the sinc function is an indeterminate form when $k = 11$. You will have to compute b_{11} yourself and include it at the proper location in the vector b_k. Compare xs_filt to the original image. In addition, plot the magnitude of the frequency response for this FIR filter.

RECONS-
TRUCT
MOVIES

C.9.3.2   Zooming for an Image   Zooming in on a section of an image is very similar to the D-to-A reconstruction process because it also requires interpolation. The three interpolation systems (zero-order hold, linear interpolation, and lowpass filtering) developed in Section C.9.3.1 can be applied to do zooming by a factor of 4.

1. Take a small patch of an image (about 50×50) where there is some interesting detail (e.g., the eye or feather of lenna). There are several ways to produce a larger image that appears zoomed. One way is to simply repeat pixel values. So, in order to zoom a 50×50 section up to 200×200, you would repeat each pixel four times in each direction. Display a zoomed portion of an image by repeating pixel values.

2. Another way to produce a larger image is to insert zeros between the existing samples. In other words, you must produce an image that is 200×200 with data values only in rows or columns whose index is a multiple of 4. Consider the following code that does this for a one-dimensional vector.

```
L = length(xx);
yy = zeros(1,4*L);
yy(4:4:4*L) = xx;
```

Generalize this idea to write a function that will insert zeros in the rows and columns of a two-dimensional matrix.

3. Now filter the image from item 2 to do the interpolation. Use both the linear and the sinc function interpolators.

4. Comment on the ability of all three zooming operators to preserve detail and edges in the image while expanding the size of details. Try to explain your observations by considering the frequency content of the "zoomed" image.

5. MATLAB has a zoom command in the Image Processing Toolbox. The function is called imzoom. If this is available on your system, try to determine what sort of interpolation scheme it is using.

C.10   LABORATORY: THE z-, n-, AND $\hat{\omega}$-DOMAINS

C.10.1   Objective

The objective for this lab is to build an intuitive understanding of the relationship between the location of poles and zeros of $H(z)$ (the z-domain), the impulse response $h[n]$ (in the n domain), and the frequency response $H(e^{j\hat{\omega}})$ (the $\hat{\omega}$ domain). A graphical user interface (GUI) called PeZ was written in MATLAB for doing interactive explorations of the three domains.[23]

C.10.2   Warm-up

pez.m

If you have the *DSP First Toolbox* installed, invoke PeZ by first typing dspfirst at the MATLAB prompt and then selecting "Pole/Zero Plotter." A control panel with a few buttons and a plot of the unit circle in the complex z-plane will pop up. You can use the PeZ controller to selectively place poles and zeros in the z-plane, and then observe how their placement affects the impulse and frequency responses. If the plots need manual updating, click on the Redo Plots button under the <Quicksize...> menu.

The Real Time Drag Plots button will put PeZ in a mode such that an individual pole or zero (pair) can be moved around, and the corresponding $H(e^{j\hat{\omega}})$ and $h[n]$ plots will be updated.

GUI: PeZ

Since exact placement of poles and zeros with the mouse is difficult, the button Edit By Co-Ord is provided for numerical entry of the real and imaginary parts, or magnitude and angle (a separate edit window will appear when you use this option). Before you can edit a pole or zero, however, you must first select it with the mouse. Removal of individual poles or zeros can also be performed by clicking on the Delete Poles & Zeros (again, a separate window will appear). Note that all poles and zeros can be easily cleared by clicking on the <Clear...> menu, and then selecting Poles, Zeros, or All. A complete set of documentation for PeZ can be found on the CD-ROM.

Play around with PeZ for a few minutes to gain some familiarity with the interface. Try implementing a 3-point running sum filter by placing its zeros (and poles) at the correct location in the z-plane.

Instructor Verification Sheet

Instructor Verification (separate page)

C.10.3   Laboratory: Relationships Between z-, n-, and $\hat{\omega}$-domains

Work through the following exercises and record your observations on the worksheet at the end of this laboratory. In general, you want to make note of the following quantities: How does $h[n]$ change with respect to oscillation period and rate of decay? How does $H(e^{j\hat{\omega}})$ change with respect to peak location and width?

---

[23] PeZ was written by Craig Ulmer.

C.10.4 Real Poles

1. Use PeZ to place a pole at $z = 0.5$. You may have to use the Edit by Co-Ord button to get the location exactly right. Use the plots for this case as the reference for answering the next four items.

2. Move the pole close to the origin (still on the real axis). You can do this by clicking the pole and dragging it to the new location. Describe the changes in the impulse response $h[n]$ and the frequency response $H(e^{j\hat{\omega}})$.

3. You can also move poles or zeros under the influence of the Real Time Drag Plots option in PeZ. When this box is checked, the impulse response and frequency response plots are immediately updated while you are moving the pole (or zero). Once this mode is set, click the pole (or zero) you want to move and start to drag it slowly. Watch for the update of the plots in the secondary window. Release the the mouse button to stop the updating. As long as the mode is set, you can move another pole or zero. Finally, the display may be jerky if you have a slower computer. Move the real pole slowly from $z = \frac{1}{2}$ to $z = 1$ and observe the changes in the impulse response $h[n]$ and the frequency response $H(e^{j\hat{\omega}})$.

4. Place the pole exactly on the unit circle. Describe the changes in $h[n]$ and $H(e^{j\hat{\omega}})$.

5. Move the pole outside the unit circle. Describe the changes in $h[n]$ and $H(e^{j\hat{\omega}})$.

6. Recall what you learned in Chapter 8 about placing poles to guarantee system stability. By stability, we mean that the system's output does not blow up (see Section 8.8).

C.10.5 Complex Poles

If the denominator polynomial $A(z)$ has a complex root, it will have a second root at the conjugate location when the polynomial coefficients are real. For example, if we place a root at $z = \frac{1}{3} + j\frac{1}{2}$, then we will also get one at $z = \frac{1}{3} - j\frac{1}{2}$.

1. What property of the polynomial coefficients of $A(z) = 1 + a_1 z^{-1} + a_2 z^{-2}$ will guarantee that the roots come in conjugate pairs?

2. Clear all the poles and zeros from PeZ. Now place a pole with magnitude 0.75 at an angle of $45°$; then two zeros at the origin. Note that PeZ automatically places a conjugate pole in the z-domain.

3. Derive the filter coefficients for the denominator $A(z)$ and numerator $B(z) = b_0 + b_1 z^{-1} + b_2 z^{-2}$. Use the following relationship:

$$H(z) = \frac{B(z)}{A(z)} = G\frac{(1 - z_1 z^{-1})(1 - z_2 z^{-1})}{(1 - p_1 z^{-1})(1 - p_2 z^{-1})} \tag{C.10.1}$$

where z_1 and z_2 are the zeros, and p_1 and p_2 are the poles from item 2. (Remember that MATLAB can multiply polynomials through its conv function.) Record the filter coefficients for later use.

4. Change the angle of the pole: Move the pole to $90°$, then $135°$. Describe the changes in $h[n]$ and $H(e^{j\hat{\omega}})$.

5. Increase the magnitude of the pole: First try 0.9, then 0.95, and then go outside the unit circle. Describe the changes in $h[n]$ and $H(e^{j\hat{\omega}})$.

C.10.6 Filter Design

In this section, we will use PeZ to place the poles and zeros to make a filter with a desirable frequency response. First of all, we will put poles at the origin and zeros on the unit circle to gain some understanding of these special cases.

1. Clear all the poles and zeros from PeZ. Now place a single pole at $z = 0$ and observe the impulse and frequency response. Now place another pole at the origin and observe carefully the changes in $h[n]$ and the magnitude and phase of $H(e^{j\hat{\omega}})$. Finally, place a third pole at $z = 0$. What can you conclude about the effect of poles at the origin on the impulse response, magnitude response, and phase response of a system?

2. Clear all poles and zeros from PeZ. Now place zeros at the following locations: $z_1 = -1$, $z_2 = 0 - j$ and $z_3 = 0 + j$ (remember that conjugate pairs such as z_2 and z_3 will be entered simultaneously). Judging from the impulse and frequency responses, what type of filter have you just implemented? Measure (as accurately as possible) the slope of the phase response.

Design of a filter involves selecting the coefficients $\{a_k\}$ and $\{b_k\}$ to accomplish a given task. The task here is to create a filter that has a very narrow "notch." This filter would be useful for removing one frequency component while leaving others undisturbed. The *notch filter* can be synthesized from the cascade of the two simpler filters shown in Fig. C.25.

3. Start the process by using PeZ to design each of the filters given in Fig. C.25. Both filters are second-order. Make sure that you enter the poles and zeros precisely. PeZ will do the conversion between root locations and polynomial coefficients, but you could also do this with the MATLAB commands roots and poly. You should check your results directly by also calculating the filter coefficients by hand. Record the coefficients of your filters in the table provided on the worksheet.

4. Use freqz( ) to verify that the frequency response of each filter is correct.

5. Now use PeZ to put the filters together in a cascade. Place the poles and zeros, and then view the frequency response. What are the filter coefficients for the cascaded filter $H(z)$?

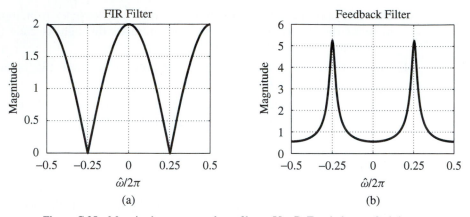

Figure C.25   Magnitude response of two filters. Use PeZ to help you find the filter coefficients that will match these frequency responses as closely as possible. (a) Second-order FIR filter; (b) second-order IIR filter.

6. Use `freqz( )` to determine the frequency response of the cascade of the two filters that you "designed" in item 5. Plot the magnitude of the overall frequency response of the cascade system for $-\frac{1}{2} < \hat{\omega}/(2\pi) < \frac{1}{2}$, and print a copy of the plot for your lab report. Explain briefly why the frequency response magnitude has a notch, and explain why the gains at $\hat{\omega} = 0$ and $\hat{\omega} = \pi$ are the same.

Worksheet for Observations

Name:_____   Date:_____

Implemented 3-point running sum with PeZ:_____

Part	Observations
C.10.4(1)	$h[n]$ decays exponentially with no oscillations, $H(e^{j\hat{\omega}})$ has a hump at $\hat{\omega} = 0$
C.10.4(2)	
C.10.4(3)	
C.10.4(4)	
C.10.4(5)	
C.10.4(6)	
C.10.5(1)	
C.10.5(2)	
C.10.5(3)	
C.10.5(4)	
C.10.5(5)	
C.10.6(1)	
C.10.6(2)	

Part C.10.6(3,4,5,6)	a_k	b_k
b_k		
Filter 1		
Filter 2		
Cascade of 1 and 2		

Explanation of cascade frequency-response:

C.11   LABORATORY: EXTRACTING FREQUENCIES OF MUSICAL TONES

This lab is built around a single project that involves the implementation of a system for automatically writing a musical score by analyzing the frequency content of a recording (a sampled signal). A primary component of such a system is the spectrogram, which produces a time-frequency representation of the recorded waveform. However, to make a working system, several other processing components are needed after the spectrogram to extract the important information related to the notes. The design of these additional blocks will lead naturally to a deeper understanding of what the spectrogram actually represents.

C.11.1 Overview

In Chapter 9 we introduced the spectrogram as an important tool for time-frequency analysis. We also gave two viewpoints that aid in understanding the spectrogram: a filter bank structure and a sliding-window FFT structure. The filter bank view helps to explain frequency resolution, which can be related to the bandwidth of the channel filters. The FFT view is more useful when discussing computational issues.

Music GUI

   For music signals, the spectrogram tends to produce an image with only a few peaks. Finding these peaks and identifying their frequencies and durations is the main issue in this lab. A MATLAB GUI for showing the spectrogram along with musical notation is available for experimentation.

   In order to make the project manageable, we will progress through several different signal types while testing. These include:

Recorded
Songs

1. Sine waves at a specific frequency.
2. Sine waves that make up a C-major scale.
3. Sinusoids that create the tune for *Twinkle, Twinkle, Little Star*.
4. A piano rendition of *Twinkle, Twinkle, Little Star*.
5. Other recorded songs are available for processing: *Jesu, Joy of Man's Desiring*; *Minuet in G*; *Beethoven's Fifth Symphony*; and *Für Elise*.

Instructor
Verification
Sheet

C.11.2 Warm-up: System Components

In the warm-up, we will investigate several M-files needed to build the complete processing chain. The instructor verification sheet can be found on the CD-ROM and printed out as needed.

C.11.2.1 Spectrogram Computation

In MATLAB, there is already a function for calculating the spectrogram, but it is not always available since it is part of the Signal Processing Toolbox. A similar function called spectgr( ) is included as part of the *DSP First Toolbox*, and we have preserved the same list of arguments. The calling format for spectgr is the following:

```
[B,T,F] = spectgr( xx, Nfft, fs, window, Noverlap )
```

which is the identical to the format used in MATLAB's specgram( ) function. The input arguments are defined as follows: xx is the input signal, Nfft is the FFT length, fs is the sampling frequency, window is a column vector containing the coefficients of the window, and Noverlap is the number of points in the overlapped part of consecutive sections.

The outputs are the spectrogram matrix, B; a vector T, containing the time locations of the windowed segments,[24] and a vector F, which contains the list of *scaled* frequencies corresponding to the spectrogram analysis. The vectors T and F are useful for labeling plots. Both are scaled by the sampling frequency so the units are seconds for T, and hertz for F. The spectrogram B contains complex values and its size is such that it has a column length (number of rows) equal to length(F) and a row length equal to length(T). In the *DSP First* implementation, the calling program must provide all of the arguments. The MATLAB function, on the other hand, allows the caller to omit arguments, but that just adds complexity in the programming.

In this section, we will present the steps in a spectrogram calculation, so that you can write your own function. The preferred viewpoint for calculation is that of a sliding-window FFT. In this implementation, we take a segment of the signal of length L (the length of the window vector), multiply the segment by a window, and then compute a zero-padded N-point FFT to form one column of the spectrogram matrix. Then the starting point of the data segment is moved over by an amount $L - N_{\text{overlap}}$ and the process is repeated. Eventually, we run out of data and the spectrogram is complete.

spectgr.m

For a MATLAB program, we need to write a while loop that tests whether any signal remains. The following example shows all the code needed for the inner loop:

```
B = zeros( Nfft/2 + 1, num_segs );      %- Pre-allocate the matrix
L = length(window);             %-- assuming a user generated window
iseg = 0;
while( )                                %<==== FILL IN THE TEST CONDITION
   nstart = 1 + iseg*(L-Noverlap);
   xsegw = window .* xx( nstart:nstart+L-1 );    %-- xx is a column
   XX = fft( xsegw, Nfft );
   iseg = iseg + 1;
   B(:,iseg) = XX(1:Nfft/2+1);
end
```

Explain how each of the steps in the spectrogram is being calculated. Explain how to calculate the number of segments num_segs ahead of time. Also, explain the purpose of the last line in the while loop. Finally, determine a test that can be used to terminate the while loop.

Instructor Verification (separate page)

---

[24] There are several conventions for defining the time: (1) start of the segment, (2) middle of the segment, or (3) end of the segment. The middle choice probably makes the most sense, but it really doesn't matter in this project because only relative times will be significant.

C.11.2.2 Generating the Window   The call to `spectgr` requires a window of length L. One possibility is the *rectangular* window consisting of all ones, but a better window is that for the Hann filter. The rectangular window corresponds to a running-sum filter. The definition of the Hann window is

$$w[n] = \tfrac{1}{2} - \tfrac{1}{2}\cos\left(\frac{2\pi n}{L+1}\right) \qquad n = 1, 2, \ldots L \qquad \text{(C.11.1)}$$

There are some variations on this definition, but the one given here omits endpoints that would be zero. Write a function that will generate the Hann window, making sure that it returns a column vector. Then you can use this when calling the `spectgr` function. Make a plot of the Hann window for $L = 64$.

Instructor Verification (separate page)

show_img.m

C.11.2.3 Display the Spectrogram   The display of the `spectgr` output can be done with the MATLAB function `imagesc` or with the *DSP First* function `show_img`. On a computer monitor, the spectrogram display can use color which tends to enhance details, but the conventional printout is a grayscale image with black indicating large values. In addition, if the gray level is proportional to the magnitude of B, small details may be lost, so it may be advantageous to convert to a logarithmic scale covering 30 or 40 dB (called "log mag"). Finally, the default orientation in MATLAB is a matrix orientation with the origin in the upper left-hand corner. To change this orientation so that the origin is in the lower left-hand corner, use `axis xy`. The following code fragment summarizes this type of display:

```
if (LOG)                    %-- assume LOG is a true/false variable
    B = 20*log10( abs(B) );        %-- ignore log(0) warnings
    dBmax = 30;
    B = B - max(abs(B(:))) + dBmax;
    B = B.*(B>0);       %-- dB range is now 0 <= B <= dBmax.
else
    B = abs(B);
end
imagesc( T, F, B ); colormap(1-gray(256)); axis xy;
```

C.11.2.4 Finding Peaks   Although it may be easy to spot peaks in the spectrogram visually, it is much harder to write a computer program to extract the same peaks reliably. This first step in the process, however, is merely to extract all the peaks. Then we can follow this up with an editing program that removes extraneous peaks. The peak-picking function needs to do only one-dimensional picking along the frequency axis because the music spectrogram has a definite horizontal bias; the tones last for a long duration along the time (horizontal) axis. If we scan each column of the $B(k, \ell)$ matrix for peaks, we can merge peaks from neighboring columns to see whether a note is present and also to determine how long it lasts.

pkpick.m

A one-dimensional peak-picker is available in the function pkpick.m, whose help comments are given below:

```
function  [peaks, locs] = pkpick( xx, thresh, number )
%PKPICK        pick out the peaks in a vector
%   Usage:  [peaks,locs] = pkpick( xx, thresh, number )
%          peaks  :  peak values
%          locs   :  location of peaks (index within a column)
%          xx     :  input data  (if complex, operate on mag)
%          thresh :  reject peaks below this level
%          number :  max number of peaks to return
%
```

Test that pkpick.m works as you expect by generating a cosine wave and finding its peaks.

Instructor Verification (separate page)

An unexpected problem with peak-picking is the quantization of the frequency axis. The peak-picking function will give an output that is on the grid of possible frequencies. If we need to estimate the peak location between these grid points, interpolation is needed. The function pkinterp.m is available for that purpose.

C.11.3 Design of the Music-Writing System

The complete system for automatically writing the music is quite complicated, so we follow the engineering practice of breaking down the system into smaller, more manageable, components.

C.11.3.1   Block Diagram for the System   Figure C.26 shows the major subsystems needed to extract enough information from a musical recording to write the sheet music for that input. Each of these should be implemented as a separate MATLAB function.

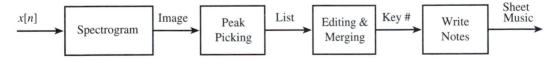

Figure C.26   Block diagram of major components in music-writing system.

C.11.3.2   Write a Spectrogram Function   Use the code fragment above as the basis for writing your own spectrogram function. Test your function by having it compute the spectrogram of a sine wave. The display should be one horizontal line at the frequency of the sinusoid.

C.11.3.3   Parameters of the Spectrogram

Window length: Derive a frequency resolution requirement for separating notes, and then use it to specify a window length. The resolution must be converted from continuous-time frequency (in Hz) to discrete-time frequency:

$$\hat{\omega} = 2\pi \frac{f}{f_s}$$

FFT Length: Use a power-of–2 FFT for efficiency. A long FFT will give more frequencies and reduce the gridding problem for peak interpolation.

Overlap: Determine the time-spacing needed to find the duration of notes. Be careful when making the time-spacing very small, because the amount of computation will increase dramatically.

C.11.3.4   Peak Picking and Editing
The peak-picking operation is relatively straightforward to implement, only the number of peaks and a threshold need to be specified. If the threshold is too low, the editing phase will have to deal will many extraneous peaks. The peak-picker should generate a list consisting of triplets (frequency, time, amplitude). The function pkpick.m provided in Section C.11.2.4 will only find peaks for one vector, so it must be modified to find peaks as a function of both time and frequency.

Editing is the crucial step and also the hardest to specify. Unlike the spectrogram which is a well-defined calculation with only a couple of parameters to adjust, the editing process can take many different forms. The editing system must take a list of frequency-time-amplitude triplets generated by the peak-picker and eliminate many of them based on rules that are derived from common sense. The following issues should be considered.

1. *Frequency:*
    (a) How close is the frequency to one of the allowable frequencies of the piano keys?
    (b) The harmonics must be eliminated, but there are cases where an octave is played, so the second harmonic may be allowed. In addition, when the song has both bass and treble sections, it is the case that frequencies can be 4 or 8 times each other.
2. *Time:*
    (a) Check the peak duration. Is it a half note, quarter note, etc.? This requires that peaks be merged and tracked along the time axis.
    (b) Consider the timing of the notes. We expect the notes to start at regular times because the music has a rate, such as 2/4 time, or 4/4 time.
    (c) In fact, an interesting sidelight project would be to extract the "beat" of the music. This might help to establish a time base for the piece, and help set some bounds on the expected durations.

(d) There is a minimum duration, unless we have a piece with lots of special effects, trills, grace notes, etc.

3. *Amplitude:*

   (a) Keep the strongest peaks—but how many?

   (b) If we also look in the time domain, the "attack" may be found. This is the sharp rise in amplitude at the beginning of a note. The attack would help in identifying the start times of the notes.

C.11.3.5   Writing the Musical Score   The output of the editing process should be a list of key numbers and durations that define the music. We have provided a function `wrinotes.m` that will create a MATLAB image that has the notes in the musical score. Consult the `help` for `wrinotes.m` to learn the data structure that is needed for its input.

C.11.4 Testing the Music Extraction Program

This project is relatively complicated, and testing will not be easy. However, several test files are available, progressing from easy cases to difficult ones. Run your program on the following four test cases:

1. Sine waves at a specific frequency.

2. Sine waves that make up a C-major scale.

3. Sinusoids that create the tune for *Twinkle, Twinkle, Little Star*.

4. A piano rendition of *Twinkle, Twinkle, Little Star*.

In each case, you know what the true answer should be, so you can assess the capabilities of your music writer.

All the piano pieces are sampled at 11.025 kHz. They have been played without a pedal to reduce interference between neighboring notes, thereby making the editing logic a little simpler. Alternate songs are: *Jesu, Joy of Man's Desiring*; *Minuet in G*; *Beethoven's Fifth Symphony*; and *Für Elise*. Each of these will be quite difficult and challenging, unless your editing logic is very sophisticated.

Remember that the objective of the lab is to make a working system containing the major components listed in Fig. C.26. Even with a few simple tests, you should learn quite a bit about the spectrogram, its strengths, and its shortcomings.

D

About the CD
Table of Contents

Chapters	Demos	Labs	Exercises	Homework
1 Introduction	■	■	■	■
2 Sinusoids	■	■	■	■
3 Spectrum Representation	■	■	■	■
4 Sampling and Aliasing	■	■	■	■
5 FIR Filters	■	■	■	■
6 Frequency Response of FIR Filters	■	■	■	■
7 Z-Transform	■	■	■	■
8 IIR Filters	■	■	■	■
9 Spectrum Analysis	■	■	■	■
A Complex Numbers	■	■	■	■

DSP First	<u>1. Introduction</u>	In chapter one, you are introduced to signal and systems.
	<u>2. Sinusoids</u>	In chapter two, you are introduced to the most basic waveform in signal processing, the cosine wave. The frequency of a cosine wave determines what you will hear in the audio experiments. Then the complex exponential is introducted. The real part of the complex exponential is a cosine, and its imaginary part is the sine function, so a plot of the complex exponential is a rotating vector with a constant length A.
	<u>3. Spectrum Representation</u>	In chapter three, the graphical representation of signals via their frequency content is treated. In addition, we will synthesize more complicated sinusoidal waveforms composed of sums of sinusoidal signals, each with a different frequency.
	<u>4. Sampling and Aliasing</u>	In chapter four, the conversion of signals between the analog and digital domains is studied. The basic ideas underlying sampling and signal reconstruction are presented.
	<u>5. FIR Filters</u>	The class of FIR (finite-impulse-response) filters is introduced. These filters use a running weighted average to form the output from the input.
	<u>6. Frequency Response of FIR Filters</u>	The frequency response function for FIR filters is introduced. The magnitude and phase versus frequency govern the response of sinusoidal input signals through the filter.
	<u>7. Z-Transform</u>	The z-Transform is introduced. This algebraic method introduces polynomials into the analysis of linear systems.
	<u>8. IIR Filters</u>	The class of feedback filters is introduced. These filters have infinite-length impulse responses, so they are usually referred to as infinite-impulse-response (IIR) filters. Their Z-transforms contain both poles and zeros, so their frequency reponse can have very sharp peaks, as well as nulls.

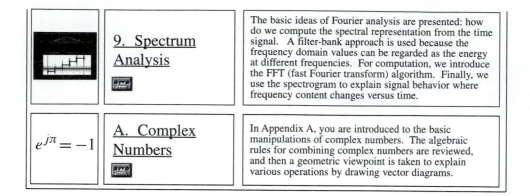

| | 9. Spectrum Analysis | The basic ideas of Fourier analysis are presented: how do we compute the spectral representation from the time signal. A filter-bank approach is used because the frequency domain values can be regarded as the energy at different frequencies. For computation, we introduce the FFT (fast Fourier transform) algorithm. Finally, we use the spectrogram to explain signal behavior where frequency content changes versus time. |
| $e^{j\pi} = -1$ | A. Complex Numbers | In Appendix A, you are introduced to the basic manipulations of complex numbers. The algebraic rules for combining complex numbers are reviewed, and then a geometric viewpoint is taken to explain various operations by drawing vector diagrams. |

Demos

 1. Introduction

2. Sinusoids

	Sinusoids	This is an introduction to plotting sinusoids (both sine and cosine waves) from equations. In addition, the tutorial reviews how to write the equation given a plot of the sinusoid.
	Rotating Phasors	Here are four movies showing rotating phasors and how the real part of the phasor traces out a sinusoid versus time. Two of the movies show how rotating phasors of different frequencies interact to produce complicated waveforms such as beat signals.
	Sine Drill	This is a MATLAB program that tests your ability to calculate the parameters of a sine wave.
	Tuning Forks	This demo shows how the size and stiffness of a tuning fork affect the tone produced by three different tuning forks.
	Clay Whistle	Here's a demo that shows the nearly sinusoidal waveforms produced by two clay whistles.

 3. Spectrum Representation

	FM Synthesis of Instrument Sounds	This demo gives the mathematical derivation of how instrument sounds can be synthesized using the principles of frequency modulation. The example sounds include a bell and a clarinet.
	Sounds and Spectrograms	This demo illustrates the connection between a variety of sounds and their spectrograms. Among the different sounds are simple tones and waveforms, real and synthesized music, and chirp signals.

4. Sampling and Aliasing 📼

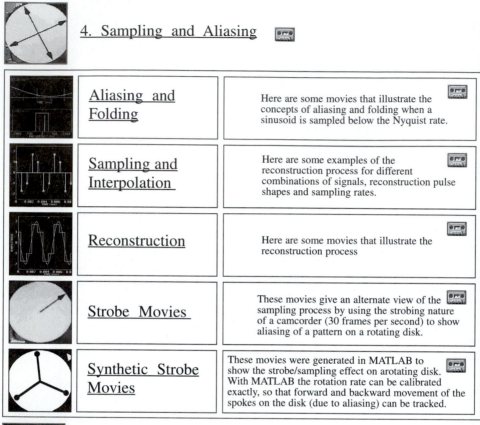 Aliasing and Folding		Here are some movies that illustrate the concepts of aliasing and folding when a sinusoid is sampled below the Nyquist rate. 📼
	Sampling and Interpolation	Here are some examples of the reconstruction process for different combinations of signals, reconstruction pulse shapes and sampling rates. 📼
	Reconstruction	Here are some movies that illustrate the reconstruction process 📼
	Strobe Movies	These movies give an alternate view of the sampling process by using the strobing nature of a camcorder (30 frames per second) to show aliasing of a pattern on a rotating disk. 📼
	Synthetic Strobe Movies	These movies were generated in MATLAB to show the strobe/sampling effect on arotating disk. With MATLAB the rotation rate can be calibrated exactly, so that forward and backward movement of the spokes on the disk (due to aliasing) can be tracked. 📼

5. FIR Filters 📼

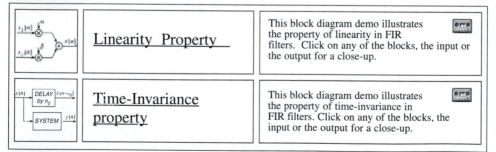 Linearity Property		This block diagram demo illustrates the property of linearity in FIR filters. Click on any of the blocks, the input or the output for a close-up. 📼
	Time-Invariance property	This block diagram demo illustrates the property of time-invariance in FIR filters. Click on any of the blocks, the input or the output for a close-up. 📼

6. Frequency Response of FIR Filters 📼

	Cascading FIR Filters	LTI FIR filters are used to process images, thus demonstrating that low-pass filtering is blurring, while high-pass filtering will sharpen edges.
	Introduction to FIR filters	A brief introduction to FIR filters and how they can change the sound of speech signals.

7. Z-Transform

	Three-Domains	The connection between the Z-transform domain of poles and zeros and the time domain, and also the frequency domain is illustrated with several movies where individual zeros or zero pairs are moved continuously.
	PeZ	A MATLAB tool for pole/zero manipulation. Poles and zeros can be placed anywhere on a map of the z-plane. The corresponding time domain (n) and frequency domain (omega) plots will be displayed. When a zero pair (or pole pair) is dragged, the impulse response and frequency response plots will be updated in real time.

8. IIR Filters

	Three - Domains	The connection between the Z-transform domain of poles and zeros and the time domain, and also the frequency domain is illustrated with several movies where individual poles, or zeros or pole pairs of IIR filters are moved continuously.
	IIR Filtering	A short tutorial on first- and second-order IIR (infinite-length impulse response) filters. This demo shows plots in the three domains for a variety of IIR filters with different filter coefficients.
	Z to Freq	A demo that illustrates the connection between the complex Z-plane and the frequency response of a system. The frequency response is obtained by evaluating H(z) on the unit circle in the complex Z-plane.

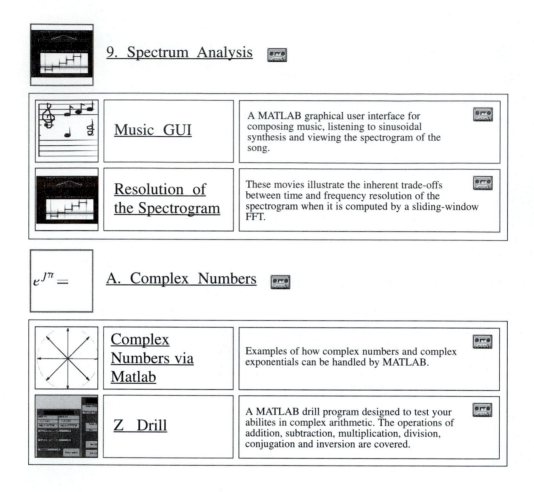

9. Spectrum Analysis

	Music GUI	A MATLAB graphical user interface for composing music, listening to sinusoidal synthesis and viewing the spectrogram of the song.
	Resolution of the Spectrogram	These movies illustrate the inherent trade-offs between time and frequency resolution of the spectrogram when it is computed by a sliding-window FFT.

A. Complex Numbers

	Complex Numbers via Matlab	Examples of how complex numbers and complex exponentials can be handled by MATLAB.
	Z Drill	A MATLAB drill program designed to test your abilites in complex arithmetic. The operations of addition, subtraction, multiplication, division, conjugation and inversion are covered.

Labs

 1. Introduction

 2. Sinusoids

Lab 1: Introduction to Matlab	In this lab we introduce the fundamentals of Matlab. Matlab is a programming environment that you will find helpful for many of the exercises in this text.
Lab2: Introduction to Complex Exponentials	Manipulating sinusoid functions using complex exponentials turns trigonometric problems into simple arithmetic and algebra. In this lab, we first review the complex exponential signal and the phasor addition property needed for adding cosine waves. Then we will use Matlab to make plots of phasor diagrams that show the vector addition needed when combining sinusoids.

3. Spectrum Representation

Lab 3: Synthesis of Sinusoidal	In this lab, we will synthesize more complicated sinusoidal waveforms composed of sums of sinusoidal signals, each with a different frequency. The sounds synthesized will one of several songs.
Lab 4: AM and FM Sinusoidal Signals	The objective of this lab is to introduce more complicated signals that are related to the basic sinusoid. These are signals which implement frequency modulation (FM) and amplitude modulation (AM) are widely used in communication systems such as radio and television, but they also can be used to create interesting sounds that mimic musical instruments.

 4. Sampling and Aliasing

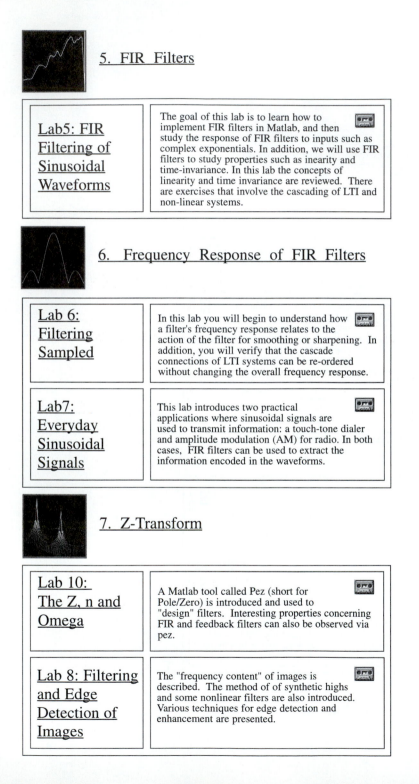

5. FIR Filters

Lab5: FIR Filtering of Sinusoidal Waveforms	The goal of this lab is to learn how to implement FIR filters in Matlab, and then study the response of FIR filters to inputs such as complex exponentials. In addition, we will use FIR filters to study properties such as inearity and time-invariance. In this lab the concepts of linearity and time invariance are reviewed. There are exercises that involve the cascading of LTI and non-linear systems.

6. Frequency Response of FIR Filters

Lab 6: Filtering Sampled	In this lab you will begin to understand how a filter's frequency response relates to the action of the filter for smoothing or sharpening. In addition, you will verify that the cascade connections of LTI systems can be re-ordered without changing the overall frequency response.
Lab7: Everyday Sinusoidal Signals	This lab introduces two practical applications where sinusoidal signals are used to transmit information: a touch-tone dialer and amplitude modulation (AM) for radio. In both cases, FIR filters can be used to extract the information encoded in the waveforms.

7. Z-Transform

Lab 10: The Z, n and Omega	A Matlab tool called Pez (short for Pole/Zero) is introduced and used to "design" filters. Interesting properties concerning FIR and feedback filters can also be observed via pez.
Lab 8: Filtering and Edge Detection of Images	The "frequency content" of images is described. The method of of synthetic highs and some nonlinear filters are also introduced. Various techniques for edge detection and enhancement are presented.

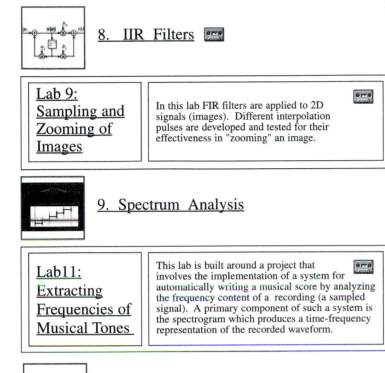

8. IIR Filters

Lab 9: Sampling and Zooming of Images

In this lab FIR filters are applied to 2D signals (images). Different interpolation pulses are developed and tested for their effectiveness in "zooming" an image.

9. Spectrum Analysis

Lab11: Extracting Frequencies of Musical Tones

This lab is built around a project that involves the implementation of a system for automatically writing a musical score by analyzing the frequency content of a recording (a sampled signal). A primary component of such a system is the spectrogram which produces a time-frequency representation of the recorded waveform.

$e^{J\pi} = -1$ A. Complex Numbers

Index

END USER LICENSE AGREEMENT

DEFECTIVE CD REPLACEMENT

E-mail Prentice Hall at **dsp_first@prenhall.com** for CD replacement. You must send us your damaged or defective CD, and we will provide you with a new one.

WEB SITE

A website exists containing the CD content in its evolving form at **http://www.ece.gatech.edu/~dspfirst**. The serial number printed on the CD is your password.